International
REVIEW OF
Neurobiology
Volume 99

International REVIEW OF *Neurobiology*

Volume 99

SERIES EDITORS

R. ADRON HARRIS
Waggoner Center for Alcohol and Drug Addiction Research
The University of Texas at Austin
Austin, Texas, USA

PETER JENNER
Division of Pharmacology and Therapeutics
GKT School of Biomedical Sciences
King's College, London, UK

EDITORIAL BOARD

Recent advances in the use of Drosophila in neurobiology and neurodegeneration

EDITED BY

NIGEL ATKINSON

The University of Texas at Austin, Texas, USA

ELSEVIER

AMSTERDAM • BOSTON • HEIDELBERG • LONDON
NEW YORK • OXFORD • PARIS • SAN DIEGO
SAN FRANCISCO • SINGAPORE • SYDNEY • TOKYO
Academic Press is an imprint of Elsevier

Academic Press is an imprint of Elsevier
32 Jamestown Road, London, NW1 7BY, UK
Radarweg 29, PO Box 211, 1000 AE Amsterdam, The Netherlands
Linacre House, Jordan Hill, Oxford OX2 8DP, UK
225 Wyman Street, Waltham, MA 02451, USA
525 B Street, Suite 1900, San Diego, CA 92101-4495, USA

First edition 2011

ISBN: 978-0-12-387003-2
ISSN: 0074-7742

For information on all Academic Press publications
visit our website at elsevierdirect.com

Printed and bound in USA

11 12 13 14 10 9 8 7 6 5 4 3 2 1

CONTENTS

The roles of *Fruitless* and *Doublesex* in the Control of Male Courtship

BRIGITTE DAUWALDER

Circadian Plasticity: from Structure to Behavior

LIA FRENKEL AND MARÍA FERNANDA CERIANI

Learning and Memory in *Drosophila*: Behavior, Genetics, and Neural Systems

LILY KAHSAI AND TROY ZARS

Studying Sensorimotor Processing with Physiology in Behaving *Drosophila*

JOHANNES D. SEELIG AND VIVEK JAYARAMAN

Modeling Human Trinucleotide Repeat Diseases in *Drosophila*

Zhenming Yu and Nancy M. Bonini

From Genetics to Structure to Function: Exploring Sleep in *Drosophila*

Daniel Bushey and Chiara Cirelli

CONTRIBUTORS

Numbers in parentheses indicate the pages on which the authors contributions begin.

Nigel S. Atkinson (23), The Waggoner Center for Alcohol and Addiction Research, Section of Neurobiology, The University of Texas at Austin, Austin, Texas, USA

Nancy M. Bonini (191), Department of Biology, University of Pennsylvania; Howard Hughes Medical Institute, Philadelphia, PA 19104, USA

Daniel Bushey (213), Department of Psychiatry, University of Wisconsin, Madison, WI, USA

María Fernanda Ceriani (107), Laboratorio de Genética del Comportamiento, Fundación Instituto Leloir, Instituto de Investigaciones Bioquímicas de Buenos Aires, CONICET, Buenos Aires, Argentina

Chiara Cirelli (213), Department of Psychiatry, University of Wisconsin, Madison, WI, USA

Brigitte Dauwalder (87), Department of Biology and Biochemistry, University of Houston, 369 SR2, Houston, TX 77204-5001, USA

Lia Frenkel (107), Laboratorio de Genética del Comportamiento, Fundación Instituto Leloir, Instituto de Investigaciones Bioquímicas de Buenos Aires, CONICET, Buenos Aires, Argentina

Alfredo Ghezzi (23), The Waggoner Center for Alcohol and Addiction Research, Section of Neurobiology, The University of Texas at Austin, Austin, Texas, USA

Iris C. Howlett (1), Department of Molecular and Cell Biology, University of California, Berkeley, CA 94720, USA

Vivek Jayaraman (169), Howard Hughes Medical Institute, Janelia Farm Research Campus, 19700 Helix Drive, Ashburn, VA 20147, USA

Lily Kahsai (139), University of Missouri, Division of Biological Sciences, 114 Lefevre Hall, Columbia, MO 65211, USA

Louise Parker (1), Department of Environmental Science, Policy and Management, University of California, Berkeley, CA 94720, USA

Zeid M. Rusan (1), Department of Molecular and Cell Biology, University of California, Berkeley, CA 94720, USA

Johannes D. Seelig (169), Howard Hughes Medical Institute, Janelia Farm Research Campus, 19700 Helix Drive, Ashburn, VA 20147, USA

Bruno van Swinderen (51), Queensland Brain Institute, The University of Queensland, Brisbane, Australia

Mark A. Tanouye (1), Department of Environmental Science, Policy and Management, University of California, Berkeley, CA 94720, USA; Department of Molecular and Cell Biology, University of California, Berkeley, CA 94720, USA

Zhenming Yu (191), Department of Biology, University of Pennsylvania

Troy Zars (139), University of Missouri, Division of Biological Sciences, 114 Lefevre Hall, Columbia, MO 65211, USA

PREFACE

Drosophila has been genetically studied for more than 100 years. Because of this legacy, it has the most sophisticated genetic toolkit of any metazoan. The original reasons for studying Drosophila, dating from the early 1900's, are still relevant today. Plain and simple, raising flies in the lab is easy and cheap. The animal is tiny and has a simple diet whose basis can be found in the grocery store, and one can successfully raise them with almost no training. In addition, flies have the same gene classes as humans without the extensive gene families that complicate mutant analysis. Originally, flies were used to glean genetic keys for understanding the nature of mutation, recombination, evolution, and development. However, the application of genetics required the drosophilist to passively acquire some understanding of this little fly's behavior. This understanding set the stage for Drosophila neurogenetics.

In the 1960s and 1970s, Drosophila neurogenetics was born of two fathers–Seymour Benzer and Martin Heisenberg. These newly self-minted neurobiologists were not limited by the common knowledge of what was possible. Because of this, they tried things that were at the time (and sometimes even today) generally believed to be impossible. Both scientists independently used unbiased genetic screens to demonstrate that single genes could be linked to a behavior.

Seymour Benzer and Martin Heisenberg had different and complementary approaches that continue today in their scientific offspring. Benzer treated flies as particles of behavior. He used behavior as a tool that magnified small genetically induced disruptions in the nervous system. This allowed him to quantify behavior from populations of animals to produce statistically meaningful results in a single afternoon, an experiment that in a mammal might have taken months or years to perform. Benzer's behavioral apparatuses are incredibly clever but usually technologically simple. His particles of behavior approach identified genes important for circadian rhythms, courtship, learning, memory, regulation of neural excitability, neurodegeneration, and aging. Martin Heisenberg took a different approach. He looked for mutations that disrupted the physical organization of the nervous system and then asked how behavior was affected. The focus of his behavioral studies tends to be the individual fly. Heisenberg's career demonstrates a remarkable ability to develop new machines and new tools. This brought us fly collars that revolutionized serial sectioning of the Drosophila brain, optomotor mazes to probe visual pattern recognition, tethered flies and flight simulators for operant and classical conditioning learning assays, and the study of attention and motivation. Many Drosophila neurobiologists consider themselves to be scientific descendants or at least relatives of these pioneers.

The topics in this book while eclectic merely reflect the origins of the field. All are knit together by the use of Drosophila genetics. Chapter 1 covers the use of Drosophila as an epilepsy model. Chapter 2 discusses the role of a BK channel gene in the production of drug tolerance and a withdrawal response. In Chapter 3, we see how electrophysiological recording in behaving tethered animals can be used to study the neurobiology of attention. Chapter 4 describes the genes and neurons of male sex-specific behavior. The rhythmic control of the outputs of the circadian clock and their effect on memory and social interactions are discussed in Chapter 5. The topics discussed in Chapter 6 are the structures and genes involved in olfactory, visual recognition, and place learning and memory. Chapter 7 returns to the recording of activity of neurons in awake behaving flies to understand sensorimotor integration. Chapter 8 discusses models of human neurodegenerative disease caused by trinucleotide repeat expansion. To conclude, we have Chapter 9, which describes mutant analysis in the study of the mechanics and function of sleep. This volume samples many exciting advancements in Drosophila neurogenetics.

SEIZURE AND EPILEPSY: STUDIES OF SEIZURE DISORDERS IN *DROSOPHILA*

Louise Parker[1], Iris C. Howlett[2], Zeid M. Rusan[2] and Mark A. Tanouye[1,2]

[1]Department of Environmental Science, Policy and Management, Helen Wills Neuroscience Institute, 131 Life Sciences Addition, University of California, Berkeley, CA 94720, USA
[2]Department of Molecular and Cell Biology, Helen Wills Neuroscience Institute, 131 Life Sciences Addition, University of California, Berkeley, CA 94720, USA

Despite the frequency of seizure disorders in the human population, the genetic and physiological basis for these defects has been difficult to resolve. Although many genetic contributions to seizure susceptibility have been identified, these involve disparate biological processes, many of which are not neural specific. The large number and heterogeneous nature of the genes involved makes it difficult to understand the complex factors underlying the etiology of seizure disorders. Examining the effect known genetic mutations have on seizure susceptibility is one approach that may prove fruitful. This approach may be helpful in both understanding how different physiological processes affect seizure susceptibility and identifying novel therapeutic treatments. We review here factors contributing

INTERNATIONAL REVIEW OF NEUROBIOLOGY, VOL. 99
DOI: 10.1016/B978-0-12-387003-2.00001-X

0074-7742/10 $35.00

to seizure susceptibility in *Drosophila*, a genetically tractable system that provides a model for human seizure disorders. Seizure-like neuronal activities and behaviors in the fruit fly are described, as well as a set of mutations that exhibit features resembling some human epilepsies and render the fly sensitive to seizures. Especially interesting are descriptions of a novel class of mutations that are second-site mutations that act as seizure suppressors. These mutations revert epilepsy phenotypes back to the wild-type range of seizure susceptibility. The genes responsible for seizure suppression are cloned with the goal of identifying targets for lead compounds that may be developed into new antiepileptic drugs.

I. Introduction

A. Studying Human Seizure Disorders with *Drosophila*

Using *Drosophila* to model human disease is an especially attractive idea. The goal is to leverage the sophisticated genetic, molecular genetic, and transgenic experimental capabilities available in the fly to uncover fundamental biological principles underlying the causes and cures of human pathology, in much the same way as was done previously for classical genetics, developmental biology, and neurobiology. Also compelling is the prospect that novel therapeutics might be identified by studying disease-causing genes followed by targeted drug development, or by establishing platforms for high-throughput drug screening. In this chapter, we discuss the progress made by us, and others, in the development of a *Drosophila* model for studying human epilepsy. The model is based mainly on a "forward genetics" approach for studying seizure disorders in the fly with electrophysiological and behavioral assays for evoking seizures, and identification of mutations that affect susceptibility to seizure: seizure-sensitive, seizure-resistant, and seizure-suppressor mutations.

The idea that fruit flies can have seizures and serve as a model for human epilepsy is somewhat surprising, yet there are important similarities at the cellular and subcellular levels between fly and mammalian nervous systems, particularly in excitable membrane components. Voltage-gated and ligand-gated signaling molecules such as Na^+, K^+, and Ca^{2+} channels, and acetylcholine, glutamate, and gamma-aminobutyric acid (GABA) transmitter receptors all have highly conserved homologues in the fruit fly. There are some features of the mammalian central nervous system that are thought to contribute to the severity of some epilepsies that are not present in the fly. For example, in flies the central nervous system is organized as a ganglionic structure with synaptic neuropilar regions, whereas the mammalian cortex is organized into synaptic layers. Nevertheless, in the fly, electrical shock of sufficient intensity delivered to the brain of adult *Drosophila* elicits neuronal spiking activity that is seizure-like in appearance (Kuebler and Tanouye, 2000; Kuebler *et al.*, 2001; Lee and Wu, 2002; Pavlidis and Tanouye, 1995; Song and Tanouye, 2006), similar to other animals with complex nervous systems, including humans.

B. Human Idiopathic Epilepsies and Channelopathies

Human seizure disorders are a serious health concern because of the large number of people affected and the inadequacy of available treatments. Seizures are abnormal, synchronous, and rhythmic firing of neurons in the central nervous system. About 10% of the population will experience a seizure during their lifetime, and about 1% of people suffer from the persistent, spontaneous seizures that define epilepsy (Shneker and Fountain, 2003). A further difficulty is intractable epilepsy: about 15 million people worldwide cannot adequately control their seizures with any available medication. Thus, overall, seizure disorders represent a pervasive class of neurological disease with unsatisfactory treatment options.

Seizures can result from a variety of brain insults including head trauma, fever, illness, and electroconvulsive shock, but a main source of seizure susceptibility appears to be genetic predisposition. More than 70 genes have been linked to epilepsies from work done on inherited disorders in humans, mice, and flies (Noebels, 2003). These genes encode a variety of products ranging from ion channel proteins to tRNAs. The number of disparate genes involved in epileptogenesis is further confounded by there frequently being no obvious functional relationships between mutation and seizure susceptibility, thereby complicating our ability to understand epilepsy on a mechanistic level (Jacobs *et al.*, 2001). The clearest cases appear to be the epilepsies that are channelopathies, that is, due to mutations within ion channel genes (Catterall *et al.*, 2008; Reid *et al.*, 2009).

An example of channelopathies is epilepsies caused by mutation of the voltage-gated Na^+ channel gene *SCN1A* (Meisler and Kearney, 2005; Mahoney *et al.*, 2009; Mulley *et al.*, 2005). In humans, such mutations are associated with three forms of epilepsy: GEFS+ (generalized epilepsy with febrile seizures plus), characterized by febrile seizures that persist beyond the age of 6 years; SMEI (severe myoclonic epilepsy in infancy), an intractable epilepsy frequently resulting in convulsive status epilepticus; and ICEGTC (intractable childhood epilepsy with generalized tonic–clonic), an atypical SMEI that does not cause myoclonic seizures. In flies, severe seizure-like behaviors and electrical abnormalities are caused by the $para^{bss}$ mutations that have been recently mapped to the *Drosophila* Na^+ channel gene *para*; however, all other *para* mutations cause hypoexcitability and are generally seizure-resistant mutations that can act as seizure suppressors.

An interesting example of an epilepsy that is not a channelopathy is a familial myoclonus-ataxia epilepsy caused by mutation of the human *PRICKLE1* gene (Bassuk *et al.*, 2008; Tao *et al.*, 2011). Mutant homologues of human *PRICKLE1* in several model organisms, *Drosophila*, mouse, and zebrafish, also caused seizure sensitivity when mutated (Tao *et al.*, 2011). The *PRICKLE1* gene product is a LIM domain protein that modulates Wnt/Fz signaling. The fly *prickle*^{sple} (*pk*^{sple}) mutant has been shown to be critical for planar polarity in imaginal discs, but had not been previously known to cause neurological phenotypes (Gubb *et al.*, 1999).

II. Seizure Studies in *Drosophila*

A. Seizure-Like Electrical Activity in Adult Flies

Electrophysiology methods were developed in flies to evoke seizures and record subsequent seizure-like electrical activity (Kuebler and Tanouye, 2000; Kuebler *et al.*, 2001; Lee and Wu, 2002; Pavlidis and Tanouye, 1995; Pavlidis *et al.*, 1994). Especially valuable is the capability of quantifying seizure susceptibility that allows comparisons between wild-type and mutant flies, and with other genetic and chemical manipulations (Kuebler and Tanouye, 2000; Kuebler *et al.*, 2001; Song and Tanouye, 2006). Seizure-like electrical activity is evoked by short wave trains of high-frequency electrical stimuli (HFS, 0.4 ms pulses at 200 Hz for 300 ms) delivered to the brain at sufficient intensity and recorded from thoracic flight muscles. This seizure-like activity reflects motoneuron action potentials on a one-to-one basis. Seizures are extensive, spreading to at least 14 different muscles, containing 60 muscle fibers innervated by 58 motoneurons. Seizures are evoked in an all-or-nothing manner: HFS stimuli below a threshold voltage never elicit seizures, whereas just above threshold, seizures are always elicited. Seizure threshold is similar for all individuals of a given genotype. Thus, each genotype possesses a characteristic seizure susceptibility that may be quantified by its threshold voltage. Normal seizure susceptibility, the wild-type range, is about 30–40 V HFS (Canton Special = 30.1, Oregon R = 39.3 V HFS; Kuebler *et al.*, 2001).

B. Seizure Disorders in Flies: Seizure Sensitivity in the Bang-Sensitive Paralytic Mutant Class

An obvious attraction of working with *Drosophila* is to take advantage of valuable mutant collections that may be screened electrophysiologically for alterations in seizure susceptibility. All of the mutants in one behavioral class, called the bang-sensitive (BS) paralytic class, were consistently found to be sensitive to seizure (Kuebler *et al.*, 2001; Pavlidis and Tanouye, 1995). The BS class contains 14 mutant alleles representing 12 genes and a variety of gene products (Table I). Three BS mutants have been used as experimental representatives of the class. (1) The *slamdance (sda)* mutant displays a weak seizure sensitivity that is fully penetrant, but is the easiest to suppress by drug or suppressor mutant. (2) The easily *shocked*

Table I
SEIZURE-SENSITIVE MUTANTS AND THEIR GENE PRODUCTS.

Seizure-Sensitive Mutant	Seizure Threshold (V HFS)	Gene Product
paralyzed (*para*^{bss1}, *para*^{bss2})	3.2, 3.7	Na^+ channel (Parker *et al.*, 2011)
easily shocked (eas)	3.4	Ethanolamine kinase (Pavlidis *et al.*, 1994)
slamdance (sda)	6.7	Aminopeptidase N (Zhang *et al.*, 2002)
bang sensitive (*bas*^{1}, *bas*^{2})	7.6, 3.8	Unknown
technical knockout (tko)	9.9	ribosomal protein S12 (Royden *et al.*, 1987)
jitterbug (jbug)	10.5	Unknown
couch potato (cpo)	11.1	RNA-binding protein (Glasscock and Tanouye, 2005)
kazachoc (kcc)	17.0	K^+ Cl^- cotransporter (Hekmat-Scafe *et al.*, 2006)
knockdown (kdn)	20.2	Citrate synthase (Fergestad *et al.*, 2006)
prickle (*pk*^{sple})	Unknown	LIM domain protein (Tao *et al.*, 2011)
stress-sensitive (sesB)	Unknown	Adenine nucleotide translocase (Zhang *et al.*, 1999)
rock-n-roll (rnr)	Unknown	Unknown

Seizure threshold is voltage of high-frequency stimulation (V HFS). For comparison the seizure threshold is 30.1 V HFS for Canton-Special wild-type flies (Kuebler *et al.*, 2001).

(eas) mutant shows moderate seizure sensitivity. (3) The *bang senseless*^{1} allele of the *paralyzed* gene (*para*^{bss1}) is a severe BS mutation: electrophysiologically and behaviorally the most sensitive to seizure with a prominent tonic–clonic-like phenotype. It is the most difficult BS mutant to suppress genetically and pharmacologically, and was presented to model human intractable epilepsy (Parker *et al.*, 2011).

The *para*^{bss1} mutation (formerly named *bss*^{1}) is presented as an example of the BS mutant class. Like most other BS mutants, *para*^{bss1} flies display unexceptional behaviors under normal conditions. Feeding, grooming mating behaviors, and phototaxis and geotaxis responses all appear normal. Overall activity levels are unaltered; flies are neither hyperactive nor sluggish. Abnormal behavior is induced in *para*^{bss1} following a mechanical shock, a tap of the culture vial or brief vortex mixing (a "bang"). The resulting behavioral phenotype is complex with six distinguishable phases (Fig. 1A).

i. An initial seizure is typical for BS mutants, lasting several seconds, and characterized by leg shaking, abdominal muscle contractions, wing flapping and scissoring, and proboscis extensions.
ii. An initial paralytic period is also typical of BS mutants; *para*^{bss1} flies are immobile and unresponsive to mechanical stimulus, as described previously (Ganetzky and Wu, 1982).
iii. Unlike other BS mutants, initial paralysis in *para*^{bss1} is followed by an extended period of tonic–clonic-like activity. During this period, the fly is

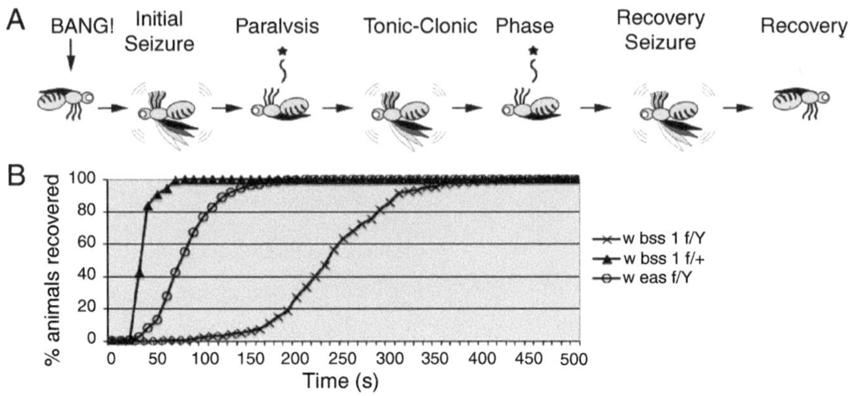

FIG. 1. ***Drosophila para^{bss1}* mutant behavior**. (A) Cartoon depicting stereotypic behavioral phenotype of *para^{bss1}* flies subjected to a mechanical shock (10 s vortex = BANG): initial seizure-like behavior, followed by complete paralysis, then a tonic–clonic-like period that is unique to *para^{bss1}* and not evident in other BS mutant genotypes. In the figure, one clonus-like event is depicted, but the number can vary, as can the duration of the period. The tonic–clonic-like period is followed by a recovery seizure and the fly then recovers. Not depicted is a quiescent period of variable duration often observed between the recovery seizure and recovery, as well as the refractory period during which flies are resistant to further seizures that occurs immediately following recovery. (B) For *para^{bss1}/Y* hemizygous males, recovery time from behavioral paralysis is substantially longer than for *para^{bss1}*/+ heterozygous females or for another BS mutant, *eas^{PC80}/Y*.

mainly quiescent, resembling a tonic phase. The quiescence is broken up by multiple bouts of clonus-like activity.

iv. Recovery seizure is also typical of all BS mutants including *para^{bss1}* flies; this seizure resembles the initial seizure and clonus-like activity.

v. A refractory period is observed in *para^{bss1}*, as in all other BS mutants during which behaviorally normal flies cannot be induced to have further seizures.

vi. Finally, there is complete recovery and *para^{bss1}* flies regain bang sensitivity.

The total paralytic recovery time for *para^{bss1}* (Fig. 1B) is longer than that for other BS mutants, mainly because of the tonic–clonic activity; mean recovery time for *para^{bss1}* is ~240 s compared with *sda or eas*, 38 and 81 s, respectively.

C. Seizure Sensitivity in *para^{BSS1}* and Electrophysiology of Tonic–Clonic-Like Spontaneous Firing

The electrophysiology phenotype of *para^{bss1}* is generally similar to that seen in other BS mutants, even though more extreme in some aspects. As was the case for behavior, electrophysiology is normally unexceptional in *para^{bss1}* mutants as determined using the adult giant fiber (GF) system neural circuit as proxy (Tanouye and Wyman, 1980). Thus, for example, single-pulse stimulation of the GF produced evoked muscle potential responses in the dorsal longitudinal muscles that were normal in appearance, threshold (2.04 ± 0.48 V), and latency (2.07 ms). Also normal was the ability to follow high-frequency stimulation and respond to short interpulse twin pulse intervals.

Seizure-like electrical activity in *para*^{bss1} mutants is evoked by HFS stimuli at exceptionally low stimulation voltages (4.4 V HFS; Fig. 2), substantially lower than

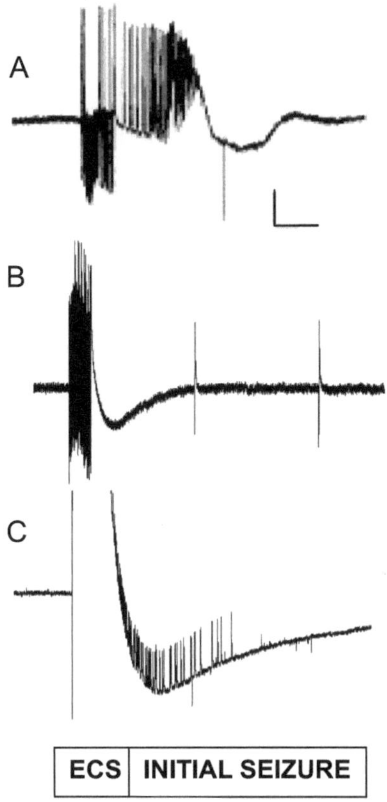

FIG. 2. **The BS mutant*sda*is more susceptible to seizures than wild type, and therefore has a lower seizure threshold.** (A) seizure-like activity (initial seizure) is elicited in a *para*^{bss1} fly by a high-frequency stimulus of low strength (4 V) and displayed at a high sweep speed. (B) A low-voltage high-frequency (HF) stimulus of 8 V fails to elicit a seizure in a wild-type Canton-Special (CS) fly because the stimulus is below the seizure threshold. Following the HF stimulus artifact, there is no seizure activity observed in this recording displayed at a high sweep speed. Note also that there is no period of synaptic failure and single-pulse stimulation of the giant fiber (GF) (0.5 Hz) continues to evoke dorsal longitudinal muscle (DLM) potentials. Two such effective single-pulse stimuli are depicted in this trace; each was effective in evoking a DLM potential. (C) Seizure-like activity is elicited in a wild-type CS fly by a high-voltage HF stimulus (32 V), which is above the threshold for seizure. The seizure in this recording begins within the large stimulus artifact and is displayed at a high sweep speed. Vertical calibration bar is 20 mV. Horizontal calibration bar is 300 ms. Adapted from Kuebler *et al.*, 2001.

for wild type and slightly lower than for *eas* and *sda* BS mutant males (5.13 and 6.2 V HFS, respectively). Thus, using the criterion of HFS threshold, *para*^{bss1} mutants are about seven times more seizure-sensitive than Canton-Special wild-type flies (Kuebler and Tanouye, 2000; Lee and Wu, 2002; Parker *et al.*, 2011). Electrophysiological features of evoked *para*^{bss1} seizures are complex (Fig. 3).

i. There is an initial "seizure" following delivery of a suprathreshold HFS stimulus consisting of aberrant high-frequency firing (>100 Hz) lasting for ~1 s, present in the dorsal longitudinal muscle and all other muscle fibers and motoneurons examined (Fig. 3A, B). This *para*^{bss1} seizure-like phenotype is qualitatively similar to that seen for other BS genotypes and normal flies, differing only in the HFS voltage required for eliciting it (i.e., seizure threshold), but otherwise indistinguishable in subsequent electrical activity.
ii. After the initial seizure, the next aspect of the electrophysiology phenotype for *para*^{bss1} mutants is the sudden failure of the GF neural circuit to drive muscle potentials in the dorsal longitudinal muscle. This is termed a "synaptic failure" period as it has been shown to be due to transmission failure at many central synapses (Fig. 3A; Pavlidis and Tanouye, 1995); it is likely the cause of behavioral paralysis in *para*^{bss1} mutants and for other BS

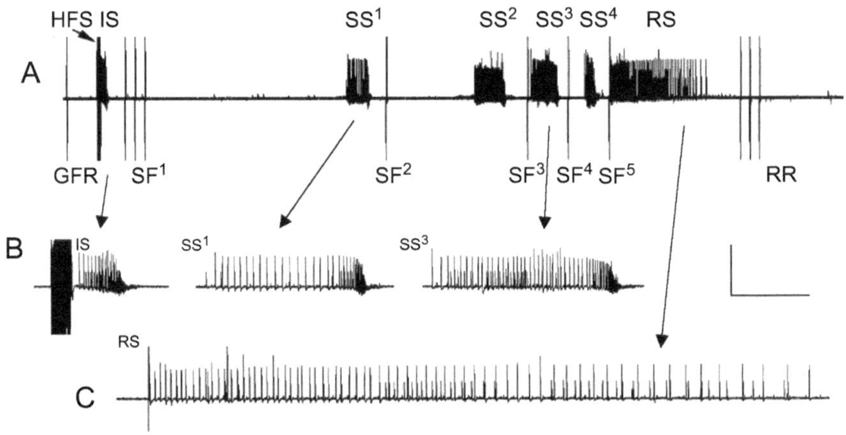

FIG. 3. **Complex seizure-like electrophysiology phenotypes in *para*^{bss1} mutants.** (A) Electrical recording of seizure-like neuronal activity and synaptic failure in a *para*^{bss1}/*Y* animal subjected to a 4-V high-frequency electrical stimulus (HFS), as well as single-pulse stimuli to trigger the giant fiber circuit, allowing assessment of synaptic function (giant fiber response, GFR). Following an initial seizure (IS) is a period of synaptic failure within which GF stimulation fails to evoke a response (SF$^{1-5}$), unlike that seen prior to the seizure. As depicted, during the period of synaptic failure, spontaneous secondary seizures are observed (SS$^{1-4}$). Although in this trace four secondary seizures are observed, the number is variable. A final recovery seizure (RS) is observed, and shortly thereafter, GF system transmission is restored (response recovery, RR). (B) Initial seizure and secondary seizures at higher sweep speed. (C) Recovery seizure at higher sweep speed. Calibration bar is 20 mV, 10 s in (A), 20 mV, 1.5 s in (B) and (C).

genotypes. However, although the period of synaptic failure is short in most BS mutants, for example about 38 s in *sda*, for *para*^{bss1}, it is longer and more complex, apparently reflecting the complexity observed behaviorally as tonic–clonic-like activity.

iii. For the purpose of describing the electrophysiology of the *para*^{bss1} tonic–clonic-like period, it is convenient to consider three seizure types: (1) initial seizure, occurring immediately after HFS; (2) secondary seizures, occurring within the tonic–clonic-like interval; and (3) recovery seizure, occurring immediately prior to long-term recovery of the GF system neural circuit response (Fig. 3). Initial and secondary seizures resemble each other in waveform and duration; recovery seizures are somewhat different in waveform and longer in duration, although the significance of this is unclear (Fig. 3C). In young *para*^{bss1} flies (1 day post-eclosion), the tonic–clonic-like period is fairly short, about 45 s, and contains just one secondary seizure. In older *para*^{bss1} flies (7 days), the tonic–clonic-like period is longer, about 75 s and typically contains three to four secondary seizures, although up to eight have been observed. This is consistent with observations of longer behavioral recovery times in older *para*^{bss1} flies.

D. *PARA*^{BSS1} IS A GAIN-OF-FUNCTION MUTATION IN THE VOLTAGE-GATED SODIUM CHANNEL GENE

The mutants in the BS are all similar in their behavioral and electrophysiology phenotypes. This suggested that their respective gene products might converge upon a common cell biological function or signaling pathway. Surprisingly, the 12-gene products encoded by the BS genes are disparate and do not immediately suggest any common function that might be shared. For example, the *eas* gene encodes ethanolamine kinase and the mutation is a two 2-base pair deletion causing a frame shift and a truncated product. This results in a defect that eliminates ethanolamine kinase activity and alters the amount of the predominant membrane lipid phosphatidylethanolamine. The *sda* gene encodes the transmembrane ectoenzyme aminopeptidase N that catalyzes the removal of neutral and basic amino acids from the N-termini of a variety of small peptide substrates. The *sda* mutation is a 2-bp insertion in the 5′ UTR that appears to reduce expression; *sda* phenotypes may be phenocopied by RNAi.

The *para*^{bss1} mutation is an unusual gain-of-function mutation in the voltage-gated Na^+ gene, described here in more detail. Although the mutation, formerly called *bss*^{1}, had been identified more than 30 years ago (Jan and Jan, 1978), molecular identification proved difficult. Recently, mapping data placed *bss*^{1} in the *para* Na^+ channel gene and the mutation was renamed as *para*^{bss1} (Parker *et al.*, 2011). This was a surprising result as, heretofore, all *para* alleles had been found to be seizure-suppressor mutations (Kuebler *et al.*, 2001; Song and Tanouye, 2007); and $para^{bss1}$ mutants do not display any of the canonical *para* loss-of-function phenotypes, such as a loss of action potentials and behavioral paralysis at elevated (29°C) temperatures in $para^{ts1}$ or $para^{ST76}$ (Siddiqi and Benzer, 1976; Suzuki *et al.*, 1971; Wu and Ganetzky, 1980). Canonical *para* mutations are lethal in double mutant combinations with the Na^{+} channel regulator mle^{napts} (Ganetzky, 1984). In contrast, mle^{napts} acts as a seizure suppressor for $para^{bss1}$ (Kuebler *et al.*, 2001). Consistent with the gain-of-function nature of the $para^{bss1}$ mutation, a *para+* RNAi construct is also a strong suppressor of $para^{bss1}$ phenotypes (Parker *et al.*, 2011).

The gain-of-function nature of $para^{bss1}$ results in unusual gene dosage relationships that may influence interpretations of gain-of-function Na_V mutations in human epilepsies, vis-à-vis loss-of-function mutations. Generally, $para^{bss1}$ is not phenotypically equivalent to a deficiency deleting the *para* gene, and extra copies of *para+* do not continuously ameliorate the effects of mutation as their dosage increases. Thus, phenotypic severity for $para^{bss1}$ is

$$\begin{aligned} para^{bss1}/Y = para^{bss1}/para^{bss1} \gg para^{bss1}/Df > para^{bss1}/+ \\ = para^{bss1}/+/+ \gg +/+ \end{aligned}$$

These observations contrast with *para* loss-of-function mutations, such as $para^{ST76}$, which behave more typically. Phenotypic severity for $para^{ST76}$ is

$$para^{ST76}/Df \geq para^{ST76}/Y = para^{ST76}/para^{ST76} \gg para^{ST76}/+ = +/+$$

The $para^{bss1}$ mutation is due to a single amino acid substitution (leu→phe) of a conserved residue at position 1699 within the hydrophobic S3b membrane-spanning segment of homology domain IV in the Na^{+} channel protein (Parker *et al.*, 2011). Na^{+} channel α-subunits, such as encoded by *para*, comprise four homologous domains (I–IV), each containing six transmembrane segments (S1–S6). The ion pore is formed centrally by the collective organization of S5–S6 segments from each domain. Surrounding the ion pore are the four voltage sensors comprising the S1–S4 segments from each domain. Crucial to the action of each voltage sensor is a modular unit termed a "paddle motif," an S3b–S4 helix–turn–helix motif that is thought to move at the protein–lipid interface, driving activation of the voltage sensors and opening and closing of the pore (Alabi *et al.*, 2007; Bosmans *et al.*, 2008; Catterall *et al.*, 2008). Current thinking on Na^{+} channel paddle domains suggests that the voltage sensor paddles of domains I–III drive channel activation, whereas the paddle of domain IV drives channel inactivation (Bosmans *et al.*, 2008). Consistent with this interpretation, in heterologous Xenopus ooctye expression experiments utilizing *para* cDNA, the functional effect of the L1699F substitution was shown to shift the voltage of fast inactivation to more positive potentials with no effect on activation voltage (Parker *et al.*, 2011).

III. Genetic Suppression of Seizure Susceptibility in *Drosophila*

A. Identification of Mutations that Suppress Seizures by Reverse Genetics

This section examines a novel class of genes called "seizure suppressors." Suppressors are modifying mutations that yield individuals phenotypically more like the wild type, that is, the mutant phenotype is "suppressed." In some cases, the gross wild-type phenotype is completely restored; in other cases, restoration is only partial. Thus, seizure suppressors are mutations that revert or partially revert the seizure-sensitive phenotype of BS mutants, but can be genetically separated, by recombination, from the mutation that they suppress. To our knowledge, the basic approach of utilizing second-site suppressor mutations had not been previously exploited for any neurological syndrome, leading us to initiate mutant searches with the most basic of questions. (i) Are there seizure-suppressor genes, and how might they lead to new therapeutics? (ii) Can we devise screens to identify seizure suppressors? (iii) What is the entire range of potential gene products that can act as seizure suppressors? (iv) Is this range limited to nervous system-specific gene products, such as signaling molecules, or does it include non-nervous system gene products as well? There are now answers to all of these questions, such that the approach is now poised to discover an antiepileptic drug (AEDs) with good seizure-suppressing capabilities and minimal anticipated side effects. Indeed, some promising candidates may have already been identified.

Existing *Drosophila* neurological mutant strains were searched for ones that are seizure-resistant, that is, those that display characteristically high thresholds to evoked seizures (Table II; Kuebler *et al.*, 2001; Song and Tanouye, 2006). Several seizure-resistant mutants are partial loss-of-function mutations of ion channel genes. Thus, the most seizure resistant of the mutants identified is *shakingB*^{2} with an evoked seizure threshold that is three times higher (*shakB*^{2} seizure threshold = 94.7 V HFS). Another seizure-resistant mutant is *para*^{ST76}, a loss-of-function allele of the *para* Na^+ channel gene that causes behavioral paralysis and a loss of action potentials at restrictive temperatures (Loughney *et al.*, 1989; Ramaswami and Tanouye, 1989; Siddiqi and Benzer, 1976; Suzuki *et al.*, 1971; Wu and Ganetzky, 1980). The *para*^{ST76} mutant has a seizure threshold of 65.0 6 ± 7.2, about two times that of wild type (Kuebler *et al.*, 2001). The *no action potential* allele of the *maleless* gene (*mle*^{napts}) is also a mutation that causes seizure resistance (*mle*^{napts} seizure threshold = 72.2 ± 7.3 V HFS) (Kuebler *et al.*, 2001). The *mle*^{napts} allele is a gain-of-function mutation in an RNA helicase-like protein (Kernan *et al.*, 1991; Reenan *et al.*, 2000) that causes a reduction in voltage-gated Na^+ channel expression, a loss of action potentials, and behavioral paralysis at elevated temperatures (Jackson *et al.*, 1984; Kauvar, 1982; Wu and Ganetzky, 1980; Wu *et al.*, 1978).

Table II
LIST OF SEIZURE-SUPPRESSOR MUTANTS AND THEIR GENE PRODUCTS.

Seizure-Suppressor Mutant	Gene Product
Reverse genetics	
paralyzed (*para*^{ST76})	Na^+ channel (Kuebler *et al.*, 2001)
male lethal (*mle*^{napts})	Na^+ channel regulator (Kuebler *et al.*, 2001)
shakingB (*shakB*^{2})	Gap junction channel (Kuebler *et al.*, 2001; Song and Tanouye, 2007)
Shaker (*Sh*^{KS133})	K channel (Kuebler *et al.*, 2001)
Forward genetics	
paralyzed (*para*^{JS1})	Na^+ channel (Song and Tanouye, 2007)
shakingB (*shakB*^{JS})	Gap junction channel (Song and Tanouye, 2007)
escargot (*esg*^{EP684} + 4 alleles)	Zn^{2+}-finger transcription factor (Hekmat-Scafe *et al.*, 2005)
meiosis-P26 (*mei-P26*^{EG16}, *mei-P26*^{1})	Ring finger B-box coiled-coil-NHL protein (Glasscock *et al.*, 2005)
topoisomerase I (*top1*^{JS} + 3 alleles)	DNA topoisomerase type I (Song *et al.*, 2007)
kazal-domain protein-1 (*kdp1*)	Kazal-type serine protease inhibitor (Hekmat-Scafe *et al.*, 2005)
kazal-domain protein-2 (*kdp2*)	Kazal-type serine protease inhibitor (Hekmat-Scafe *et al.*, 2005)
suppressor of eas7 (*su(eas7)*)	Unknown (Glasscock *et al.*, 2005)
suppressor of eas13 (*su(eas13)*)	Unknown (Glasscock *et al.*, 2005)

This table lists separately suppressors identified by reverse genetics and by forward genetics.

Because seizure-resistant mutants are much less susceptible to seizures than wild type, they were tested by reverse genetics to determine whether they are capable of suppressing seizures, that is if they functioned as seizure-suppressor mutations. Indeed, several were found to revert phenotypes of seizure-sensitive mutants in homozygous double mutant combinations, including *para*^{ST76}, *mle*^{napts}, and *shakB*^{2} (Table III). Of these, *shakB*^{2} was chosen for a more detailed analysis as representative of the class (Song and Tanouye, 2006). The *shakB* gene encodes a gap junction channel, mutation in which perturbs electrical synaptic transmission (Crompton *et al.*, 1995; Krishnan *et al.*, 1993; Phelan *et al.*, 1998; Thomas and Wyman, 1984). Gap junction communication is an important target for AEDs as drugs that reduce electrical transmission diminish seizures, and enhanced electrical transmission increases the frequency and severity of seizures (Carlen *et al.*, 2000; Jahromi *et al.*, 2002). Seizure suppression by *shakB*^{2} differs in double mutant combination with different BS mutations (Song and Tanouye, 2006; Table III). Some BS mutants show full suppression (Class III) or partial suppression (Class II) of seizure susceptibility; however, for the mutants with the most severe phenotypes, *shakB*^{2} has no effect (Class I). Loss of *shakB* function is responsible for this suppression by driving expression in *shakB*^{2}*; sda* double mutants. The resulting flies regained seizure sensitivity (i.e., they lost their seizure-suppression). Thus, *shakB*^{2} may act to suppress seizures in less severe BS mutants by a mechanism similar to

Table III
THE *SHAKB*^{2} GAP JUNCTION CHANNEL MUTATION ACTS AS A SEIZURE SUPPRESSOR FOR SOME EPILEPSY MUTATIONS.

	BS Mutation	Seizure Threshold (No Suppressor) (V HFS)	Seizure Threshold (+ *shakB*^{2} suppressor) (V HFS)
Class I	*para*^{bss1}	3.2	3.6
(no suppression)	*para*^{bss2}	3.7	5.0
	bas^2	3.8	5.8
Class II	*eas*	3.4	15.3
(partial suppression)	*tko*	5.0	26.8
Class III	*sda*	6.7	32.8
(full suppression)	*jbug*	10.5	38.2
	kdn	20.2	41.6

Seizure susceptibility for some BS mutants are completely reverted to wild-type levels (Class III). Some mutations (Class II) show partial suppression. Some mutations show no change in seizure threshold (Class I). All values are in volts HFS ($n \geq 10$ for each genotype).

that proposed for drugs such as carbenoxolone that block gap junction activity (Szente *et al.*, 2002). Observations presented here are generally consistent with such a mechanism; indeed feeding carbenoxolone to BS flies also ameliorated seizures (Song and Tanouye, 2006). These results are supportive of an electrical synaptic failure mechanism for *shakB*^{2} seizure suppression.

B. IDENTIFICATION OF SEIZURE-SUPPRESSOR MUTATIONS BY FORWARD GENETIC SCREENS: SUPPRESSION BY LOSS-OF-FUNCTION MUTATIONS IN THE *PARA* NA$^+$ CHANNEL GENE

Forward genetic screens for new seizure-suppressor mutations are performed by mutagenesis in a seizure-sensitive genetic background. For example, Song and Tanouye (2007) utilized mutagenesis by P-element transposons in an *eas* genetic background. Primary screening is behavioral, selecting for exceptional bang-resistant *eas* flies that indicate a reversion of the *eas* BS paralytic phenotype (Glasscock and Tanouye, 2005; Hekmat-Scafe *et al.*, 2005; Song and Tanouye, 2007; Song *et al.*, 2007). Suppressor mutations (second-site suppressors) are separated from *eas* genetically, then mapped, cloned, and characterized (Table II). A total of nine seizure-suppressor mutations have been identified, to date by forward genetics screens, including two that validated screening procedures: a new allele of the Na^+ channel gene (*para*^{JS1}) and a new allele of the gap junction channel gene (*shakB*^{JS}). Several novel seizure-suppressor mutants were also identified including DNA topoisomerase I (*top1*^{JS}), a Zn^{2+}-finger transcription factor (*esg*), and a ring finger

B-box coiled-coil-NHL protein (*mei-P26*^{EG16}) (Glasscock and Tanouye, 2005; Hekmat-Scafe *et al.*, 2005; Song and Tanouye, 2007; Song *et al.*, 2007).

Identification of new *para* and *shakB* alleles as seizure suppressors (Song *et al.*, 2007) is taken as validation for screening methodology, since we had previously shown by reverse genetics that extant *para* and *mle*^{napts} Na^+ channel mutations and the *shakB*^{2} gap junction mutation are potent seizure suppressors (Kuebler *et al.*, 2001). For *para*^{JS1}, the P-element insertion is in the 3′ UTR of *para*, resulting in a 45% reduction in transcription, which apparently accounts for seizure suppression. The *para*^{JS1} mutation is a highly effective seizure suppressor; we found the behavioral phenotype of *eas* is suppressed by 87%, whereas the seizure threshold of *eas* is increased from 3.4 ± 0.5 to 26.4 ± 3.0 V, close to the wild-type threshold (Song and Tanouye, 2007). The *para*^{JS1} mutation is a general seizure suppressor in that it suppresses *sda* and *para*^{bss1} mutant phenotypes, as well as *eas*. In a wild-type background, the *para*^{JS1} mutation causes a weak seizure-resistant phenotype. The seizure threshold of *para*^{JS1} flies is 51.2 ± 5.2 V HFS, about 1.5 times greater than that of wild-type flies. The *para*^{JS1} mutation causes no other apparent phenotypes: electrophysiology, behavior, and morphology are all wild type; *para*^{JS1} is not a temperature-sensitive paralytic mutant (Song and Tanouye, 2007). These results are consistent with our previous observations from reverse genetics that *para*^{ST76} and the *mle*^{napts} mutant that results in a dramatic reduction of *para* translation also function as a seizure suppressor (Kuebler *et al.*, 2001).

C. A Mutation of DNA Topoisomerase I is a Novel and Unexpected Seizure Suppressor

A variety of different gene products have been found to contribute to seizure sensitivity in humans, rodents, and flies. This suggests that similarly, there may be a diversity of products contributing to seizure-suppression. Unexpected classes of seizure-suppressor genes are especially intriguing for study as they may yield novel insight into mechanisms underlying seizure disorder. They additionally have the potential to open the door to new ideas for therapeutics. Seizure suppressors may suggest novel targets that may lead to development of new classes of AEDs. Targets that do not otherwise compromise nervous system function may lead to drugs with minimal side effects. As example, the discovery and characterization of the *DNA topoisomerase 1* allele *top1*^{JS} is described in greater detail. The *top1*^{JS} allele was identified in a P-element transposon screen for *eas* suppression (Song *et al.*, 2007). The identification of *top1*^{JS} as a seizure suppressor was surprising as DNA topoisomerases have not previously been associated with seizure, seizure control, or any other electrical excitability functions. Rather type I DNA topoisomerase is thought to resolve the torsional tension associated with DNA replication or gene transcription (Champoux, 2001).

Seizure suppression is caused by insertion of the P-element in the 5′ UTR of *top1*, 257 bp upstream of the translation start site resulting in a 12.5-fold reduction of transcription. The *top1* gene is an essential gene of several known *top1* alleles; *top1^{JS}* is the only viable, all others are homozygous lethal (Lee *et al.*, 1993; Zhang *et al.*, 2000). The *top1^{JS}* mutation is a general seizure suppressor ameliorating phenotypes of *sda*, *eas*, and *bss* (Song *et al.*, 2007). As example, *top1^{JS}* suppresses *eas* behavioral paralysis phenotypes in 63% of animals and seizure threshold is raised about 3.5-fold. One possible explanation for suppression is that *top1^{JS}* causes increased neuronal apoptosis. Transcription in active neurons generates supercoiled DNA that must be continuously relaxed to sustain high levels of RNA synthesis. The binding of *top1* to DNA ordinarily forms a cleavable complex leading to relaxation of the DNA supercoil (Champoux, 2001). Reduced levels of *top1* activity in mutants may lead to damage of DNA and cell death. The *top1^{JS}* mutant shows high levels of neuronal apoptosis and expression of the *DIAP1* inhibitor of apoptosis, blocks *top1^{JS}* suppression of *eas* seizures and this could, in part, account for seizure suppression.

IV. Chemical Suppression of Seizure Susceptibility in *Drosophila*

A. AEDs Reduce Seizure Sensitivity in BS Flies

Investigations of human AED effects on *Drosophila* BS mutant phenotypes provide several insights. They provide an additional link between the fly model and the human pathology: strengths and weaknesses of the model may be inferred from effectiveness and ineffectiveness of different AEDs. We may further infer that novel drugs developed because they are effective against *Drosophila* seizures may become effective treatments for human seizure disorders. The ability of known AEDs to suppress seizures in *Drosophila* would demonstrate the utility of this model to identify and evaluate novel anticonvulsant compounds. Finally, identifying specific AEDs that are able to suppress seizures in the BS mutants may shed light on the mechanisms causing the seizure-sensitive defect in these strains. In particular, as some AEDs show specificity to certain human epileptic syndromes, mutants may be identified from the BS mutant collection that might most closely resemble those based on drug profile. Reynolds *et al.* (2003) showed that phenytoin and gabapentin are effective at suppressing BS mutant behavioral phenotypes; carbamazepine, ethosuximide, and vigabactrin were not. Additional studies have shown that valproate, potassium bromide, and carbenoxolone are effective at suppressing BS phenotypes (Kuebler and Tanouye, 2002; Song and Tanouye, 2006; Song *et al.*, 2008; Tan *et al.*, 2004). Taken together, these combined studies indicate that many, but not all, anticonvulsants used to treat human seizure disorders are effective against *Drosophila* seizures.

As an example, the treatment of BS mutants with valproate is described in more detail. Valproate is a wide-spectrum AED used to treat generalized and partial seizures. It is also used increasingly to treat nonepileptic neurological syndromes such as anxiety, schizophrenia, and bipolar disorder (Greenhill and Jones, 2010; Loscher, 2002; Landmark, 2008; White *et al.*, 2007). The broad antiepileptic efficacy of valproate is thought to reflect a combination of neurochemical and neurophysiological mechanisms directed against multiple molecular targets (Loscher, 2002). Valproate has been found to inhibit voltage-gated Na^+ channels and block repetitive firing of action potentials, to block T-type Ca^{2+} channels, and to elevate GABA levels and potentiate GABA responses (Landmark, 2007; White *et al.*, 2007). Injection of valproate into the brain or the heart suppresses seizures in BS (Howlett and Tanouye, 2011; Kuebler and Tanouye, 2002). For example, a 25-mM injection in *sda* mutants increased seizure threshold by about a factor of seven from 9 to 62 V HFS (Howlett and Tanouye, 2011). In contrast, for $para^{bss1}$ mutants valproate effect was more modest, increasing seizure threshold by about a factor of 2.5, from 5 to 13 V HFS. The effect on *eas* was intermediate, increasing seizure threshold to the wild-type range, from 5 to 33 V HFS. In contrast to injected-drug, valproate-feeding methods have been less effective in suppressing BS mutant seizures and show mixed results. For example, in pk^{sple1} homozygous mutant flies, feeding of 10 mM valproate was lethal. However, valproate significantly suppressed seizure-like behaviors in pk^{sple1} heterozygotes (Tao *et al.*, 2011). In $para^{bss1}$ and *eas* mutants, valproate feeding was also ineffective at ameliorating seizure phenotypes (Song *et al.*, 2008). For example, $para^{bss1}$ flies fed valproate (10 mM) for 3 days responded especially poorly. A number of flies (11%) died after bang stimulation. For those that survived, all showed paralysis and actually took significantly longer to recover from paralysis than sucrose-fed control flies (Song *et al.*, 2008). These results on valproate are in contrast to feeding experiments with potassium bromide, which significantly ameliorated seizure phenotypes (Song *et al.*, 2008; Tan *et al.*, 2004).

B. AED Development From the $top1^{JS}$ Seizure Suppressor: Camptothecin and Other Top1 Inhibitors

A major attraction of *Drosophila*-based models is the idea that mutants may be used as a platform to facilitate the development of novel therapeutics, for example, AED development inspired by seizure-suppressor mutants (Song and Tanouye, 2008, 2009). The goal would to define new drug targets, beyond Na^+ channels, Ca^{2+} channels, and GABA-related proteins that are the main targets of current AEDs. Top1 provides an interesting test of this notion as compounds have been identified that act as potent Top1 inhibitors, and these can be compared with seizure suppression by the $top1^{JS}$ mutation. The involvement of *top1* in seizures is surprising because DNA topoisomerases have not previously been associated with seizures or any other electrical excitability neuronal functions. Top1 is an essential nuclear enzyme that acts to relieve the torsional stress that DNA encounters during activities such as replication, transcription, and chromatin condensation (Champoux, 2001). Top1 enzyme binds to DNA and causes a single-strand break. It remains bound in an enzyme–DNA cleavage complex, facilitating DNA unwinding, and then mediating re-ligation.

Top1 inhibitors are compounds such as camptothecin (CPT), a quinoline-based alkaloid, which is a phytochemical derived from the Chinese Happy Tree (*Camptotheca acuminata*). CPT interferes with enzyme function by covalently binding to the top1–DNA complex, thereby blocking re-ligation (Boege *et al.*, 1996; Pommier *et al.*, 1998, 1999). If this occurs during replication, it is thought that there is a cell cycle checkpoint failure leading to apoptosis. In the developing brain, we would expect that Top1 functions mainly during chromosome replication and for apoptosis to be evident. In the adult brain, we might expect less replication and for Top1 function to be most important in neurons with changing transcriptional needs. The mechanism of action for CPT inhibition of Top1 is completely different from that of the $top1^{JS}$ mutation, which causes a 12 times transcript reduction. Nevertheless, CPT has been found to ameliorate BS seizure phenotypes in drug-feeding experiments. For example, $para^{bss1}$-fed CPT recover from paralysis about two-thirds faster than control flies, and tonic–clonic-like activity is almost completely suppressed (Song *et al.*, 2007). Drug treatment also causes a modest, even though, significant increase in seizure threshold with a great decrease in synaptic failure time. Similar results are observed with the Top1 inhibitors kaempferol and apigenin. In addition, Top1 inhibitors are tolerated much better than valproate with considerably less toxicity (Song *et al.*, 2007).

V. Concluding Remarks. Seizure Studies in Flies: Insights From Mutant Analyses, and Comparisons With Human Seizures

Several findings from seizure investigations in flies support the utility of these studies in modeling seizure disorders in humans. Other findings that are possible in flies because of advanced genetic capabilities cannot be replicated in humans, but can provide rich insight into the genetics of seizures and seizure susceptibility. Among these observations are

(a) All individual flies have a seizure threshold indicating a characteristic seizure susceptibility.

(b) Seizure susceptibility can be modulated by mutation and by genetic background in characteristic and predictable ways.
(c) Some mutations that cause seizure sensitivity in flies such as $para^{bss1}$ and pk^{sple} are similar to those responsible for some human epilepsies.
(d) Seizure-like activity in flies spreads through the central nervous system along particularly pathways that depend on functional synaptic connections and recent electrical activity.
(e) Seizure-like activity in flies can be spatially segregated into particular regions of the central nervous system.
(f) Electroconvulsive shock treatment in flies raises the threshold for subsequent seizure-like activity.
(g) Drosophila seizure phenotypes can be ameliorated by the human AEDs valproate, phenytoin, gabapentin, and potassium bromide (KBr).
(h) Na^+ channel mutations are excellent seizure suppressors, consistent with the notion that many AEDs are targeting Na channels.
(i) Several mutations that cause seizure sensitivity or seizure suppression are surprising because their gene products had not previously been implicated in seizures or any other neurological function.

Acknowledgments

This study was supported by awards from the McKnight Foundation and the NIH (NS31231) to M.A.T. We thank the members of the Tanouye laboratory for helpful discussions throughout the project.

References

Alabi, A.A., Bahamonde, M.I., Jung, H.J., Kim, J.I., and Swartz, K.J. (2007). Portability of paddle motif function and pharmacology in voltage sensors. *Nature* **450**, 370–376.

Bassuk, A.G., Wallace, R.H., Buhr, A., Buller, A.R., Afawi, Z., Shimojo, M., Miyata, S., Chen, S., Gonzalez-Alegre, P., and Griesbach, H.L et al., (2008). A homozygous mutation in human PRICKLE1 causes an autosomal-recessive progressive myoclonus epilepsy-ataxia syndrome. *Am. J. Hum. Genet.* **83**, 572–581.

Boege, F., Straub, T., Kehr, A., Boesenberg, C., Christiansen, K., Anderson, A., Jacob, F., and Kohrle, J. (1996). Selected novel flavones inhibit the DNA binding or the DNA religation step of eukaryotic topoisomerase I. *J. Biol. Chem.* **271**, 2262–2270.

Bosmans, F., Martin-Eauclair, M.F., and Swartz, K.J. (2008). Deconstructing voltage sensor function and pharmacology in sodium channels. *Nature* **456**, 202–209.

Carlen, P.L., Frances, S., Zhang, L., Naus, C., Kushnir, M., and Velazquez, J.L. (2000). The role of gap junctions in seizures. *Brain Res. Rev.* **32**, 235–241.

Catterall, W.A., Dib-Hajj, S., Meisler, M.H., and Pietrobon, D. (2008). Inherited neuronal ion channelopathies: new windows on complex neurological diseases. *J. Neurosci.* **28**, 11768–11777.

Champoux, J.J. (2001). DNA topoisomerases: structure, function and mechanism. *Annu. Rev. Biochem.* **70**, 369–413.

Crompton, D., Todman, M., Wilkin, M., Ji, S., and Davies, J. (1995). Essential and neural transcripts from the *Drosophila shaking-B* locus are differentially expressed in the embryonic mesoderm and pupal nervous system. *Dev. Biol.* **170**, 142–158.

Fergestad, T., Bostwick, B., and Ganetzky, B. (2006). Metabolic disruption in Drosophila bang-sensitive seizure mutants. *Genetics* **173**, 1357–1364.

Ganetzky, B. (1984). Genetic studies of membrane excitability in Drosophila: lethal interaction between two temperature-sensitive paralytic mutations. *Genetics* **108**, 897–911.

Ganetzky, B., and Wu, C.F. (1982). Indirect suppression involving behavioral mutants with altered nerve excitability in Drosophila melanogaster. *Genetics* **100**, 597–614.

Glasscock, E., Singhania, A., and Tanouye, M.A. (2005). The mei-p26 gene encodes an RBCC-NHL protein that regulates seizure susceptibility in Drosophila. *Genetics* **170**, 1677–1689.

Glasscock, E., and Tanouye, M.A. (2005). Drosophila couch potato mutants exhibit complex neurological abnormalities including epilepsy phenotypes. *Genetics* **169**, 2137–2149.

Greenhill, S.D., and Jones, R.S.G. (2010). Diverse antiepileptic drugs increase the ratio of background synaptic inhibition to excitation and decrease neuronal excitability in neurons of the rat entorhinal cortex in vitro. *Neuroscience* **167**, 456–474.

Gubb, D., Green, C., Huen, D., Coulson, D., Johnson, G., Tree, D., Collier, S., and Roote, J. (1999). The balance between isoforms of the prickle LIM domain protein is critical for planar polarity in Drosophila imaginal discs. *Genes Dev.* **13**, 2315–2327.

Hekmat-Scafe, D.S., Dang, K.N., and Tanouye, M.A. (2005). Seizure suppression by gain-of-function escargot mutations. *Genetics* **169**, 1477–1493.

Hekmat-Scafe, D.S., Lundy, M.Y., Ranga, R., and Tanouye, M.A. (2006). Mutations in the K^+/Cl^- cotransporter gene kazachoc (kcc) increase seizure susceptibility in Drosophila. *J. Neurosci.* **26**, 8943–8954.

Howlett, I.C., and Tanouye, M.A. (2011). Seizure-sensitivity in Drosophila is ameliorated by dorsal vessel injection of the antiepileptic drug valproate. *J. Neurogenet.* (in press)

Jackson, F.R., Wilson, S.D., Strichartz, G.R., and Hall, L.M. (1984). Two types of mutants affecting voltage-sensitive sodium channels in Drosophila melanogaster. *Nature* **308**, 189–191.

Jacobs, M.P., Fischbach, G.D., Davis, M.R., Dichter, M.A., Dingledine, R., Lowenstein, D.H., Morrell, M.J., Noebels, J.L., Rogawski, M.A., Spencer, S.S., and Theodore, W.H. (2001). Future directions for epilepsy research. *Neurology* **57**, 1536–1542.

Jahromi, S.S., Wentlandt, K., Piran, S., and Carlen, P. (2002). Anticonvulsant actions of gap junctional blockers in an *in vitro* seizure model. *J. Neurophysiol.* **88**, 1893–1902.

Jan, Y.N., and Jan, L.Y. (1978). Genetic dissection of short-term and long-term facilitation at the Drosophila neuromuscular junction. *Proc. Natl. Acad. Sci. U S A* **75**, 515–519.

Kauvar, L.M. (1982). Reduced [3H]-tetrodotoxin binding in the *napts* paralytic mutant of *Drosophila. Mol. Gen. Genet.* **187**, 172–173.

Kernan, M.J., Kuroda, M.I., Kreber, R., Baker, B.S., and Ganetzky, B. (1991). nap^{ts}, a mutation affecting sodium channel activity in Drosophila, is an allele of mle, a regulator of X chromosome transcription. *Cell* **66**, 949–959.

Krishnan, S.N., Frei, E., Swain, G., and Wyman, R.J. (1993). Passover, a gene required for synaptic connectivity in the giant fiber system of Drosophila. *Cell* **73**, 967–977.

Kuebler, D., and Tanouye, M.A. (2000). Modifications of seizure susceptibility in Drosophila. *J. Neurophysiol.* **83**, 998–1009.

Kuebler, D., and Tanouye, M.A. (2002). The anticonvulsant sodium valproate reduces seizure-susceptibility in mutant Drosophila. *Brain Res.* **958**, 36–42.

Kuebler, D., Zhang, H.G., Ren, X., and Tanouye, M.A. (2001). Genetic suppression of seizure susceptibility in *Drosophila. J. Neurophysiol.* **86**, 1211–1225.

Landmark, C.J. (2007). Targets for antiepileptic drugs in the synapse. *Med. Sci. Monit.* **13**, RA1–RA7.

Landmark, C.J. (2008). Antiepileptic drugs in non-epilepsy disorders: relations between mechanisms of action and clinical efficacy. *CNS Drugs* **22**, 27–47.

Lee, J., and Wu, C.F. (2002). Electroconvulsive seizure behavior in *Drosophila*: analysis of the physiological repertoire underlying a stereotyped action pattern in bang-sensitive mutants. *J. Neurosci.* **22**, 11065–11079.

Lee, M.P., Brown, S.D., Chen, A., and Hsieh, T.S. (1993). DNA topoisomerase I is essential in *Drosophila melanogaster. Proc. Natl. Acad. Sci. U S A* **90**, 6656–6660.

Loscher, W. (2002). Basic pharmacology of valproate: a review after 35 years of clinical use for the treatment of epilepsy. *CNS Drugs* **16**, 669–694.

Loughney, K., Kreber, R., and Ganetzky, B. (1989). Molecular analysis of the para locus, a sodium channel gene in Drosophila. *Cell* **58**, 1143–1154.

Mahoney, K., Moore, S.J., Buckley, D., Alam, M., Parfrey, P., Penney, S., Merner, N., Hodgkinson, K., and Young, T.L. (2009). Variable neurologic phenotype in a GEFS+ family with a novel mutation in SCN1A. *Seizure* **18**, 492–497.

Meisler, M.H., and Kearney, J.A. (2005). Sodium channel mutations in epilepsy and other neurological disorders. *J. Clin. Invest.* **115**, 2010–2017.

Mulley, J.C., Scheffer, I.E., Petrou, S., Dibbens, L.M., Berkovic, S.F., and Harkin, L.A. (2005). SCN1A mutations and epilepsy. *Hum. Mutat.* **25**, 535–542.

Noebels, J.L. (2003). The biology of epilepsy genes. *Annu. Rev. Neurosci.* **26**, 599–625.

Parker, L., Padilla, M., Du, Y., Dong, K., and Tanouye, M.A. (2011). Drosophila as a model for epilepsy: bss is a gain-of-function mutation in the Para sodium channel gene that leads to seizures. *Genetics* **187**, 523–534.

Pavlidis, P., Ramaswami, M., and Tanouye, M.A. (1994). The Drosophila easily shocked gene: a mutation in a phospholipid synthetic pathway causes seizure, neuronal failure, and paralysis. *Cell* **79**, 23–33.

Pavlidis, P., and Tanouye, M.A. (1995). Seizures and failures in the giant fiber pathway of Drosophila bang-sensitive paralytic mutants. *J. Neurosci.* **15**, 5810–5819.

Phelan, P., Stebbings, L.A., Baines, R.A., Bacon, J.P., Davies, J.A., and Ford, C. (1998). Drosophila shaking-B protein forms gap junctions in paired Xenopus oocytes. *Nature* **391**, 181–184.

Pommier, Y., Pourquier, P., Fan, Y., and Strumberg, D. (1998). Mechanism of action of eukaryotic DNA topoisomerase I and drugs targeted to the enzyme. *Biochim. Biophys. Acta* **1400**, 83–105.

Pommier, Y., Pourquier, P., Urasaki, Y., Wu, J., and Laco, G.S. (1999). Topoisomerase I inhibitors: selectivity and cellular resistance. *Drug Resist. Update* **2**, 307–318.

Ramaswami, M., and Tanouye, M.A. (1989). Two sodium channel genes in *Drosophila*: implications for channel diversity. *Proc. Natl. Acad. Sci. U S A* **86**, 2079–2082.

Reenan, R.A., Hanrahan, C.J., and Ganetzky, B. (2000). The mle(napts) RNA helicase mutation in Drosophila results in a splicing catastrophe of the para Na^+ channel transcript in a region of RNA editing. *Neuron* **25**, 139–149.

Reid, C.A., Berkovic, S.F., and Petrou, S. (2009). Mechanisms of human inherited epilepsies. *Prog. Neurobiol.* **87**, 41–57.

Reynolds, E.R., Stauffer, E.A., Feeney, L., Rojahn, E., Jacobs, B., and McKeever, C. (2003). Treatment with the antiepileptic drugs phenytoin and gabapentin ameliorates seizure and paralysis of *Drosophila* bang-sensitive mutants. *J. Neurobiol.* **58**, 503–513.

Royden, C.S., Pirrotta, V., and Jan, L.Y. (1987). The *tko* locus, site of a behavioral mutation in *D. melanogaster*, codes for a protein homologous to prokaryotic ribosomal protein S12. *Cell* **51**, 165–173.

Shneker, B.F., and Fountain, N.B. (2003). Epilepsy. *Dis. Mon.* **49**, 426–478.

Siddiqi, O., and Benzer, S. (1976). Neurophysiological defects in temperature-sensitive paralytic mutants of *Drosophila melanogaster*. *Proc. Natl. Acad. Sci. U S A* **73**, 3253–3257.

Song, J., Hu, J., and Tanouye, M.A. (2007). Seizure suppression by *top1* mutations in Drosophila. *J. Neurosci.* **27**, 2927–2937.

Song, J., Parker, L., Hormozi, L., and Tanouye, M.A. (2008). DNA topoisomerase I inhibitors ameliorate seizure-like behaviors and paralysis in a Drosophila model of epilepsy. *Neuroscience* **156**, 722–728.

Song, J., and Tanouye, M.A. (2006). Seizure suppression by *shakB2*, a gap junction mutation in Drosophila. *J. Neurophysiol.* **95**, 627–635.

Song, J., and Tanouye, M.A. (2007). A role for *para* sodium channel gene 3' UTR in the modification of Drosophila seizure susceptibility. *Dev. Neurobiol.* **67**, 1944–1956.

Song, J., and Tanouye, M.A. (2008). From bench to drug: human seizure modeling using Drosophila. *Prog. Neurobiol.* **84**, 182–191.

Song, J., and Tanouye, M.A. (2009). The genetics and molecular biology of seizure-susceptibility in Drosophila. In: Baraban, S. (Ed.), Animal Models of Epilepsy: Methods and Innovation. Springer-Verlag, New York, LLC, pp. 27–44.

Suzuki, D., Grigliatti, T., and Williamson, R. (1971). Temperature-sensitive mutations in *Drosophila melanogaster*, VII. A mutation (*para^{ts}*) causing reversible adult paralysis. *Proc. Natl. Acad. Sci. U S A* **68**, 890–893.

Szente, M., Gajda, Z., Said,,Said Ali, K., and Hermesz, E. (2002). Involvement of electrical coupling in the *in vivo* ictal epileptiform activity induced by 4-aminopyridine in the neocortex. *Neuroscience* **115**, 1067–1078.

Tan, J.S., Lin, F., and Tanouye, M.A. (2004). Potassium bromide, an anticonvulsant, is effective at alleviating seizures in the Drosophila bang-sensitive mutant bang senseless. *Brain Res.* **1020**, 45–52.

Tanouye, M.A., and Wyman, R.J. (1980). Motor outputs of giant nerve fiber in Drosophila. *J. Neurophysiol.* **44**, 405–421.

Tao, H., Manak, J.R., Sowers, L., Mei, X., Kiyonari, H., and Abe, T et al., (2011). Mutations in Prickle orthologs cause seizures in flies, mice, and humans. *Am. J. Hum. Genet.* **88**, 138–149.

Thomas, J.B., and Wyman, R.J. (1984). Mutations altering synaptic connectivity between identified neurons in *Drosophila melanogaster*. *J. Neurosci.* **4**, 530–538.

White, H.S., Smith, M.D., and Wilcox, K.S. (2007). Mechanisms of action of antiepileptic drugs. *Int. Rev. Neurobiol.* **81**, 85–110.

Wu, C.F., and Ganetzky, B. (1980). Genetic alteration of nerve membrane excitability in temperature-sensitive paralytic mutants of Drosophila melanogaster. *Nature* **286**, 814–816.

Wu, C.F., Ganetzky, B., Jan, L.Y., and Jan, Y.N. (1978). A *Drosophila* mutant with a temperature-sensitive block in nerve conduction. *Proc. Natl. Acad. Sci. U S A* **75**, 4047–4051.

Zhang, C.X., Chen, A.D., Gettel, N.J., and Hsieh, T.S. (2000). Essential functions of DNA Topoisomerase I in *Drosophila melanogaster*. *Dev. Biol.* **222**, 27–40.

Zhang, H., Tan, J., Reynolds, E., Kuebler, D., Faulhaber, S., and Tanouye, M.A. (2002). The Drosophila slamdance gene: a mutation in an aminopeptidase can cause seizure, paralysis and neuronal failure. *Genetics* **162**, 1283–1299.

Zhang, Y.Q., Roote, J., Brogna, S., Davis, A.W., Barbash, D.A., Nash, D., and Ashburner, M. (1999). Stress sensitive B encodes an adenine nucleotide translocase in Drosophila melanogaster. *Genetics* **153**, 891–903.

HOMEOSTATIC CONTROL OF NEURAL ACTIVITY: A DROSOPHILA MODEL FOR DRUG TOLERANCE AND DEPENDENCE

Alfredo Ghezzi and Nigel S. Atkinson

Waggoner Center for Alcohol and Addiction Research, Section of Neurobiology, The University of Texas at Austin, 1 University Station C0920, Austin Texas 78712-0248, USA

Drug addiction is a complex condition of compulsive drug use that results in devastating physical and social consequences. *Drosophila melanogaster* has recently emerged as a valuable genetic model for investigating the mechanisms of addiction. Drug tolerance is a measurable endophenotype of addiction that can be easily generated and detected in animal models. The counteradaptive theory for drug dependence postulates that the homeostatic adaptations that produce drug tolerance become counteradaptive after drug clearance, resulting in symptoms of dependence. In flies, a single sedation with ethanol or with an organic solvent anesthetic (benzyl alcohol) induces functional tolerance, an adaptation of the nervous system that reduces the effect of these neural depressants. Here we review the role of the BK channel gene (*slo*) and genes that encode other synaptic proteins in the process of producing functional tolerance. These proteins are predicted to be part of an orchestrated response that involves specific interactions across a highly complex synaptic protein network. The response of the *slo* gene to drug exposure

INTERNATIONAL REVIEW OF NEUROBIOLOGY, VOL. 99
DOI: 10.1016/B978-0-12-387003-2.00002-1

and the consequence of induced *slo* expression fit nicely the tenets of the counteradaptive theory for drug tolerance and dependence. Induction of *slo* expression represents an adaptive process that generates tolerance because it enhances neuronal excitability, which counters the sedative effects of the drugs. After drug clearance, however, the increase in *slo* expression leads to an allostatic withdrawal state that is characterized by an increase in the susceptibility for seizure. Together, these results demonstrate a common origin for development of drug tolerance and withdrawal hyperexcitability in Drosophila.

I. Introduction

Drug addiction is a complex neurobiological condition that is characterized by compulsive and escalating drug use in spite of the social and physical harm that it causes. Although addiction is a direct product of repeated drug use, its development and expression are strongly influenced by a number of genetic, psychosocial, and environmental factors. Historically, addiction has been defined in psychological terms with an emphasis placed on the motivational and behavioral aspects of the symptoms. More recently, it has begun to be defined in physiological terms as adaptations to the effects of a drug (Koob and Bloom, 1988; Littleton and Little, 1994). These adaptations persist following drug clearance and are thought to lead to withdrawal phenotypes. The diagnosis of addiction or alcoholism in humans includes negative changes in social behavior and status, which complicates its modeling in experimental systems. Although the constellation of symptoms that represent the addicted state has not been captured in any model organism, the endophenotypes of addiction do lend themselves to study. An endophenotype of addiction is a drug response that has a clear genetic or physiological basis. It is believed that the molecular mechanisms that generate the endophenotype overlap with those that produce the addicted state. By describing the molecular basis of the endophenotype, we come to understand a portion of the molecular basis of the addicted state. A current challenge for neurobiological research is to identify and understand the neuroadaptive mechanisms that at the molecular, cellular, and systems levels generate the behavioral manifestations of addiction.

A. Tolerance, Dependence, and Withdrawal: A Recipe for Drug Addiction

Two common manifestations of the physiological neuroadaptations to an addictive drug are the endophenotypes of tolerance and physical dependence. Tolerance refers to a diminishing effect of a drug as a consequence of prior drug exposure, while physical dependence is a product of the physical necessity for a drug. Withdrawal refers to the symptoms of dependence that appear during abstinence following drug use (Koob and Bloom, 1988).

The concept of physical drug dependence dates back to the early 1940s when C. K. Himmelsbach (1941) first described the morphine abstinence syndrome. Himmelsbach defined physical drug dependence as "a state in which certain physiological processes have become so conditioned to the effects of a drug that its presence is requisite to the maintenance of homeostasis." In his studies with morphine users, Himmelsbach also noted that a state of dependency could only be diagnosed after removal of the drug in which patients exhibit characteristic withdrawal symptoms. These symptoms, he describes, "possibly reflect the degree to which mechanisms for the maintenance of homeostasis have been affected by the drug." Under this definition, the manifestation of drug dependence is equated with the appearance of withdrawal symptoms.

This homeostatic view of dependence has since expanded into a comprehensive view of addiction and has become a theoretical framework for understanding the relationship between physical dependence and drug tolerance. In the homeostatic and redundancy theory of tolerance and dependence, W. R. Martin (1968) proposed that tolerance and dependence are part of the same phenomena. He postulated that the neuroadaptive mechanisms that counteract an effect of the drug to produce drug tolerance persist even after the drug has been cleared. Once uncovered by the extinction of the drug effects, the counteradaptive changes translate into withdrawal symptoms—the physical manifestation of dependence (Fig. 1).

The psychological drive for drug use is believed to have a direct connection to the physiological processes that underlie tolerance and dependence. The "opponent process theory" of acquired motivation provides a context for understanding the interrelationship between drug tolerance and dependence. This is an independent inception of the counteradaptive theory couched in the syntax of psychology (Solomon and Corbit, 1974). This theory posits that homeostatic processes (the opponent processes) that counter the effects of a drug continue during abstinence to produce a state of dysregulation that can only be appeased by further drug use (Koob and Le Moal, 2006; Littleton, 1998; Solomon, 1980). Thus, the opponent process is responsible for both the tolerance to the pleasurable effects of a drug and the production of withdrawal symptoms that lead to a negative affective state and produce the feed-forward behavioral changes that promote addiction (Koob, 1996).

Animal models are crucial for understanding the neuroadaptive mechanisms responsible for the development of addiction. Unfortunately, the difficulty of measuring affective states in animals makes it impossible to generate complete models of the disease. However, tolerance and withdrawal are measurable

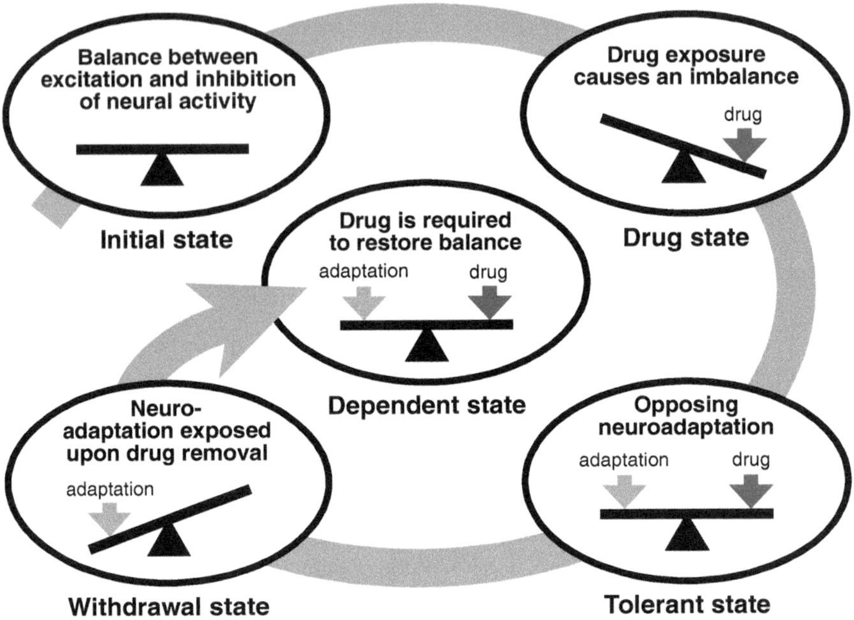

FIG. 1. **Homeostatic counteradaptive model of drug tolerance and dependence.** In the counteradaptive model of drug tolerance and dependence, the carefully controlled balance between excitation and inhibition of neural activity in the brain (Initial state) is significantly altered by exposure to a psychoactive drug, creating a state of activity imbalance (Drug state). In an attempt to restore the balance of neural activity, homeostatic neuroadaptive mechanisms are activated (Tolerant state). After drug clearance, the homeostatic neuroadaptation is exposed, resulting in an opposing unbalanced state (Withdrawal state), leading back to the requirement of the drug to restore balance (Dependent state). Continuing use of the drug leads to further adaptation and an intensified requirement for the drug, resulting in a spiraling-negative feed-forward cascade. Figure adapted from Littleton (1998).

endophenotypes that define both physiological and behavioral aspects of addiction and can be easily quantified in animal models.

B. Drug Tolerance Versus Drug Resistance

In common parlance, tolerance, and resistance are often considered to be synonymous. However, it is important to note that in a pharmacological context, drug tolerance differs significantly from drug resistance. Resistance is the term that describes the innate or baseline magnitude of drug responsivity. Therefore, resistance should be reported only for drug-naive animals. Drug tolerance is synonymous with inducible resistance triggered by prior drug exposure and implies a change from the preexisting state. Sensitization, on the contrary, is the opposite of tolerance. Sensitization describes a drug-induced reduction in resistance to an effect of the drug as a consequence of prior drug exposure. Both tolerance and sensitization are plastic phenomena.

A popular strategy for investigating the genetics of drug abuse is to measure the innate resistance of an organism to the effects of a particular drug. The rationale for this approach originated from the observation that humans with elevated behavioral resistance to ethanol have a fourfold increased probability of future alcoholism (Schuckit, 1994). Resistance is a risk factor for alcoholism because while these individuals are resistant to alcohol intoxication, they are not necessarily resistant to the addictive effects of the drug, and these individuals drink more to achieve the desired degree of intoxication. The mechanism of resistance is believed to arise at least in part from differences in the responsivity of the direct targets of ethanol (Mayfield *et al.*, 2008). Thus, the identification of allelic variants or mutations that enhance resistance should identify medically relevant targets of ethanol. The identification of mutations that decrease resistance is also potentially valuable, with the caveat that mutations that merely reduce vigor might also decrease drug resistance. Tolerance is a drug-induced change that precedes and overlaps the genesis of addiction to many drugs. The changes that produce tolerance are believed to be a subset of those directly responsible for the addicted state. Therefore, a genetic study of tolerance offers a tractable approach for identifying genes that underlie the changes that cause addiction.

In mammals, there are two mechanistically distinct types of tolerance: metabolic tolerance and functional tolerance. Metabolic tolerance, also called pharmacokinetic tolerance, results from increased clearance of a drug and/or reduced drug uptake. These changes reduce the drug concentration or the time of exposure that cells experience. Tolerance caused by a reduction in the cellular response to a given concentration of drug is called functional tolerance or pharmacodynamic tolerance (Julien, 2004). For the nervous system, functional tolerance has been conceptualized in terms of homeostatic adaptations that attempt to restore more-normal neuronal excitability or activity and in terms of changes that resemble those involved in learning and memory (Hyman *et al.*, 2006; Koob and Le Moal, 2006; Young and Goudie, 1995). It is not clear that these events would necessarily involve different mechanisms.

Tolerance is also categorized based on how it was induced. Acute tolerance refers to a decrease in sensitivity to the effect of a drug that develops during the course of a single drug exposure. The other extreme is chronic tolerance, which refers to a decrease in sensitivity that results from repetitive or sustained exposure to a drug. Finally, rapid tolerance describes the tolerance that exists after a single dose of the drug has been metabolically cleared (Young and Goudie, 1995).

C. Modeling Addiction in Drosophila

Evidence gathered from familial studies of drug addiction points toward a significant genetic component associated with the risk of substance abuse (Goldman *et al.*, 2005; Kendler *et al.*, 2003). Identification of the genes behind the risk factors linked to addictive behaviors should uncover the mechanisms underlying the physiological neuroadaptions that ultimately lead to addiction. Because the ability to study genetics in humans is limited, the use of a genetically tractable organism is crucial for dissecting the endophenotypes of addiction. With an extraordinarily malleable genome and with a high degree of gene homology with mammalian genomes, the fruit fly *Drosophila melanogaster* has become a powerful genetic model system for studying addiction. After more than a century of genetic progress, Drosophila sports the most sophisticated genetic toolbox of any metazoan (Bellen *et al.*, 2010; Duffy, 2002).

In spite of the perceived simplicity of the fly's anatomy and physiology, the fly maintains a surprising conservation of genetic architecture with its mammalian counterparts. Most important for the neurobiologist is that most mammalian genes associated with synaptic transmission and neural plasticity have homologs in Drosophila. These include the genes encoding ion channels, synaptic proteins, and the receptors for the major neurotransmitters used in mammals (Littleton and Ganetzky, 2000; Lloyd *et al.*, 2000). At the behavioral level, Drosophila also show a remarkable degree of homology with higher organisms. Flies not only display basic behaviors—such as foraging, courtship, sleeping, and fighting—that are similar to those of mammals (Hendricks *et al.*, 2000; Chen *et al.*, 2002; Sokolowski, 1980; Spieth, 1974), but they can also remember, learn from experience, and modify their behavior in a goal-oriented manner (Pick and Strauss, 2005; Quinn *et al.*, 1974). For more than two decades, the fruit fly has been the subject of a large number of studies modeling behavioral responses to a variety of potentially addictive psychoactive drugs. As a model organism, genetic studies with Drosophila have made substantial contributions toward the understanding of responses to psychostimulants, organic solvent anesthetics, and ethanol.

As in mammals, in flies psychostimulant drugs such as cocaine, nicotine, caffeine, and amphetamines suppress sleep and promote hyperactive behavior characterized by increased arousal and increased locomotion. Acute exposure to volatilized cocaine or nicotine produce clear stereotypic behaviors such as increased grooming, aberrant walking patterns, and bursts of hyperactivity that are characterized by fast uncontrolled movements. In mammals, the action of these drugs on dopaminergic signaling is central to their effects on arousal and reinforcement. In flies, these drugs enhance dopaminergic signaling, and this signaling has been shown to directly contribute to the drug-induced behavior (Atkinson, 2009; Rothenfluh and Heberlein, 2002; Wolf and Heberlein, 2003).

In flies, volatile organic solvent anesthetics, such as halothane and isoflurane, induce a state of sedation characterized by complete immobility and the loss of the ability to maintain postural control or respond to stimuli (Allada and Nash, 1993; Krishnan and Nash, 1990). Less-volatile organic solvents, such as ethanol and benzyl alcohol, have been shown to induce a biphasic response characterized by a brief initial increase in locomotor activity followed by sedation (Cowmeadow *et al.*, 2005; Ghezzi *et al.*, 2004; Moore *et al.*, 1998; Parr *et al.*, 2001; Scholz *et al.*, 2000). Similarly, in humans, organic solvents elicit a wide range of behavioral and physiological responses that range from hyperactivity to profound sedation. Low or moderate exposure to volatile solvents can have excitatory effects, whereas higher levels of exposure produce sedation and anesthesia (Flanagan and Ives, 1994).

Volatile organic solvent anesthetics, inhalants, and ethanol are intriguing in that they are "dirty drugs" that alter the properties of many different molecules that participate in electrical signaling (Harris, 1999; Kopp Lugli *et al.*, 2009). There is clear evidence that organic solvents have cellular and behavioral consequences related to those produced by more-conventional drugs of abuse. These consequences include modulating specific neurotransmitter receptor subtypes and triggering dopamine release in the nucleus accumbens to generate behavioral reinforcement (Lubman *et al.*, 2008). In the rat hippocampus, volatile anesthetics have been shown to modulate both excitatory and inhibitory synaptic transmission culminating in a depression of synaptic transmission (Harris *et al.*, 1995; Wakasugi *et al.*, 1999). Gamma-aminobutyric acid (GABA) receptors are the best-established candidate for the drug target that mediates the sedative action of ethanol and volatile anesthetics in mammals (Beckstead *et al.*, 2000; Harris, 1999). In flies, the inhibitory $GABA_B$ receptor has been implicated in the sedative effects of alcohol (Dzitoyeva *et al.*, 2003), whereas the initial stimulating effects of ethanol have been linked to dopaminergic signaling (Bainton *et al.*, 2000; Kong *et al.*, 2010). The BK Ca^{2+}-activated K^+ channel has also been shown to be involved in organic solvent/ethanol responses in Caenorhabditis, Drosophila, and mammals (discussed below).

The conservation of drug-induced behaviors and targets from Drosophila to mammals is indeed encouraging. In light of these similarities, it is reasonable to expect that Drosophila genetics can be used to identify the genes that underlie previously undescribed drug responses and neuroadaptations that contribute to addiction. Ethanol tolerance was first described in Drosophila by Scholz and colleagues (2000). These authors show that adult flies acquire functional tolerance but not metabolic tolerance. Since then, several laboratories have used flies for investigating the mechanisms underlying this adaptation (reviewed in Atkinson, 2009 and Rodan and Rothenfluh, 2010). A focus of this review will be the role of the BK-type Ca^{2+}-activated K^+ channel and other synaptic proteins in the production of anesthetic and ethanol tolerance and dependence in *D. melanogaster*.

II. Behavioral Analysis of Rapid Drug Tolerance in Drosophila

In flies, inhalation of benzyl alcohol vapor or ethanol vapor results in similar behavioral responses. At low, nonsedating doses, both drugs induce a gradual increase in locomotor activity accompanied by erratic movements. At higher doses that culminate in sedation, the hyperactive phase is observed only in the initial minutes of exposure and decays gradually, transitioning into a phase marked by low-mobility levels, ataxia, and the loss of postural control. Eventually, flies will completely sedate, showing no signs of movement and remaining immobile on their backs. After the flies are moved to a drug-free vial, recovery from sedation is also gradual and can span between ~5 and 60 min, depending on the dose. While still on their backs, flies first begin to move their legs in a twitchy, spasmodic manner. These spasms last noticeably longer in ethanol treated flies. Subsequently, flies regain postural control and then pass through a period of standing immobility followed by ever more successful attempts to walk or climb (Cowmeadow *et al.*, 2005; Ghezzi *et al.*, 2004; Rothenfluh and Heberlein, 2002; Singh and Heberlein, 2000).

A. Benzyl Alcohol Tolerance

Our introduction to benzyl alcohol came while we were examining the *slo* gene for homeostatic responses to the insecticide imidacloprid. Imidacloprid is a partial nicotinic-acetylcholine receptor agonist (Matsuda *et al.*, 1998). A sample of imidacloprid, dissolved in an undisclosed solvent, was provided to us by a drug company. We thought that the *slo* gene responded to this mixture in an important way, but when we requested the name of the solvent we were denied by a rather protective company representative. A brief analysis by mass spectrometry and nuclear magnetic resonance revealed that the solvent was benzyl alcohol. To our surprise, it was this carrier solvent that produced most of the effects that originally intrigued us (Bohm, 2000). Benzyl alcohol proved to be a near-ideal organic solvent for sedating flies. We continued to study it to help describe the neuronal responses to organic solvent sedation.

Intentional exposure to benzyl alcohol in humans arises primarily from its use as an injectable local anesthetic (Fleisher and Ludwig, 2010) and because it is the first breakdown product of toluene (IARC Working Group on the Evaluation of Carcinogenic Risks to Humans, 1999). Toluene is both an abused inhalant and a solvent whose commercial use is widespread (Dinwiddie, 1994). The metabolism of toluene probably represents the primary source of human exposure to benzyl alcohol.

Benzyl alcohol is an organic solvent anesthetic that is easy to administer and is extremely well tolerated by flies. Analysis of behavior showed that low doses of benzyl alcohol acted as a stimulant, whereas large doses produced sedation. Our standard animal for behavioral analysis is a female fly, ages 3–5 days old, with a Canton S wild-type genetic background. To sedate flies with benzyl alcohol, we coat vials with benzyl alcohol in acetone (usually 0.4% benzyl alcohol). Because of the difference in vapor pressures of the solvents, the acetone rapidly evaporates and leaves behind a fine coat of benzyl alcohol. Flies placed in the vial quickly succumb to the intoxicating effects of benzyl alcohol and are sedated within 10 to 15 min. Recently, we have used the vapor from small cups of benzyl alcohol positioned below the vials to sedate the flies. A vapor-permeable KimWipe barrier is used to keep the flies from falling into the solvent. The emanating vapor sedates the flies with a similar time course and produces the same responses as the original method (Ghezzi *et al.*, 2004, 2010).

Because flies are negatively geotactic, they spend much of their time climbing the walls of their vial. During recovery from benzyl alcohol sedation, flies return to this pastime shortly after the return of the ability to stand. The return of wall climbing is a reliable indicator that flies have recovered from sedation with benzyl alcohol. This behavior lends itself to computer monitoring. We use a simple image subtraction program paired with consumer digital cameras to record the recovery from benzyl alcohol sedation (Ramazani *et al.*, 2007).

In response to a single sedation, flies acquire functional tolerance that can be easily scored in an assay that measures the period of sedation. For tolerance experiments, flies are placed in the vials until sedated, moved to a benzyl alcohol-free vial to recover, and then incubated with food for 24 h. Control animals are mock sedated in vials that were handled in the same manner but with the omission of the benzyl alcohol. The next day both groups are sedated simultaneously with benzyl alcohol vapor, switched to a fresh-air environment, and the duration of sedation is monitored. Flies recover more rapidly from their second benzyl alcohol sedation than from their first sedation; that is, they acquire tolerance to the drug. Examples of this behavior are shown in Figs. 2 and 3.

B. Ethanol Tolerance and Cross-Tolerance

Flies also acquire tolerance to ethanol sedation. The ethanol tolerance assay is also a two-day protocol. On the first day, a population of 3- to 5-day-old females is divided into two groups. One group is sedated with a saturated ethanol vapor stream produced by ethanol bubblers, and the other group is mock sedated. After the flies recover from sedation, both groups are returned to food. Twenty-four hours after the first sedation, both groups are sedated, moved to a fresh-air environment for recovery, and the rate of recovery is recorded. A major difference between the ethanol and benzyl alcohol responses is that flies delay climbing after ethanol sedation, even though they are obviously performing other types of normal

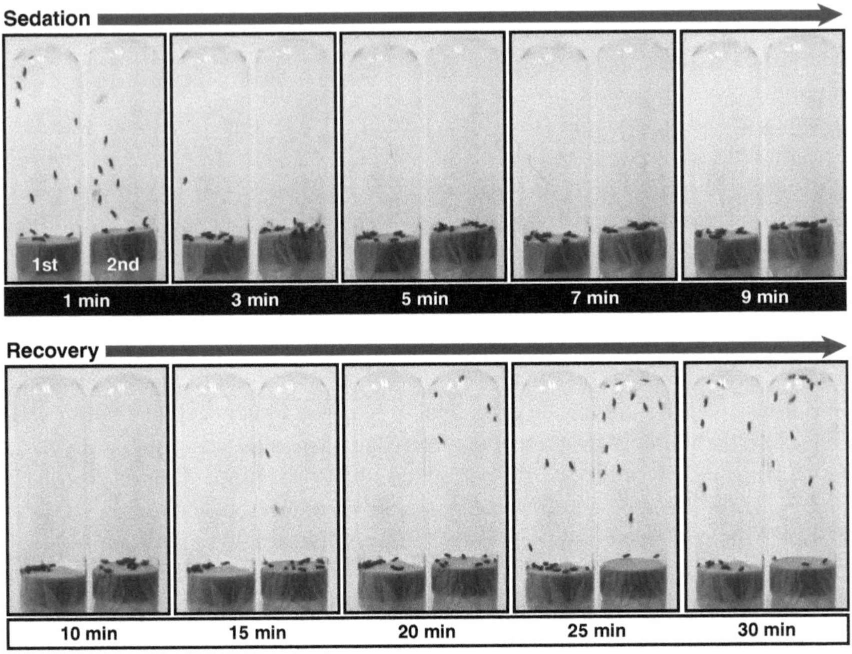

FIG. 2. **Benzyl alcohol tolerance assay.** Shown are sequential pictures of two vials of flies taken at different time points of sedation and recovery from 0.4% benzyl alcohol exposure. The vial on the left in every picture contains naive flies, which have never been treated before; this is their first exposure (1st). The vial on the right contains flies that were previously sedated (24 h earlier) with a similar dose of benzyl alcohol; this is their second exposure (2nd). The time point at which each picture was taken is indicated under each picture in minutes after start of the treatment. Time points from 1 to 9 min (white text over black) are in the presence of the solvent. The solvent has been removed at 10 min. Time points from 10 to 30 min (black text over white) are during the recovery in a solvent-free tube. Although both groups knock down simultaneously, flies recovering from the 2nd sedation recover negative geotaxis at earlier time points than flies recovering from their 1st sedation do.

behavior. The population returns to climbing rather asynchronously. Therefore, for the ethanol tolerance assays we visually monitor the return of postural control (standing) as a sign of recovery from sedation. Flies recover from a second ethanol sedation (delivered 24 h after the first sedation) more rapidly than from their first ethanol sedation (Cowmeadow *et al.*, 2005).

The molecular mechanism that produces ethanol tolerance appears to be completely or substantially the same as the mechanism that produces benzyl alcohol tolerance. This conclusion is based on the fact that both drugs induce mutual cross-tolerance and that the capacity to acquire tolerance to either drug is blocked by the same genetic mutations (Al-Hasan *et al.*, 2011; Cowmeadow *et al.*, 2005; Ghezzi *et al.*, 2004; Krishnan *et al.*, in press).

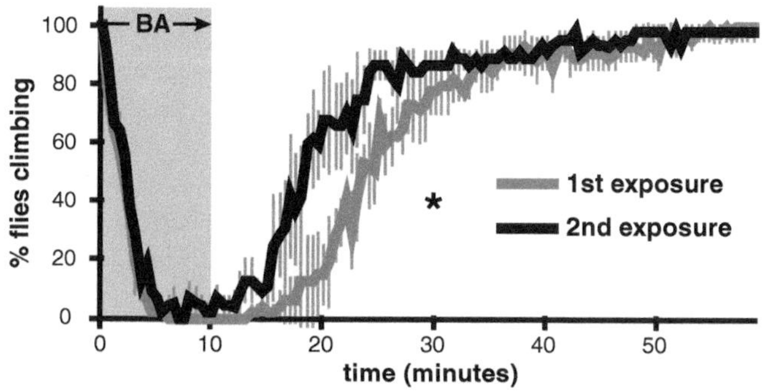

FIG. 3. **Quantification of benzyl alcohol tolerance.** Shown are knockdown and recovery curves of wild-type flies after one (1st exposure) or two (2nd exposure) exposures to 0.4% benzyl alcohol. Values are plotted as percentage of flies climbing during sedation with benzyl alcohol (gray background) and during recovery from sedation (white background). Time between exposures is 24 h. Significant difference is determined by the log-rank test (error bars are SEM for each data point).

C. Statistical Analysis of Tolerance

When describing these tolerance assays, we have not mentioned the magnitude of tolerance but only whether it occurs. When we began work on this topic, we rationalized that the magnitude of tolerance would show day-to-day variations caused by events that we could not control or could not imagine. For example, Drosophila show changes in behavior in response to barometric pressure (a factor that also affects solvent vaporization) (Ankney, 1984). Variability of this kind can be avoided if we treat tolerance as a binary phenomenon that either occurs or does not. Therefore, we compare a population's recovery from first sedation to its recovery from second sedation and ask only if the rates differ significantly from one another. Certainly, tolerance in a wild-type fly must be a quantitative trait that is produced by the concerted contributions of a number of gene products. Viewing tolerance as a binary phenomenon means that one can identify only mutations that have such a strong effect on the production of tolerance that the separation of the first and second sedation recovery curves is no longer statistically significant. Although this simplification makes the assay very robust, it is limiting in that it does not detect the effect of mutations that make smaller incremental contributions to the production of tolerance. Fortunately, other investigators have screened for mutations that reduce the magnitude of tolerance, and therefore such genes will not be overlooked.

To determine whether the first-sedation and second-sedation groups recover at a significantly different rate, we use the log-rank test for equality of survival.

Survival statistics are ideal for evaluating time to a specific event (Hosmer *et al.*, 2002, 2008). Although this statistical test is most commonly applied to test for changes in the rate of deaths, disease, marriage, divorce, or failure of a mechanical part, it works well for determining whether the two groups recover from sedation at significantly different rates. Using this statistical test for tolerance, we observe that benzyl alcohol tolerance and ethanol tolerance each persist for about 10 days (Cowmeadow *et al.*, 2005; Ghezzi and Atkinson, 2011). Thus, functional tolerance in flies is a long-term neuroadaptive response.

III. BK Channels and the Production of Rapid Tolerance

The mechanisms of action of ethanol and volatile anesthetics on the nervous system are well characterized. In both mammals and Drosophila, substantial evidence suggests that these drugs suppress neural function at least in part through interactions with voltage- and ligand-gated ion channels (Covarrubias and Rubin, 1993; Evans and Balster, 1991; Harris, 1999; Leibovitch *et al.*, 1995; Nash, 2002). In general, organic solvent sedation is mediated by the inhibition of excitatory ion channels and the potentiation of inhibitory ion channels to produce a generalized depression of the nervous system. From a homeostatic perspective, it is expected that the adaptive mechanisms that lead to tolerance involve changes in neural activity that strive to restore normal neural excitability. In neurons, electrical excitability is an emergent property arising from the activity of and electrical interaction between an assortment of ion channel proteins. Therefore, likely regulators of a homeostatic adaptation are the ion channel proteins themselves.

A. Homeostatic Modulation of BK Channel Gene Expression Mediates Drug Tolerance

The *slo* gene encodes BK-type Ca^{2+}-activated K^{+} channels and is widely expressed throughout excitable tissues. BK channel activity plays a major role in shaping neuronal excitability, regulating synaptic activity, modulating smooth muscle tone, and controlling neuroendocrine secretion (Gribkoff *et al.*, 2001). Our data suggested that *slo* expression was homeostatically regulated in response to drug exposure and that rapid functional tolerance to benzyl alcohol or ethanol sedation was dependent on *slo* expression in the nervous system. We observed that a single benzyl alcohol or ethanol sedation caused a slight increase in neuronal *slo* mRNA abundance concomitant with the appearance of tolerance. Conversely, low doses of benzyl alcohol that acted as a stimulant caused a reduction in neuronal *slo* mRNA abundance and produced behavioral sensitization. Animals homozygous for slo^4 are healthy but do not acquire either benzyl alcohol tolerance or ethanol tolerance in our duration-of-sedation assays. The slo^4 mutant allele had been shown to be a null mutation by behavioral, molecular, and electrophysiological analysis. In addition, we used a *slo* allele ($ash2^{18}$) that eliminates only *slo* expression from the nervous system to show that *slo*-dependent tolerance is a neural response. Finally, tolerance could be phenocopied by transgenic induction of *slo* expression (Cowmeadow *et al.*, 2005, 2006; Ghezzi *et al.*, 2004).

The observation that increased *slo* expression countered the sedating properties of two drugs led to the hypothesis that increased BK channel activity must act as a neural excitant. At the time that this hypothesis was proposed (Ghezzi *et al.*, 2004), the dogmatic belief that K^+ channel activity must always be inhibitory was still widely held, in spite of evidence accumulating to the contrary. How can increased expression of BK channels enhance neural excitability? In some cells, BK channel activity has been shown to shorten the refractory period, thereby increasing neural excitability in the form of an increased capacity for repetitive firing. The refractory period is the period that must elapse before the neuron can fire again. The refractory period is shortened because the high-conductance BK channels cause the rapid repolarization of the synapse, thereby preventing the activation of other classes of K^+ channels that would otherwise occur. If these secondary K^+ channels are activated they produce a long-lasting hyperpolarization that reduces the maximum firing rate (Brenner *et al.*, 2005; Lovell and McCobb, 2001; Pattillo *et al.*, 2001; Van Goor *et al.*, 2001).

To determine whether increased *slo* expression reduced the neural refractory rate, we examined the response of the giant fiber visual escape response pathway to benzyl alcohol sedation and to increased *slo* expression. In the giant fiber preparation, two stimulating tungsten electrodes are placed in the eyes, and a recording electrode is placed in an indirect flight muscle. A depolarizing stimulus is delivered to the eyes. From there the stimulus propagates through the brain and activates the giant fiber. The giant fiber axons extend from the brain to the thoracic ganglion where they are electrically coupled to an interneuron that activates a motor neuron that terminates on a thoracic flight muscle. At high stimulation potentials, the giant fiber is directly stimulated (in essence, the entire brain is depolarized by the stimulus), and the flight muscles report the firing pattern of the giant fiber pathway. This is called the short latency visual escape response pathway. A total of three neurons are involved in transmitting the signal to the flight muscles. Recordings from flight muscle report the firing pattern of this three-neuron circuit (Tanouye and Wyman, 1980).

In the fly giant fiber preparation, medical inhalation anesthetics had been shown to produce a dose-dependent increase in the neural refractory period that culminated in a failure of the neuron to respond to stimulation (Lin and Nash, 1996). Once the drug is removed, the firing capacity gradually recovers, in parallel with behavioral recovery. It was an exciting moment when we realized that the proposed reduction of the refractory period by increased BK channel activity could directly counter an effect of the anesthetic.

To determine whether *slo* induction enhances the neural capacity for repetitive firing, we compared the following-frequency profile of the giant fiber pathway before and after induction of *slo* gene expression. In one experiment, a single 15–20 min benzyl alcohol sedation was used to induce *slo* expression. When tested 24 h after sedation, the capacity of the giant fiber to follow high-frequency stimulation in the absence of the drug was significantly enhanced (reduced refractory period) compared with the capacity of the untreated control, indicative of an adaptive process that overlaps with the development of tolerance (Fig. 4). As is the case with tolerance, the enhancement in following frequency is *slo* dependent, in that mutations that block only neural expression of *slo* prevent it. Most important, the increase in following frequency can be produced in the absence of drugs by activating an inducible *slo* transgene. These results show that an increase in *slo* neural expression is both necessary and sufficient for the increase in the capacity for repetitive firing (Ghezzi *et al.*, 2010).

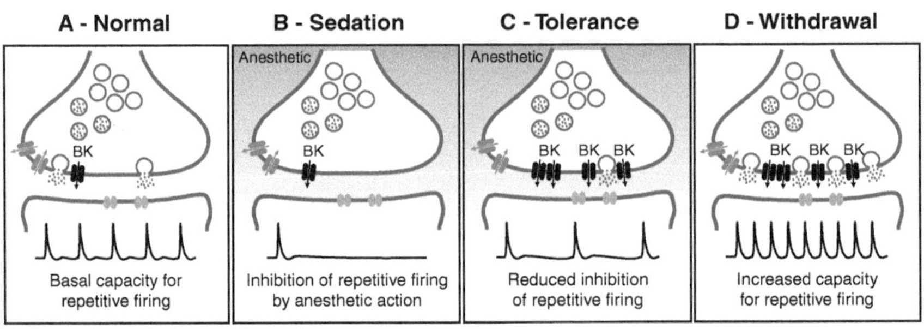

FIG. 4. **Model for how homeostatic regulation of BK channels contributes to both drug tolerance and withdrawal.** (A) In a naive, no drug state, neurons of the giant fiber pathway in Drosophila exhibit a basal capacity for repetitive firing when evoked by high-frequency stimulation (Normal). (B) The capacity for repetitive firing is significantly inhibited by exposure to sedative drugs (Sedation), leading to the overall depression of neural activity characteristic of sedation. (C) Drug exposure induces expression of presynaptic BK channels as part of a homeostatic response to sedation. Increased BK channel activity enhances the capacity for repetitive firing, leading to a reduced effect of the anesthetic on neural firing during a subsequent exposure (Tolerance). (D) After drug clearance, however, another effect of increased BK channel expression is unmasked, resulting in an enhanced neural excitability in the form of increased basal firing capacity and an increased susceptibility for seizures (Withdrawal). The electrophysiological traces shown here are schematic representations of hypothetical data, not real traces.

B. Modulation of the BK Channel Gene Produces a Withdrawal Symptom

The stereotypical image of a person in alcohol withdrawal is that of a person who has shaking hands, cannot sleep, suffers bouts of delirium, and may have spontaneous seizures. At least three of these symptoms are probably indicative of an increased baseline of neural excitability. In humans, the symptoms of alcohol withdrawal were experimentally documented with prisoners who were allowed to drink large amounts alcohol (averaging 256–489 mL of 95% ethanol per day) for long periods (weeks to months) (Isbell *et al.*, 1955). The connection between alcohol withdrawal and alcohol withdrawal seizures has been well reviewed in Porter *et al.* (1990) and Rogawski (2005). The counteradaptive theory of tolerance/dependence nicely account for these symptoms in that the adaptation (presumed increased excitability) that counters an effect of a neural depressant is uncovered during abstinence and could lead to these symptoms.

As described above, increased BK channel activity enhances the capacity for repetitive neural firing—a hallmark of seizure. Recent studies have tightly linked enhanced BK channel activity with specific forms of epilepsy. A human mutation in the pore-forming subunit of the BK channel that increases channel open probability is the cause of generalized epilepsy and paroxysmal dyskinesia (Du *et al.*, 2005). Increased activity of BK channels caused by a mutation in the β4 BK channel auxiliary subunit increases high-frequency firing and leads to temporal lobe seizures (Brenner *et al.*, 2005).

Flies have been used as a model organism for studying epilepsy for many years (see Chapter 1). We appropriated an electrophysiological method used to study epileptic-like seizures in flies to determine whether tolerance to benzyl alcohol sedation came at the cost of a reduction in the seizure threshold. Stimulation of the giant fiber pathway at high frequency (200 Hz) is electroconvulsive and induces a seizure with a characteristic pattern. The magnitude of the voltage required to elicit seizure activity represents an accurate measure of seizure susceptibility (Kuebler and Tanouye, 2000). We demonstrated that flies that have become tolerant to benzyl alcohol by prior exposure and have increased *slo* expression, displayed seizure activity at lower voltages than control flies did. Similarly, artificial induction of *slo* from a transgene also reduced the threshold voltage for eliciting seizures (Ghezzi *et al.*, 2010). Finally Kuebler *et al.* (2001) have demonstrated that null mutations in *slo* had the opposite effect and increased the seizure threshold.

C. How Does the *slo* Gene Sense Drug Sedation?

We wanted to understand how the *slo* gene sensed drug sedation. To do so requires an understanding of how the *slo* gene is regulated. The *slo* transcriptional control region is large and complex. The 7-kb transcriptional control region contains at least five tissue-specific transcription start sites (core promoters). This control region includes the first intron after the translation start site, which has been shown to contain important regulatory elements (Brenner and Atkinson, 1996). The two most 5′ promoters appear to be responsible for all neural expression, the next two promoters have been shown to be active in epithelial cells of the digestive system, and the 3′-most promoter is responsible for muscle and tracheal cell expression (Fig. 5A). Each transcription start site begins expression with a unique 5′ UTR that will be spliced to a common exon (Bohm *et al.*, 2000; Brenner and Atkinson, 1997; Brenner *et al.*, 1996; Chang *et al.*, 2000).

One of the first steps in gene activation is generally held to be the alteration of chromatin structure. Transcription factor-induced chemical modification of histones can influence the accessibility of the underlying DNA and alter the affinity of the transcription preinitiation complex for the chromatin (Berger, 2007). Most activiting transcription factors recruit histone acetyl transferases to the transcriptional control region to acetylate a variety of histones. Histone acetylation is tightly correlated with making the underlying DNA more accessible and with transcription activation. To identify *slo* DNA enhancer elements that respond to drug sedation, we elected to map drug-induced histone acetylation footprints left behind by transcription factors. We began our survey by monitoring histone H4 acetylation. In general, histone H4 acetylation relaxes the association between nucleosomes and DNA, prevents the cross-linking of consecutive histones, and is correlated with increased gene expression (Allis *et al.*, 2007; Lee *et al.*, 1993).

An antibody that recognizes all forms of histone H4 lysine acetylation (K5, K8, K12, K16) was used in the chromatin immunoprecipitation (ChIP) assay to generate a profile of histone H4 acetylation across the *slo* transcriptional control region at different times following sedation. The H4 acetylation spikes were mapped across the transcriptional control region and correlated with the tissue-specific transcription start sites and various evolutionarily conserved elements (Fig. 5B). Four hours after sedation, an acetylation spike appeared centered over conserved element 55b. This spike preceded detectable gene induction that was first detectable at 6 h after sedation. At 6 h after sedation, the region encompassing both neural promoters was hyperacetylated. This broad acetylation peak is probably a side effect of increased transcription from the two neural promoters. At 24 h after sedation, *slo* expression remained elevated above baseline but appeared to be in decline. At this time, only the histones localized over conserved element 6b (~60 n) remained hyperacetylated. Finally, at 48 h after sedation *slo* gene expression and the acetylation state of the transcription control region had returned to baseline (Fig. 5B) (Wang *et al.*, 2007).

Histone acetylation is the product of transcription factor activity. It was suspected that the CREB (cAMP response element-binding) transcription factor contributed to this process because CREB is known to recruit a histone acetylase and because the location of the first acetylation spike, the 55b element, is flanked

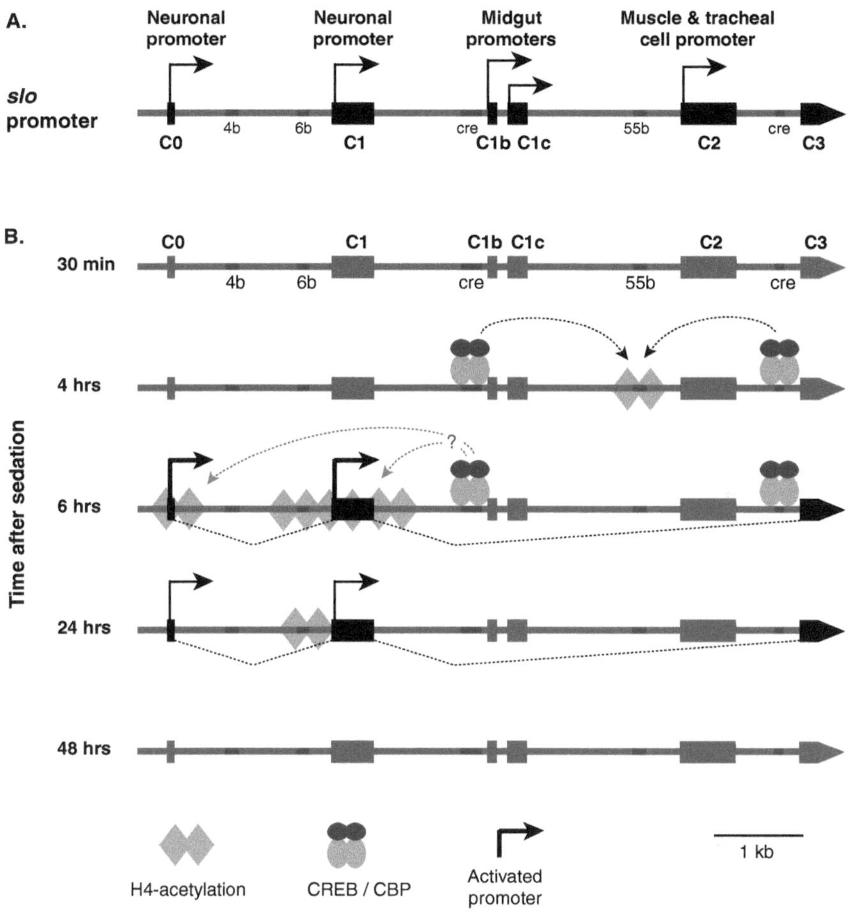

FIG. 5. **Drug-induced chromatin remodeling at the *slo* transcriptional control region.** (A) Shown is the 7 kb transcriptional control region of the *slo* gene. This gene has at least five alternative tissue-specific promoters: two neuronal promoters (C0 and C1), two midgut promoters (C1b and C1c), and one muscle cell and tracheal cell-specific promoter (C2). In addition, this region contains several DNA elements that are highly conserved across different Drosophila species (4b, 6b, 55b) and two CREB response elements (cre). (B) Dynamic histone H4 acetylation changes across the *slo* transcriptional control region after benzyl alcohol sedation. No change in acetylation is detected 30 min after sedation. At 4 h, an increase in acetylation is detected with a peak centered over the 55b element. This event is dependent on binding of phospho-CREB at the two CRE sites. After 6 h, the acetylation peak relocates to the neural promoter area with a small peak around C0 and a broad peak around C1 and mRNA expression from these promoters become evident. After 24 h, the histone acetylation peak becomes focused at the 6b element, and mRNA expression decays. By 48 h, histone acetylation and mRNA expression return to nonsedated control level.

by two CRE consensus sites. Furthermore, the 55b element carried a motif that differed in one base from a consensus CRE site. Using antibodies specific for the CREB DNA-binding domain and for phospho-CREB, we showed that benzyl alcohol sedation increased phospho-CREB occupancy at all three sites: the two CRE consensus sites and CRE-like element in 55b. In Drosophila, there are two CREB genes that are idiosyncratically most often referred to as *dCrebA* and *dCREB2* (also known as *CrebB-17A* or *CREB-B*) (Smolik *et al.*, 1992; Usui *et al.*, 1993). Prior benzyl alcohol sedation activated expression from *dCrebA* and altered the splicing of *dCREB2* transcripts to reduce expression of a repressor isoform. These changes were predicted to induce transcriptional activation by CREB. A CRE-luciferase reporter gene confirmed that benzyl alcohol sedation activated the CREB-signaling pathway. Further implicating CREB was the demonstration that expression of an inducible dominant-negative *dCREB2b* transgene blocked histone acetylation across the *slo* promoter region, blocked *slo* induction, and blocked the development of tolerance. The *dCREB2*^{S182} hypomorphic mutation was also shown to prevent *slo* induction and to block the acquisition of tolerance (Wang *et al.*, 2007, 2009). It has not yet been determined whether *dCrebA* mutations also disrupt the capacity to acquire tolerance.

D. Role of BK Channels in Mammalian Responses to Ethanol

The rat hypothalamic-neurohypophysial model system has also been used to study the role of BK channels in alcohol tolerance. The magnocellular neurons (MCNs) of this system release arginine-vasopressin and oxytocin, triggered by the influx of Ca^{2+} through voltage-gated Ca^{2+} channels and effectively terminated by BK channel activity. Acute ethanol exposure decreases the open probability of L-type Ca^{2+} channels and enhances the open probability of the BK channels of the neurohypophysial terminals, which suppresses the firing of these cells and the resultant release of the hormones. Chronic ethanol exposure produces tolerance that is manifested as a reduction in the capacity of ethanol to inhibit hormone release. Tolerance is the product of reduction in sensitivity of these channels to ethanol (Knott *et al.*, 2002). Ethanol-mediated regulation of BK channel activity has been extensively studied in this system. At pharmacologically meaningful concentrations, ethanol enhances the activity of BK channels in neurohypophysial terminals within minutes by enhancing the open probability of BK channels without affecting ion selectivity or unitary conductance (Pietrzykowski *et al.*, 2004). Additionally, BK channel variants from human brain continue to show potentiation by ethanol when incorporated into planar lipid bilayers lacking other cellular components (Crowley *et al.*, 2003). This demonstrates that ethanol potentiation of channel activity is an intrinsic response of at least some neuronally expressed BK channels.

In the hypothalamic-neurohypophysial model system there is strong evidence that BK channel activity is modulated by ethanol exposure in ways that could underlie chronic tolerance. Twenty-four hours of ethanol exposure results in internalization of some preexisting BK channels in the nerve terminal and a declustering of those that remain (Pietrzykowski *et al.*, 2004). The reorganization of channels is coupled with a change in gene expression that produces replacement channels that are refractory to ethanol. Within minutes after ethanol exposure, expression of the miRNA miR-9 is induced, causing the destruction of *slo* mRNA splice variants encoding ethanol-sensitive BK channels. Splice variants that encode more ethanol-resistant BK channels persist to produce replacement channels. These changes produce a reduction in BK channel current density and ethanol sensitivity. The regulation of BK channel expression by miR-9 is not limited to this model system but has also been observed in medium spiny neurons (Pietrzykowski *et al.*, 2008). These changes in BK channel activity, expression, and localization nicely correlate with cellular tolerance. Although other gene products may make important contributions, it appears that the BK channels play a major role in chronic tolerance.

Acute tolerance has also been physiologically demonstrated for BK channels in the hypothalamic-neurohypophysial neurons, although a cellular correlate of the changed channel activity has not been reported. Twelve minutes after exposure to ethanol, the capacity of BK channels to be potentiated by ethanol is dramatically reduced (Pietrzykowski *et al.*, 2004). This fits the definition of acute tolerance. This reduction may occur through the dephosphoryaltion of a single CaMKII site in the α subunit of the channel. This was demonstrated by the observation that the progressive phosphorylation of bovine BK channel subunits by CaMKII converts them from channels that can be potentiated by ethanol into channels that are inhibited by ethanol (Liu *et al.*, 2006). Acute molecular tolerance to the effects of ethanol has also been shown to be influenced by accessory channel subunits that associate with the pore-forming subunit. In mice, channels that include the $\beta 4$ subunit do not show acute tolerance to ethanol exposure (Martin *et al.*, 2008).

In this rat model system, only posttranscriptional events have been documented; changes in transcription have not been observed. It is possible that transcriptional regulation does not play a role in the mammalian system and that posttranscriptional regulation does not play a role in the Drosophila system. However, this dichotomy is more likely to be a product of the specific properties or experimental advantages of each model system. Important regulatory responses tend to be controlled at many levels—as exemplified in the MCN system in which ethanol regulation of the channel activity occurs by phosphorylation, miRNA degradation of specific splice variants, channel internalization, and declustering.

IV. The Synaptic Connection

The efficient control of synaptic vesicle availability is crucial for the regulation of synaptic excitability. In the synaptic vesicle cycle, newly formed or recycled vesicles are filled with neurotransmitters and stored in a reserve pool as clusters that reversibly attach to the actin cytoskeleton through Synapsins. Upon phosphorylation of Synapsins, the filled vesicles move from the reserve pool to a readily accessible pool where they are available for docking at the active zone. The vesicles in the readily releasable pool fuse with the presynaptic cell membrane in a Ca^{2+}-dependent process to release neurotransmitter into the intersynaptic space. Released neurotransmitters activate ligand-gated receptor ion channels at the postsynaptic cell to trigger electrical and second-messenger responses in the postsynaptic cell. After fusion, presynaptic vesicles must be recycled by endocytosis for reuse (Sudhof, 2004) (Fig. 6).

BK channels are the largest conductance ion channels that a neuron encodes, and in the presynaptic terminal a small change in their activity can dictate the membrane potential of the terminal and the influx of Ca^{2+} through voltage-gated Ca^{2+} channels. This one-two punch of controlling membrane potential and Ca^{2+} influx enables BK channels to have a strong influence on synaptic release. In addition to BK channels, other synaptic proteins have also been implicated in the response to ethanol and organic solvent anesthetics (Fig. 6). These include the presynaptic proteins Dynamin, Syntaxin 1A, and Synapsin (Al-Hasan *et al.*, 2011; Godenschwege *et al.*, 2004; Krishnan *et al.*, in press); the transmembrane cell adhesion integrin subunit βPS and αPS3 (Bhandari *et al.*, 2009); the postsynaptic $GABA_B$ receptor (Dzitoyeva *et al.*, 2003); and the postsynaptic scaffolding protein Homer (Urizar *et al.*, 2007). BK channels have recently been shown to be in intimate physical contact with some of these proteins. Both mass spectrometry and co-immunoprecipitation have shown that BK channels are physically associated with both Syntaxin 1A and Dynamin 1 (Cibulsky *et al.*, 2005; Gorini *et al.*, 2010). Syntaxin 1A is involved in synaptic release, whereas Dynamin is a large GTPase that performs a key step in vesicle recycling.

Syntaxin 1A is a t-SNARE (soluble NSF attachment receptor) protein that mediates targeted vesicular fusion at the synapse. In *Caenorhabditis elegans*, a mutation in the *Syntaxin 1A* ortholog causes strong resistance to the volatile anesthetics isoflurane and halothane (van Swinderen *et al.*, 1999), and in rats these anesthetic compounds have been shown to bind to Syntaxin 1A resulting in structural alterations of the protein (Nagele *et al.*, 2005). In Drosophila, a mutation in the *Syntaxin 1A* gene disrupts the capacity to acquire ethanol tolerance (Krishnan *et al.*, in press).

In mammals, the Dynamins comprise a family of three different genes, two of which (Dynamin 1 and Dynamin 3) are expressed widely in the brain and localized to the synapse. Although Dynamin 1 is well known for its function in rapid vesicle

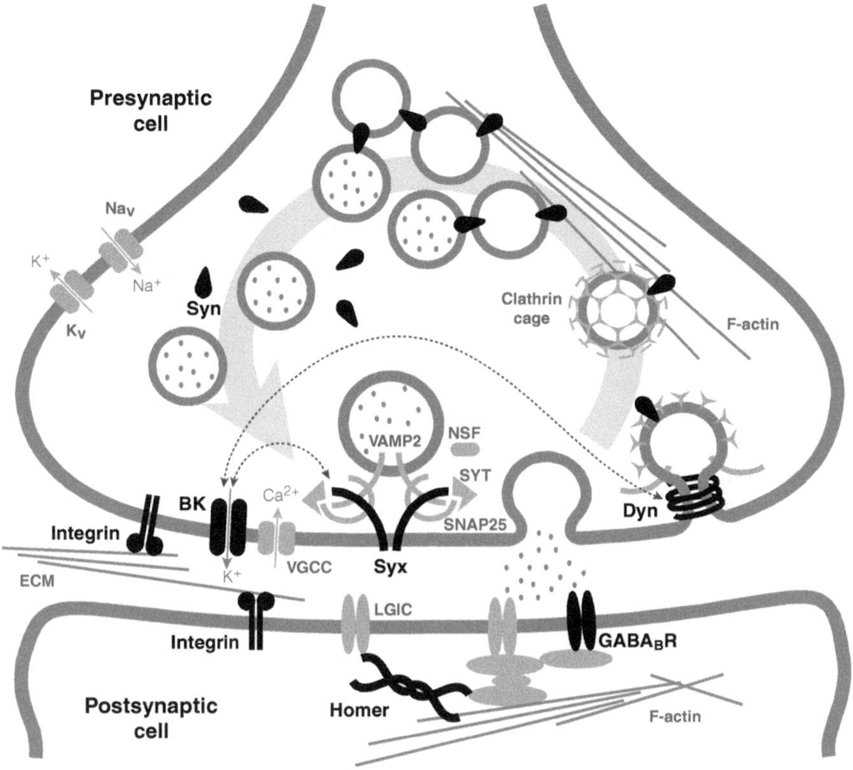

FIG. 6. **Synaptic model of drug tolerance.** Shown is a schematic representation of a synapse based on current knowledge of putative interactions between synaptic proteins. Synaptic proteins that have been implicated in the development of tolerance to alcohol and anesthetics are displayed and labeled in black. These proteins include Synapsins (Syn), Dynamin (Dyn), Homer, Integrins, BK channels (BK), Syntaxin 1A (Syx), and the $GABA_B$ receptor ($GABA_BR$). Proteins and structures in gray are included to provide context. Thin dotted lines with arrows denote known protein–protein interactions. Figure adapted from Gorini *et al.* (2010).

recycling and concentrates primarily in the presynaptic compartment (Gray *et al.*, 2003; Takei *et al.*, 1996), Dynamin 3 is best known for mediating clathrin-mediated uptake of surface receptors and concentrates in the brain within the postsynaptic density of dendrites (Gray *et al.*, 2003).

In Drosophila, the multiple forms of Dynamin are all thought to be encoded by a single gene called *shibire* (Chen *et al.*, 1991). The shi^{ts1} and shi^{ts2} are temperature-sensitive mutant alleles that cause temperature-dependent paralysis. At the restrictive temperature (~29°C), the mutant Shibire protein stops functioning, leading to a rapid and reversible cessation of synaptic vesicle recycling and synaptic signaling (Kosaka and Ikeda, 1983; van der Bliek and Meyerowitz, 1991). At the permissive temperature (~22°C), the mutant animals have been described as being essentially normal.

In a recent study, it has been shown that animals carrying the *shi*^{ts1} or the *shi*^{ts2} mutant alleles were unable to acquire benzyl alcohol tolerance at the permissive temperature. That is— with regard to benzyl alcohol tolerance—these mutant alleles did not behave as conditional mutations (Al-Hasan *et al.*, 2011). With regard to ethanol tolerance, the story was more involved. The *shi*^{ts1} mutation again blocked the acquisition of tolerance in a nonconditional way. The *shi*^{ts2} mutation, however, blocked tolerance only at the restrictive temperature and only if the temperature-induced blockade in neural signaling overlapped the ethanol exposure. If the *shibire* blockade was evoked 5 or 23 h after ethanol exposure tolerance, then the functional tolerance assayed at 24 h after the exposure appeared normal. This indicates that at the restrictive temperature, the *shi*^{ts2} mutation interferes with the acquisition but not the maintenance of functional tolerance. The difference between benzyl alcohol and ethanol responses probably indicates that, with regard to the induction of tolerance, ethanol is a more effective drug.

One might expect that any event producing a neural blockade would also block the acquisition of tolerance. However, this does not appear to be the case. A temperature-sensitive mutation affecting the neuronal voltage-gated sodium channel was also tested. This mutation, *para*^{ts1}, blocks the production of neuronal action potentials and therefore produces a neural blockade by a different method. Even though it effectively blocked neuronal activity, the *para*^{ts1} mutation did not block tolerance (Krishnan *et al.*, in press). This may mean that the *shibire*-encoded Dynamin protein is a trigger for the production of tolerance. However, at this time, one cannot rule out that when Dynamin changes conformation at the restrictive temperature, it disturbs another closely associated protein(s) and that this change blocks the acquisition of tolerance. Because it has been shown that Dynamin touches the BK channel, it is possible that this hypothetical protein is the BK channel.

The close physical interaction between two entities, BK channels and Dynamin, that have been genetically shown to be important for the acquisition of tolerance is particularly interesting since in mammalian systems Pietrzykowski and colleagues (Pietrzykowski *et al.*, 2004) have shown that ethanol tolerance initiates with the declustering and internalization of BK channels. It is thus possible that the process underlying the repositioning of BK channels from the synaptic membrane is facilitated by an interaction of the channel with the endocytic Dynamin and that the same process occurs in Drosophila.

V. Closing Remarks

Drosophila has emerged as a valuable biological model for understanding the molecular mechanisms underlying drug abuse endophenotypes. Genes that have been identified are involved in second-messenger signaling, stress responses, olfaction, metabolism, transcriptional regulation, cytoskeletal organization, and memory formation and include proteins such as transcription factors, proteases, ion channels, synaptic proteins, neurotransmitter receptors, and biosynthetic enzymes (reviewed in Atkinson, 2009 and Rodan and Rothenfluh, 2010).

Tolerance and withdrawal are two key ingredients in the recipe for addiction. Together they provide a motivation for increased drug consumption. In Drosophila, aspects of both endophenotypes originate with the same gene. Whether the genes that underlie these endophenotypes directly contribute to pathological and compulsive drug use is not known. However, the recent development of Drosophila voluntary alcohol drinking assays may enable Drosophila geneticists to help resolve this question (Devineni and Heberlein, 2009; Ja *et al.*, 2007).

In the field of developmental biology, Drosophila has provided many of the molecular keys for understanding mammalian development. Drug abuse research using Drosophila is a relatively young field that has already made important contributions toward understanding drug responses. In a sense, the experimental limitations of any given model system force one to have a distinct experimental perspective that makes possible the discovery of different facets of drug addiction. We believe that drug abuse research using Drosophila has well-breached the tipping point of discovery, making it a major contributor to our understanding of the addictive process.

Acknowledgments

We thank Jascha Pohl, Rudi Bohm, Brooks Robinson, Ben Troutwine, Rosie Robles, Harish Krishnan, and Jane Kirschman for critical reading of the manuscript. This work was supported by the National Institutes of Health grants RO1 DA022219 and R01 AA018037 to NSA.

References

Al-Hasan, Y.M., Krishnan, H.R., Ghezzi, A., Prado 3rd., F.J., Robles, R.B., and Atkinson, N.S. (2011). Tolerance to anesthesia depends on synaptic proteins. *Behav. Genet.*. doi: 10.1007/s10519-011-9451-8.

Allada, R., and Nash, H.A. (1993). Drosophila melanogaster as a model for study of general anesthesia: the quantitative response to clinical anesthetics and alkanes. *Anesth. Analg.* **77**, 19–26.

Allis, C.D., Jenuwein, T., Reinberg, D., and Caparros, M.-L. (2007). *Epigenetics*. Cold Spring Harbor Laboratory Press, Woodbury, NY.

Ankney, P.F. (1984). A note on barometric pressure and behavior in Drosophila pseudoobscura. *Behav. Genet.* **14**, 315–317.

Atkinson, N.S. (2009). Tolerance in Drosophila. *J. Neurogenet.* **23**, 293–302.

Bainton, R.J., Tsai, L.T., Singh, C.M., Moore, M.S., Neckameyer, W.S., and Heberlein, U. (2000). Dopamine modulates acute responses to cocaine, nicotine and ethanol in Drosophila. *Curr. Biol.* **10**, 187–194.

Beckstead, M.J., Weiner, J.L., Eger 2nd., E.I., Gong, D.H., and Mihic, S.J. (2000). Glycine and gamma-aminobutyric acid(A) receptor function is enhanced by inhaled drugs of abuse. *Mol. Pharmacol.* **57**, 1199–1205.

Bellen, H.J., Tong, C., and Tsuda, H. (2010). 100 years of Drosophila research and its impact on vertebrate neuroscience: a history lesson for the future. *Nat. Rev. Neurosci.* **11**, 514–522.

Berger, S.L. (2007). The complex language of chromatin regulation during transcription. *Nature* **447**, 407–412.

Bhandari, P., Kendler, K.S., Bettinger, J.C., Davies, A.G., and Grotewiel, M. (2009). An assay for evoked locomotor behavior in Drosophila reveals a role for integrins in ethanol sensitivity and rapid ethanol tolerance. *Alcohol. Clin. Exp. Res.* **33**, 1794–1805.

Bohm, R. A. (2000). *Transcriptional Control of Slowpoke, a Calcium Activated Potassium Channel Gene.* PhD thesis, The University of Texas at Austin, Austin, TX.

Bohm, R.A., Wang, B., Brenner, R., and Atkinson, N.S. (2000). Transcriptional control of Ca(2 +)-activated K(+) channel expression: identification of a second, evolutionarily conserved, neuronal promoter. *J. Exp. Biol.* **203**, 693–704.

Brenner, R., and Atkinson, N. (1996). Developmental and eye-specific transcriptional control elements in an intronic region of a Ca2+-activated K+ channel gene. *Dev. Biol.* **177**, 536–543.

Brenner, R., and Atkinson, N.S. (1997). Calcium-activated potassium channel gene expression in the midgut of Drosophila. *Comp. Biochem. Physiol. B Biochem. Mol. Biol.* **118**, 411–420.

Brenner, R., Chen, Q.H., Vilaythong, A., Toney, G.M., Noebels, J.L., and Aldrich, R.W. (2005). BK channel beta4 subunit reduces dentate gyrus excitability and protects against temporal lobe seizures. *Nat. Neurosci.* **8**, 1752–1759.

Brenner, R., Thomas, T.O., Becker, M.N., and Atkinson, N.S. (1996). Tissue-specific expression of a Ca(2+)-activated K+ channel is controlled by multiple upstream regulatory elements. *J. Neurosci.* **16**, 1827–1835.

Chang, W.M., Bohm, R.A., Strauss, J.C., Kwan, T., Thomas, T., Cowmeadow, R.B., and Atkinson, N. S. (2000). Muscle-specific transcriptional regulation of the slowpoke Ca(2+)-activated K(+) channel gene. *J. Biol. Chem.* **275**, 3991–3998.

Chen, M.S., Obar, R.A., Schroeder, C.C., Austin, T.W., Poodry, C.A., Wadsworth, S.C., and Vallee, R.B. (1991). Multiple forms of dynamin are encoded by shibire, a Drosophila gene involved in endocytosis. *Nature* **351**, 583–586.

Chen, S., Lee, A.Y., Bowens, N.M., Huber, R., and Kravitz, E.A. (2002). Fighting fruit flies: a model system for the study of aggression. *Proc. Natl. Acad. Sci. U S A* **99**, 5664–5668.

Cibulsky, S.M., Fei, H., and Levitan, I.B. (2005). Syntaxin-1A binds to and modulates the Slo calcium-activated potassium channel via an interaction that excludes syntaxin binding to calcium channels. *J. Neurophysiol.* **93**, 1393–1405.

Covarrubias, M., and Rubin, E. (1993). Ethanol selectively blocks a noninactivating K+ current expressed in Xenopus oocytes. *Proc. Natl. Acad. Sci. U S A* **90**, 6957–6960.

Cowmeadow, R.B., Krishnan, H.R., and Atkinson, N.S. (2005). The slowpoke gene is necessary for rapid ethanol tolerance in Drosophila. *Alcohol. Clin. Exp. Res.* **29**, 1777–1786.

Cowmeadow, R.B., Krishnan, H.R., Ghezzi, A., Al'Hasan, Y.M., Wang, Y.Z., and Atkinson, N.S. (2006). Ethanol tolerance caused by slowpoke induction in Drosophila. *Alcohol. Clin. Exp. Res.* **30**, 745–753.

Crowley, J.J., Treistman, S.N., and Dopico, A.M. (2003). Cholesterol antagonizes ethanol potentiation of human brain BKCa channels reconstituted into phospholipid bilayers. *Mol. Pharmacol.* **64**, 365–372.

Devineni, A.V., and Heberlein, U. (2009). Preferential ethanol consumption in Drosophila models features of addiction. *Curr. Biol.* **19**, 2126–2132.

Dinwiddie, S.H. (1994). Abuse of inhalants: a review. *Addiction* **89**, 925–939.

Du, W., Bautista, J.F., Yang, H., Diez-Sampedro, A., You, S.A., Wang, L., Kotagal, P., Luders, H.O., Shi, J., Cui, J., Richerson, G.B., and Wang, Q.K. (2005). Calcium-sensitive potassium channelopathy in human epilepsy and paroxysmal movement disorder. *Nat. Genet.* **37**, 733–738.

Duffy, J.B. (2002). GAL4 system in Drosophila: a fly geneticist's Swiss army knife. *Genesis* **34**, 1–15.

Dzitoyeva, S., Dimitrijevic, N., and Manev, H. (2003). gamma-Aminobutyric acid B receptor 1 mediates behavior-impairing actions of alcohol in Drosophila: adult RNA interference and pharmacological evidence. *Proc. Natl. Acad. Sci. U S A* **100**, 5485–5490.

Evans, E.B., and Balster, R.L. (1991). CNS depressant effects of volatile organic solvents. *Neurosci. Biobehav. Rev.* **15**, 233–241.

Flanagan, R.J., and Ives, R.J. (1994). Volatile substance abuse. *Bull. Narc.* **46**, 49–78.

Fleisher, G.R., and Ludwig, S. (2010). *Textbook of Pediatric Emergency Medicine.* Wolters Kluwer/Lippincott Williams & Wilkins Health, Philadelphia, PA.

Ghezzi, A., Al-Hasan, Y.M., Larios, L.E., Bohm, R.A., and Atkinson, N.S. (2004). slo K(+) channel gene regulation mediates rapid drug tolerance. *Proc. Natl. Acad. Sci. U S A* **101**, 17276–17281.

Ghezzi, A., Pohl, J.B., Wang, Y., and Atkinson, N.S. (2010). BK channels play a counter-adaptive role in drug tolerance and dependence. *Proc. Natl. Acad. Sci. U S A* **107**, 16360–16365.

Ghezzi, A., Atkinson, N.S. (2011). Unpublished observations.

Godenschwege, T.A., Reisch, D., Diegelmann, S., Eberle, K., Funk, N., Heisenberg, M., Hoppe, V., Hoppe, J., Klagges, B.R., Martin, J.R., Nikitina, E.A., Putz, G., Reifegerste, R., Reisch, N., Rister, J., Schaupp, M., Scholz, H., Schwarzel, M., Werner, U., Zars, T.D., Buchner, S., and Buchner, E. (2004). Flies lacking all synapsins are unexpectedly healthy but are impaired in complex behaviour. *Eur. J. Neurosci.* **20**, 611–622.

Goldman, D., Oroszi, G., and Ducci, F. (2005). The genetics of addictions: uncovering the genes. *Nat. Rev. Genet.* **6**, 521–532.

Gorini, G., Ponomareva, O., Shores, K.S., Person, M.D., Harris, R.A., and Mayfield, R.D. (2010). Dynamin-1 co-associates with native mouse brain BKCa channels: proteomics analysis of synaptic protein complexes. *FEBS Lett.* **584**, 845–851.

Gray, N.W., Fourgeaud, L., Huang, B., Chen, J., Cao, H., Oswald, B.J., Hemar, A., and McNiven, M. A. (2003). Dynamin 3 is a component of the postsynapse, where it interacts with mGluR5 and Homer. *Curr. Biol.* **13**, 510–515.

Gribkoff, V.K., Starrett Jr., J.E., and Dworetzky, S.I. (2001). Maxi-K potassium channels: form, function, and modulation of a class of endogenous regulators of intracellular calcium. *Neuroscientist* **7**, 166–177.

Harris, R.A. (1999). Ethanol actions on multiple ion channels: which are important? *Alcohol. Clin. Exp. Res.* **23**, 1563–1570.

Harris, R.A., Mihic, S.J., Dildy-Mayfield, J.E., and Machu, T.K. (1995). Actions of anesthetics on ligand-gated ion channels: role of receptor subunit composition. *FASEB J.* **9**, 1454–1462.

Hendricks, J.C., Finn, S.M., Panckeri, K.A., Chavkin, J., Williams, J.A., Sehgal, A., and Pack, A.I. (2000). Rest in Drosophila is a sleep-like state. *Neuron* **25**, 129–138.

Himmelsbach, C.K. (1941). The morphine abstinence syndrome, its nature and treatment. *Ann. Intern. Med.* **15**, 829–843.

Hosmer, D., Lemeshow, S., and May, S. (2008). *Applied Survival Analysis: Regression Modeling of Time to Event Data.* Wiley-Interscience, New York, NY.

Hosmer, D.W., Lemeshow, S., and Sunny, K. (2002). *Applied Survival Analysis: Regression Modeling of Time to Event Data (Solutions Manual)*. Wiley-Interscience, New York, NY.

Hyman, S.E., Malenka, R.C., and Nestler, E.J. (2006). Neural mechanisms of addiction: the role of reward-related learning and memory. *Annu. Rev. Neurosci.* **29**, 565–598.

IARC Working Group on the Evaluation of Carcinogenic Risks to Humans(1999). *Re-Evaluation of Some Organic Chemicals, Hydrazine and Hydrogen Peroxide*. World Health Organization, International Agency for Research on Cancer, Lyon, France.

Isbell, H., Fraser, H.F., Wikler, A., Belleville, R.E., and Eisenman, A.J. (1955). An experimental study of the etiology of rum fits and delirium tremens. *Q. J. Stud. Alcohol* **16**, 1–33.

Ja, W.W., Carvalho, G.B., Mak, E.M., de la Rosa, N.N., Fang, A.Y., Liong, J.C., Brummel, T., and Benzer, S. (2007). Prandiology of Drosophila and the CAFE assay. *Proc. Natl. Acad. Sci. U S A* **104**, 8253–8256.

Julien, R.M. (2004). *A Primer of Drug Action: A Concise, Nontechnical Guide to the Actions, Uses, and Side Effects of Psychoactive Drugs*. Worth Publishers Inc., New York, NY.

Kendler, K.S., Jacobson, K.C., Prescott, C.A., and Neale, M.C. (2003). Specificity of genetic and environmental risk factors for use and abuse/dependence of cannabis, cocaine, hallucinogens, sedatives, stimulants, and opiates in male twins. *Am. J. Psychiatry* **160**, 687–695.

Knott, T.K., Dopico, A.M., Dayanithi, G., Lemos, J., and Treistman, S.N. (2002). Integrated channel plasticity contributes to alcohol tolerance in neurohypophysial terminals. *Mol. Pharmacol.* **62**, 135–142.

Kong, E.C., Woo, K., Li, H., Lebestky, T., Mayer, N., Sniffen, M.R., Heberlein, U., Bainton, R.J., Hirsh, J., and Wolf, F.W. (2010). A pair of dopamine neurons target the D1-like dopamine receptor DopR in the central complex to promote ethanol-stimulated locomotion in Drosophila. *PLoS ONE* **5**, e9954.

Koob, G.F. (1996). Drug addiction: the yin and yang of hedonic homeostasis. *Neuron* **16**, 893–896.

Koob, G.F., and Bloom, F.E. (1988). Cellular and molecular mechanisms of drug dependence. *Science* **242**, 715–723.

Koob, G.F., and Le Moal, M. (2006). *Neurobiology of Addiction*. Elsevier/Academic Press, Boston, MA.

Kopp Lugli, A., Yost, C.S., and Kindler, C.H. (2009). Anaesthetic mechanisms: update on the challenge of unravelling the mystery of anaesthesia. *Eur. J. Anaesthesiol.* **26**, 807–820.

Kosaka, T., and Ikeda, K. (1983). Possible temperature-dependent blockage of synaptic vesicle recycling induced by a single gene mutation in Drosophila. *J. Neurobiol.* **14**, 207–225.

Krishnan, H. R., Al-Hasan, Y.M., Pohl, J.B., Ghezzi, A., and Atkinson, N.S. (2011). A role for dynamin in triggering ethanol tolerance. *Alcohol. Clin. Exp. Res*, in press.

Krishnan, K.S., and Nash, H.A. (1990). A genetic study of the anesthetic response: mutants of Drosophila melanogaster altered in sensitivity to halothane. *Proc. Natl. Acad. Sci. U S A* **87**, 8632–8636.

Kuebler, D., and Tanouye, M.A. (2000). Modifications of seizure susceptibility in Drosophila. *J. Neurophysiol.* **83**, 998–1009.

Kuebler, D., Zhang, H., Ren, X., and Tanouye, M.A. (2001). Genetic suppression of seizure susceptibility in Drosophila. *J. Neurophysiol.* **86**, 1211–1225.

Lee, D.Y., Hayes, J.J., Pruss, D., and Wolffe, A.P. (1993). A positive role for histone acetylation in transcription factor access to nucleosomal DNA. *Cell* **72**, 73–84.

Leibovitch, B.A., Campbell, D.B., Krishnan, K.S., and Nash, H.A. (1995). Mutations that affect ion channels change the sensitivity of Drosophila melanogaster to volatile anesthetics. *J. Neurogenet.* **10**, 1–13.

Lin, M., and Nash, H.A. (1996). Influence of general anesthetics on a specific neural pathway in Drosophila melanogaster. *Proc. Natl. Acad. Sci. U S A* **93**, 10446–10451.

Littleton, J. (1998). Neurochemical mechanisms underlying alcohol withdrawal. *Alcohol Health Res. World* **22**, 13–24.

Littleton, J., and Little, H. (1994). Current concepts of ethanol dependence. *Addiction* **89**, 1397–1412.

Littleton, J.T., and Ganetzky, B. (2000). Ion channels and synaptic organization: analysis of the Drosophila genome. *Neuron* **26**, 35–43.

Liu, J., Asuncion-Chin, M., Liu, P., and Dopico, A.M. (2006). CaM kinase II phosphorylation of slo Thr107 regulates activity and ethanol responses of BK channels. *Nat. Neurosci.* **9**, 41–49.

Lloyd, T.E., Verstreken, P., Ostrin, E.J., Phillippi, A., Lichtarge, O., and Bellen, H.J. (2000). A genome-wide search for synaptic vesicle cycle proteins in Drosophila. *Neuron* **26**, 45–50.

Lovell, P.V., and McCobb, D.P. (2001). Pituitary control of BK potassium channel function and intrinsic firing properties of adrenal chromaffin cells. *J. Neurosci.* **21**, 3429–3442.

Lubman, D.I., Yucel, M., and Lawrence, A.J. (2008). Inhalant abuse among adolescents: neurobiological considerations. *Br. J. Pharmacol.* **154**, 316–326.

Martin, W.R. (1968). XVI. A homeostatic and redundancy theory of tolerance to and dependence on narcotic analgesics. *Res. Publ. Assoc. Res. Nerv. Ment. Dis.* **46**, 206–225.

Martin, G.E., Hendrickson, L.M., Penta, K.L., Friesen, R.M., Pietrzykowski, A.Z., Tapper, A.R., and Treistman, S.N. (2008). Identification of a BK channel auxiliary protein controlling molecular and behavioral tolerance to alcohol. *Proc. Natl. Acad. Sci. U S A* **105**, 17543–17548.

Matsuda, K., Buckingham, S.D., Freeman, J.C., Squire, M.D., Baylis, H.A., and Sattelle, D.B. (1998). Effects of the alpha subunit on imidacloprid sensitivity of recombinant nicotinic acetylcholine receptors. *Br. J. Pharmacol.* **123**, 518–524.

Mayfield, R.D., Harris, R.A., and Schuckit, M.A. (2008). Genetic factors influencing alcohol dependence. *Br. J. Pharmacol.* **154**, 275–287.

Moore, M.S., DeZazzo, J., Luk, A.Y., Tully, T., Singh, C.M., and Heberlein, U. (1998). Ethanol intoxication in Drosophila: genetic and pharmacological evidence for regulation by the cAMP signaling pathway. *Cell* **93**, 997–1007.

Nagele, P., Mendel, J.B., Placzek, W.J., Scott, B.A., D'Avignon, D.A., and Crowder, C.M. (2005). Volatile anesthetics bind rat synaptic snare proteins. *Anesthesiology* **103**, 768–778.

Nash, H.A. (2002). In vivo genetics of anaesthetic action. *Br. J. Anaesth.* **89**, 143–155.

Parr, J., Large, A., Wang, X., Fowler, S.C., Ratzlaff, K.L., and Ruden, D.M. (2001). The inebriactometer: a device for measuring the locomotor activity of Drosophila exposed to ethanol vapor. *J. Neurosci. Methods* **107**, 93–99.

Pattillo, J.M., Yazejian, B., DiGregorio, D.A., Vergara, J.L., Grinnell, A.D., and Meriney, S.D. (2001). Contribution of presynaptic calcium-activated potassium currents to transmitter release regulation in cultured Xenopus nerve-muscle synapses. *Neuroscience* **102**, 229–240.

Pick, S., and Strauss, R. (2005). Goal-driven behavioral adaptations in gap-climbing Drosophila. *Curr. Biol.* **15**, 1473–1478.

Pietrzykowski, A.Z., Friesen, R.M., Martin, G.E., Puig, S.I., Nowak, C.L., Wynne, P.M., Siegelmann, H.T., and Treistman, S.N. (2008). Posttranscriptional regulation of BK channel splice variant stability by miR-9 underlies neuroadaptation to alcohol. *Neuron* **59**, 274–287.

Pietrzykowski, A.Z., Martin, G.E., Puig, S.I., Knott, T.K., Lemos, J.R., and Treistman, S.N. (2004). Alcohol tolerance in large-conductance, calcium-activated potassium channels of CNS terminals is intrinsic and includes two components: decreased ethanol potentiation and decreased channel density. *J. Neurosci.* **24**, 8322–8332.

Porter, R.J., Mattson, R.H., Porter, R.J., Mattson, R.H., and Cramer, J.A. (1990). *Alcohol and Seizures: Basic Mechanisms and Clinical Concepts.* F. A. Davis, Philadelphia, PA.

Quinn, W.G., Harris, W.A., and Benzer, S. (1974). Conditioned behavior in Drosophila melanogaster. *Proc. Natl. Acad. Sci. U S A* **71**, 708–712.

Ramazani, R.B., Krishnan, H.R., Bergeson, S.E., and Atkinson, N.S. (2007). Computer automated movement detection for the analysis of behavior. *J. Neurosci. Methods* **162**, 171–179.

Rodan, A.R., and Rothenfluh, A. (2010). The genetics of behavioral alcohol responses in Drosophila. *Int. Rev. Neurobiol.* **91**, 25–51.

Rogawski, M.A. (2005). Update on the neurobiology of alcohol withdrawal seizures. *Epilepsy Curr.* **5**, 225–230.

Rothenfluh, A., and Heberlein, U. (2002). Drugs, flies, and videotape: the effects of ethanol and cocaine on Drosophila locomotion. *Curr. Opin. Neurobiol.* **12**, 639–645.

Scholz, H., Ramond, J., Singh, C.M., and Heberlein, U. (2000). Functional ethanol tolerance in Drosophila. *Neuron* **28**, 261–271.

Schuckit, M.A. (1994). Low level of response to alcohol as a predictor of future alcoholism. *Am. J. Psychiatry* **151**, 184–189.

Singh, C.M., and Heberlein, U. (2000). Genetic control of acute ethanol-induced behaviors in Drosophila. *Alcohol. Clin. Exp. Res.* **24**, 1127–1136.

Smolik, S.M., Rose, R.E., and Goodman, R.H. (1992). A cyclic AMP-responsive element-binding transcriptional activator in Drosophila melanogaster, dCREB-A, is a member of the leucine zipper family. *Mol. Cell. Biol.* **12**, 4123–4131.

Sokolowski, M.B. (1980). Foraging strategies of Drosophila melanogaster: a chromosomal analysis. *Behav. Genet.* **10**, 291–302.

Solomon, R.L. (1980). The opponent-process theory of acquired motivation: the costs of pleasure and the benefits of pain. *Am. Psychol.* **35**, 691–712.

Solomon, R.L., and Corbit, J.D. (1974). An opponent-process theory of motivation. I. Temporal dynamics of affect. *Psychol. Rev.* **81**, 119–145.

Spieth, H.T. (1974). Courtship behavior in Drosophila. *Annu. Rev. Entomol.* **19**, 385–405.

Sudhof, T.C. (2004). The synaptic vesicle cycle. *Annu. Rev. Neurosci.* **27**, 509–547.

Takei, K., Mundigl, O., Daniell, L., and De Camilli, P. (1996). The synaptic vesicle cycle: a single vesicle budding step involving clathrin and dynamin. *J. Cell Biol.* **133**, 1237–1250.

Tanouye, M.A., and Wyman, R.J. (1980). Motor outputs of giant nerve fiber in Drosophila. *J. Neurophysiol.* **44**, 405–421.

Urizar, N.L., Yang, Z., Edenberg, H.J., and Davis, R.L. (2007). Drosophila homer is required in a small set of neurons including the ellipsoid body for normal ethanol sensitivity and tolerance. *J. Neurosci.* **27**, 4541–4551.

Usui, T., Smolik, S.M., and Goodman, R.H. (1993). Isolation of Drosophila CREB-B: a novel CRE-binding protein. *DNA Cell Biol.* **12**, 589–595.

van der Bliek, A.M., and Meyerowitz, E.M. (1991). Dynamin-like protein encoded by the Drosophila shibire gene associated with vesicular traffic. *Nature* **351**, 411–414.

Van Goor, F., Li, Y.X., and Stojilkovic, S.S. (2001). Paradoxical role of large-conductance calcium-activated K+ (BK) channels in controlling action potential-driven Ca2+ entry in anterior pituitary cells. *J. Neurosci.* **21**, 5902–5915.

van Swinderen, B., Saifee, O., Shebester, L., Roberson, R., Nonet, M.L., and Crowder, C.M. (1999). A neomorphic syntaxin mutation blocks volatile-anesthetic action in Caenorhabditis elegans. *Proc. Natl. Acad. Sci. U S A* **96**, 2479–2484.

Wakasugi, M., Hirota, K., Roth, S.H., and Ito, Y. (1999). The effects of general anesthetics on excitatory and inhibitory synaptic transmission in area CA1 of the rat hippocampus in vitro. *Anesth. Analg.* **88**, 676–680.

Wang, Y., Ghezzi, A., Yin, J.C., and Atkinson, N.S. (2009). CREB regulation of BK channel gene expression underlies rapid drug tolerance. *Genes Brain Behav.* **8**, 369–376.

Wang, Y., Krishnan, H.R., Ghezzi, A., Yin, J.C., and Atkinson, N.S. (2007). Drug-induced epigenetic changes produce drug tolerance. *PLoS Biol.* **5**, 2342–2353.

Wolf, F.W., and Heberlein, U. (2003). Invertebrate models of drug abuse. *J. Neurobiol.* **54**, 161–178.

Young, A.M., and Goudie, A.J. (1995). Adaptive processes regulating tolerance to behavioral effects of drugs. In: Bloom, F.E., Kupfer, D.J. (Eds.), *Psychopharmacology: Fourth Generation of Progress*. Raven Press, New York, NY, pp. 733–742.

ATTENTION IN *DROSOPHILA*

Bruno van Swinderen

Queensland Brain Institute, The University of Queensland, Brisbane, QLD 4072 Australia

As bluntly summarized by a psychologist over a century ago, everyone knows what attention is [James (1890). *The Principles of Psychology*]. Attention describes our capacity to focus perception on one or a group of related stimuli while filtering out irrelevant stimuli. The ease we have in recognizing this astounding capacity in ourselves is matched by a surprising difficulty in identifying it in others, and this is especially the case for measuring attention in other animals. Identifying and measuring attention-like processes in simple animals such as flies requires, to some extent, even more rigor than asking the same question for our closer animal relatives, such as apes and monkeys. This is because flies have completely different brains than humans do, so to study attention in these creatures one must rely purely on operational or behavioral measures rather than comparative neuroanatomy. There is a long history of using sophisticated behavioral paradigms to study visual responses in *Drosophila melanogaster*, and these studies have often provided early evidence of attention-like processes in flies. More recently, these fly paradigms have been applied to measuring visual attention directly, and the combination of electrophysiology with these preparations has provided insight into how a fly might pay attention. Together with more efficient methods for measuring some aspects of attention, such as stimulus suppression, these approaches should begin to uncover how visual attention might work in a small brain.

INTERNATIONAL REVIEW OF
NEUROBIOLOGY, VOL. 99
DOI: 10.1016/B978-0-12-387003-2.00003-3

I. Introduction: the Problem of Attention in Nonhuman Animals

To study attention in a fly is somewhat analogous to studying attention in a human baby: without a verbal report, attention processes have to be inferred from behavioral or physiological clues. The human infant, however, has an advantage compared to flies: we know that attention-like processes must be operating in a child because they soon become able to describe them. For flies, and all other animals, we only have behavior and physiology to study attention. In larger animals such as monkeys, measured behaviors may involve button presses or eye movements, whereas in smaller animals such as mice and flies, entire body movements may be followed to infer attention states. In either case, behavior alone can never be entirely convincing about attention because it is difficult to know whether the behavior reflects a simple motor response to a stimulus or whether it is the outcome of a perceptual choice made in the brain, where an alternate choice was deliberately suppressed. For this reason, brain recordings are often used in animals to bolster the behavioral data, by showing neural correlates of attention associated with competing stimuli. Together, electrophysiology and behavior can be applied to satisfy the criteria required for identifying attention in an animal. Of course, combining these tools in animals gets more complicated as animals get smaller, culminating in fruit flies with a brain the size of a poppy seed. Identifying attention in a fly seems an impossible task, but in fact the genetic model *Drosophila melanogaster* has provided insight on fly attention for the past 30 years, and research in the past few years have yielded some of the most convincing evidence that flies do have a selective attention.

Attention is sometimes equated with consciousness. The problem of understanding the mystery of consciousness has therefore created some confusion surrounding the study of attention in animals, namely that attention requires some rudimentary consciousness. A subjective association between these phenomena in humans seems to require the same in other animals, but this need not be the case, and, indeed, a proper study of attention can be separated from the mystery of consciousness (van Swinderen, 2005). As a first step in this direction, researchers discuss two different kinds of attention: top-down and bottom-up (Itti and Koch, 2000; Treisman and Gelade, 1980). Both are concerned with the allocation of perceptual resources among competing stimuli, but in bottom-up processes this facility is initially achieved by a stimulus-driven process (loud noises or bright objects attract attention), whereas top-down processes are volitional and depend to some extent on past experience (Wolfe *et al.*, 2000). Bottom-up attention can be parallel, with feature salience determining whether objects are perceived, whereas top-down attention is a serial process where perceptual resources need to be divided among potential targets and distracters (Bichot and Desimone, 2006; Treue, 2003). Although somewhat artificial, this separation narrows the discussion with regard to the problem of attention in animals: when we wonder whether animals have selective attention, we are really asking whether they have top-down processes, that is, whether they have an experience-dependent stimulus-selection/-suppression mechanism. However, it is easier to address bottom-up processes in simple animals such as insects. By thus framing the problem, we are then in a better position to devise experiments to study it.

Any study of attention in animals needs to demonstrate suppression of competing stimuli in addition to selection of the one object which might eventually lead to a behavioral response. This is actually harder to demonstrate than might be expected because it is not always obvious whether an animal has even registered a competing stimulus after it has made a behavioral choice. Brain recordings in restrained animals often come to the rescue here, by revealing neural responses that are suppressed during an attention task. Fine-grained behavioral paradigms, such as continuous tracking of optokinetic behavior (by tracking eye movements, for example), can deliver similar information about stimulus suppression (Rubinstein and Abel, 2011). As an alternative to this rather complex approach, attention studies in animals have traditionally relied on measuring reaction latencies in distracter paradigms. This can be tested in working memory paradigms, such as the delayed match to sample task (DMST), where a choice must be made following a cue (Fig. 1A), usually for a reward. Distracters presented before or during the choice phase of the experiment can test the subject's allocation of attention to the stimuli. Another paradigm, the five-choice serial reaction time task (5CSRTT), is an operant chamber where rodents (for example) collect food pellets by responding to competing stimuli in a pattern set by the experimenter (Robbins, 2002). In either scenario, the problem is similar: distracters draw attentional resources and increase reaction times to target objects, but if they are bound in some way with targets then reaction time for a behavioral task can be decreased. The readout in all cases is whether an animal performs a task when distracters are present, and if so, how long it takes to do it. Finally, although experience or training can guide animal behavior in these tasks, it should not alter the fact that top-down attention is a serial process; increasing distracter numbers should increase the time it takes to find a target.

Studies of attention in animals are often implemented as part of a classical conditioning paradigm. Thus, behavioral choice in a DMST or 5CSRTT paradigm requires considerable prior training before animals are able to provide useful data for an attention experiment. Training for an attention experiment can be problematic, though. Top-down attention by definition implies certain flexibility in being able to choose what to attend to, but the extensive training often required to perform attention experiments in rats or monkeys may instead promote habit formation, especially when rewards are involved (Maunsell, 2004). Attention resources may then even be required to *override* a habit, so training presents a real conundrum for attention studies in animals, namely that the training procedure used to test performance in an attention paradigm may itself compromise attention

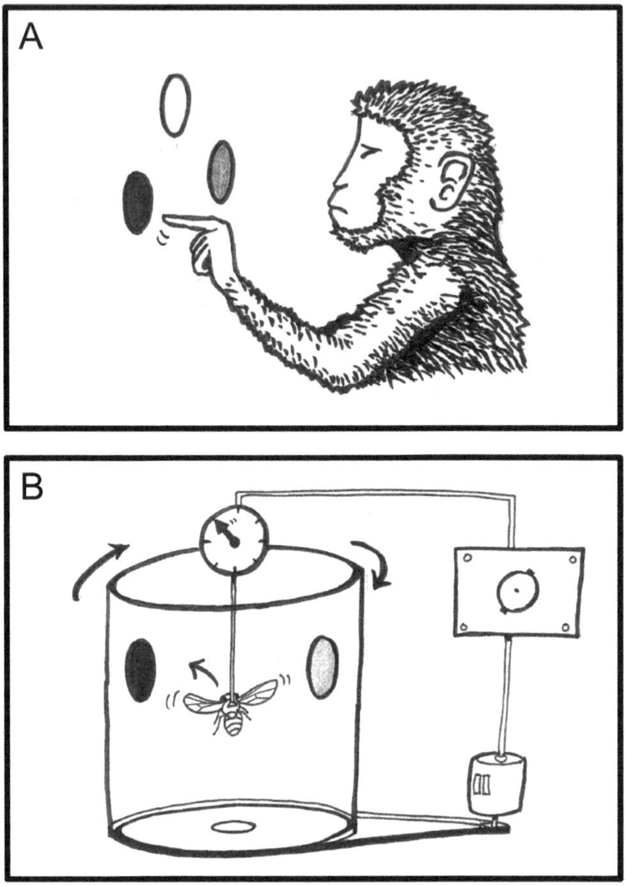

FIG. 1. Attention paradigms in monkeys and flies. (A) A delayed match to sample task (DMST), where choice is provided (black and gray circles) following a cue (white circle). (B) In the flight arena, tethered flies display choices by flying toward a visual object. A torque meter (dial) detects behavior, and a closed-loop circuit provides biofeedback for the fly (Heisenberg and Wolf, 1984).

processes. This problem applies to any animal studies, whether in a monkey, rat, or fly. The connection, if any, between attention processes and habit formation is unknown, just as it is not known whether top-down or bottom-up attention represents a continuum in the brain rather than discrete processes.

To summarize, there are a number of criteria that need to be satisfied for identifying attention-like processes in any animal, and it often appears that these criteria become stricter the further away from humans we explore for attention.

First, the original operational definition of attention put forth by Posner *et al.* (1980) still holds: it is the differentiation of objects of interest from other objects in the environment. To test this requires measuring responses to competing objects in space and time, such as targets and distracters. Ideally, some demonstration of stimulus suppression should be shown, and, finally, a thorough study of attention should also demonstrate neural correlates of these effects in the animal's brain. Research on attention in *D. melanogaster* has fulfilled all of these criteria, as will be reviewed here.

II. Classical Behavioral Approaches: Tethered Flight

Studies of attention in *Drosophila* have for the most part been an offshoot of studies of fly vision. This is for several reasons pertaining to the unique qualities of vision. First, the visual world is extremely cluttered, even for flies, and attention-like qualities would appear to be required for effective navigation through a barrage of competing cues. Early work on hoverfly flight, for example, showed how flies can track a singularity (e.g., another fly) while ignoring other motion cues (Collett, 1980). In more controlled experimental contexts, we find vision to be an ideal modality for the study of attention because visual stimuli can be partitioned in time and space more effectively than olfactory or auditory stimuli, for example. Such effective segregation of competing cues is crucial, as discussed above, for assigning behavioral responsivess to one or another competing stimulus, and by the same token, for demonstrating suppression of certain stimuli. Several decades ago, flight arenas were engineered to study visual responses in tethered flies (Poggio and Reichardt, 1976; Gotz, 1980), and these could be easily adapted to questions relevant to visual attention, such as issues of perceptual load and response latency. Using these devices (Fig. 1B), it was soon evident that optomotor behavior did not always follow the predictions of a classic input–output system (i.e., the Reichardt detector for motion processing (Reichardt, 1969)). For example, flies typically respond in open loop to a horizontally moving bar by a torque displacement in the same direction, a classic optomotor response (Heisenberg and Wolf, 1984). Introducing a second moving bar, in the opposing hemi field for example, creates unpredictable behavior assigned to one or the other bar, rather than a summation of both effects. This suggested early on that flies were ignoring entire percepts rather than integrating the whole visual scene into a predictable output. A number of experiments performed by Reinhard Wolf and Martin Heisenberg in the early 1980s solidified this anecdotal observation into real data.

The first visual attention experiments in the tethered flight arena were aimed at quantifying the effect on the optomotor response of introducing a competing visual stimulus. In the first set of experiments, the classic optomotor response was decreased by approximately 50% by introducing a static distracter (Wolf and Heisenberg, 1980). This suggested that the addition of a competing visual object increased perceptual load, and that half the time the fly's attention "spotlight" was not directed to the object that eventually moved. In a second similar set of experiments, opposing vertical bars (one to either side of the tethered fly), were oscillated front-to-back at different phases (Heisenberg and Wolf, 1984). Presented singly, each object would evoke compensatory optomotor responses in phase with the visual motion. Presented together, fly behavior was observed at times to track one or the other phase rather than a summation of both (Fig. 2A). This suggested

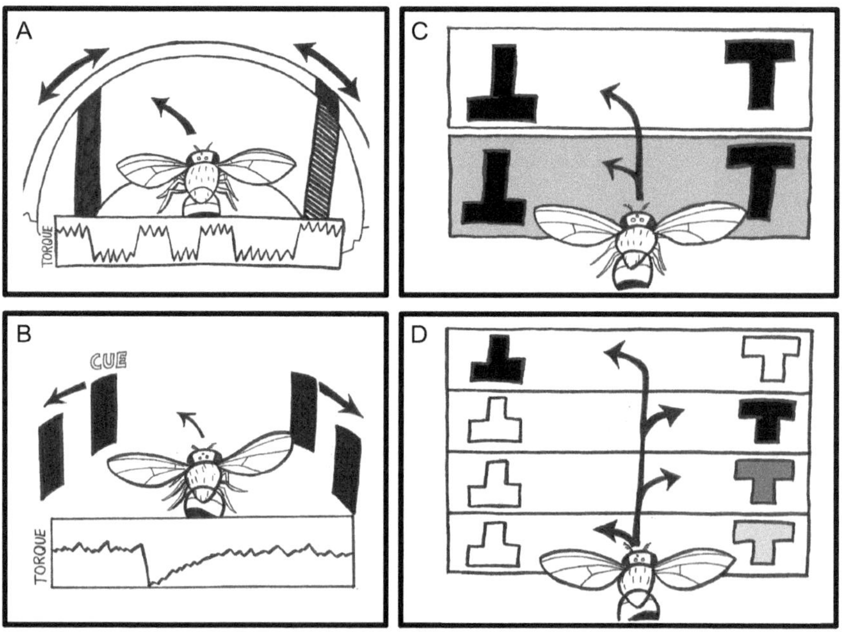

FIG. 2. Flight arena experiments. (A) Vertical bars oscillating out of phase in front of either eye can evoke responses (torque) locked to one or the other object in alternation (Wolf and Heisenberg, 1980). (B) A transient cue on the left (e.g., an oscillating bar) will bias fly responsiveness (torque) toward subsequent moving objects on the left, and suppress responses to simultaneous movement on the right (Sareen *et al.*, 2011). (C) Flies can be trained (by using heat as a negative re-enforcer) to discriminate visual objects (upper panel), and such learning is preserved through a change in background context (lower panel) (Liu *et al.*, 1999). (D) Flies can be trained to discriminate compound stimuli (Brembs and Heisenberg, 2001) (color and shape defines each object in the top panel), but will attend to individual features (e.g., color) when faced with contradictory cues (second panel). Attention will switch to the alternate feature (e.g., shape, bottom panel) at a defined threshold (Tang and Guo, 2001; Zhang *et al.*, 2007).

again that an attention spotlight was directed at only one percept while the other was suppressed. In addition to these stimulus-response experiments, closed-loop experiments (where the torque behavior of the fly actually controls the angular position of visual objects) showed that flies were able to fixate on targets while ignoring background motion (Wolf and Heisenberg, 1991).

These first forays into *Drosophila* attention-like behavior were performed on mechanical flight arenas, which involved motor-controlled drums rotating around the fly. When multiple objects were presented under separate control, as in the case of adding distracters or background motion, above, this meant adding successive layers of transparent drums. This arrangement was not ideal for fine control of visual stimuli, which is crucial for testing attention. For example, physical stimuli painted on rotating drums cannot be made to suddenly appear or disappear, so cuing or distraction effects are difficult to test. The development of virtual arenas using light emitting diodes (LEDs) solved this problem, but first considerable legwork was needed to show that optomotor responses to virtual motion using LEDs were comparable to previous experiments with physical objects (Dickinson *et al.*, 1993; Tammero *et al.*, 2004).

Thirty years after performing their seminal attention experiments in *Drosophila*, Wolf, Heisenberg, and colleagues have shown that responses of flies in the newly designed LED arenas correspond to behaviors conforming to an operational definition of attention (Sareen *et al.*, 2011). Taking advantage of the fine temporal control provided by the LEDs, the authors first tested a cueing effect on optomotor competition. This was done by oscillating one of two competing vertical bars (at 5 Hz) for 5 s prior to the displacement of either object. It was found that cueing significantly biased subsequent optomotor responses to the cued object (Fig. 2B). Importantly, the cueing effect could be dissociated in time (otherwise a direct interaction between the cue and test stimuli, such as mutual inhibition of central pattern generators, might be invoked). The study found that the salient cue attracted fly attention to its side even after a delay of 2 s, but not for longer delays such as 5 s. Further explorations using this paradigm found that the cue could attract attention to one side even if it appeared within a window 20° around the object, but not any larger, thereby defining a "window" size for the attention spotlight. Interestingly, the cueing was only effective in the lower visual field (LVF) of the fly, and this may be explained by the possibility that objects below the fly, such as landing and food sources, need to be attended more carefully. Finally, very salient distracters could override the system, as is to be expected for bottom-up attention effects. Together, this recent study provides the best behavioral evidence to date that fly attention is measureable, and that cueing effects within a narrow spatial and temporal range can modulate the choices made by a tethered fly.

In the interim between these two attention studies, the mechanical flight arena paradigm was nevertheless put to good use to study visual learning in *Drosophila* (Heisenberg *et al.*, 2001). Exploiting the closed-loop possibilities of the setup, the same group and colleagues employed an operant conditioning paradigm (Wolf and Heisenberg, 1991) to unveil ever-increasing evidence of the sophistication of *Drosophila* visual behavior. First, it was found that flies could be trained to discriminate objects following classical conditioning with heat (the unconditioned stimulus), and that such closed-loop fixation was focused on the center of the object rather than its boundaries (where temperature changes occurred during training) (Wolf and Heisenberg, 1997). This implied that flies had indeed made a classical association between the unconditioned stimulus (US) and a visual object, and were able to discriminate among distinct objects by fixating preferentially on one object while ignoring the competing visual stimuli still within the fly's visual field.

By being able to measure fly choice following training, the flight arena next provided an obvious paradigm testing visual flexibility and perception in flies, by changing the stimuli after training. A first series of experiments asked whether learned objects were position-invariant, meaning that an object is perceived as the same even if it impacts a different latitude on the fly's retina (this is possible to test because the tethered flies cannot move up or down). If learning were retinotopic, then flies would not recognize a displaced object as being the same. Initial studies suggested that fly visual learning was retinotopic (Dill *et al.*, 1993), but subsequent studies found that flies did display learning even if the conditioned objects were displaced on their vertical or horizontal visual field (Tang *et al.*, 2004). It appears that these contradictory results stem from a peculiarity of fly vision: it is extremely sensitive to context (Brembs and Wiener, 2006). In the original retinotopic studies, modifying the visuals after training (actually exchanging physical sheets on the drum) probably provided substantial contextual contamination for the conditioned stimulus.

Flies can nevertheless extract features from their context. This was demonstrated in a study where visual training and testing for learning of distinct objects (upright and inverted Ts) was performed in different color backgrounds (blue or green). Wild-type flies could still learn following a change in context (Fig. 2C), suggesting that they can suppress the background and extract the feature to be learned (Liu *et al.*, 1999). This simple experiment has relevance for studies of attention: context generalization can be considered to be a form of selective attention because the animal has to suppress one visual stream while selecting another. Subsequent studies by Bjoern Brembs, a student of Heisenberg's, confirmed that context generalization is a feature rather than a failure of the fly brain (Brembs and Heisenberg, 2001; Brembs and Hempel de Ibarra, 2006; Brembs and Wiener, 2006). Indeed, the context need not be constrained to a background, but can be an aspect of the feature itself, such as its shape, position, or color. Not surprisingly, flies are thus able to perform some rather complex feats of visual learning, such as compound learning (Brembs and Heisenberg, 2001).

Following the success of contextual experiments in the arena, the path was set for increasing the sophistication of visual learning studies in flies, eventually rivaling the kinds of experiments previously thought to be reserved for higher animals such as primates. These experiments were mostly conducted in the laboratory of Aike Guo, who also studied with Martin Heisenberg and Reinhard Wolf. By working with compound stimuli (upright or inverted Ts of either green or blue), a first study presented contradictory cues following training (upright green became blue) to test which feature of the stimulus (color or shape) was selected (color was). The color intensity was then gradually decreased in different experiments to determine the level when flies might switch to the shape feature (Fig. 2D). Interestingly, this switch happened within a rather narrow range, and the data were described by a sigmoidal function, which is often characteristic of attention-like processes in higher animals (Tang and Guo, 2001). The same conclusion, including the sigmoidal shape of decision making in flies, was reached in a later study combining color and center-of-gravity as the compound features (Zhang *et al.*, 2007). Again, although these were not attention studies *per se*, the work supported the notion that flies are able to select a feature of their visual environment while suppressing others, and that changes are subject to experience-dependent thresholds set in the fly brain. Further evidence of the "sigmoidal" nature of decision making in flies was found more recently using LED arenas, where within a single experiment, a visual could be gradually changed from vertical bar (which is attractive) to a square (which is repulsive): flies maintained fixation on a diminishing bar up to a point, whereupon they switched to antifixation, putting the now repulsive square behind them (Maimon *et al.*, 2008). The connection with this result and work from the Guo laboratory lies in the nonlinear dynamics of the behavior, which resembles attention-like effects in higher animals.

Salience is a key feature of attention. This refers to the conspicuousness of a feature, such as how luminous it is, and typically salient objects attract more attention. Predictions relating object salience to fly fixation behavior were tested in another study from the Guo laboratory (Xi *et al.*, 2008). In this study, the conspicuousness of a feature (a vertical bar) relative to background noise (contrast or spatial) was gradually decreased (in different experiments), showing that fixation behavior was correspondingly weakened. Olfactory distracters could similarly weaken fixation behavior on an object, as if perceptual resources were momentarily shifted to another modality. Finally, visual selection among three competing objects was biased toward the stronger (higher contrast) stimulus. These results were not necessarily surprising, except that ablation of an olfactory learning and memory center, the mushroom bodies, altered responsiveness thresholds for salience. Indeed, the mushroom bodies were found to modulate various aspects of complex visual learning, from context generalization (Liu *et al.*, 1999) to resolving contradictory cues (Tang and Guo, 2001; Zhang *et al.*, 2007), to establishing salience thresholds (Xi *et al.*, 2008). On the other hand, mushroom body neurons do not appear to be required for simple visual learning (Wolf *et al.*, 1998); instead, neurons of the central complex are required for elemental visual learning (Liu *et al.*, 2006;

Pan *et al.*, 2009). The neuroanatomy of visual learning, simple and nonelemental, will be revisited below in a discussion of possible attention systems in the fly brain.

There are limitations to purely behavioral approaches for studying attention, especially in animals lacking a behavioral report—but this applies to all nonhuman animals, and also some humans such as infants or paraplegic individuals. How does one separate bottom-up "reflexes" from top-down attention? Especially in flies, it is difficult to show that modulation of an optomotor reflex can be a form of attention rather than a failure (or inhibition) of a simple circuit response. If an animal responds to one stimulus and not another presented in competition, it is difficult to be certain that the animal has seen the competitor and is actually suppressing it. Ironically, the *Drosophila* flight arena presents one of the best paradigms for tackling this problem because the tethered fly cannot move away and cannot close its eyes; therefore, we are more confident about its tightly controlled visual world. In comparison, there is much less contextual control for rats or mice walking around a 5CSRTT chamber, the preferred apparatus for studying attention in rodents. Having taken fly behavior as far as possible for measuring various operational definitions of attention, it is clear that the next course needs to be insight into what is happening in the fly brain while the fly performs attention-like responses. In humans and other mammals, selective attention is associated with distinct neural correlates, such as γ-band (30–80 Hz) synchrony (Buzsaki, 2006). One might expect that if flies are really paying attention to their visual world, their brain activity might also display associated neural signatures.

III. Adaptations to Electrophysiology

Visual attention should be represented by neural correlates in the brain, for example, by increased activity of selected objects and/or decreased activity of ignored objects (Fries *et al.*, 2001). In mammalian preparations, brain activity (recorded either as field potentials of neuronal populations or as single neuron activity) informs the discussion of mechanisms of attention. For example, an ongoing debate in the field is whether attention involves increased gain of selected objects, suppression of ignored objects, or some mixture of both types of gain control (Maunsell and Treue, 2006). When multiple recording sites are considered, attention has often been found to be associated with neural synchrony or coherence within specific frequency domains. Clearly, the addition of brain recording electrodes to an attention experiment can provide substantial supporting evidence to determine whether a simple animal such as a fly is actually paying attention.

There is a long history of recording from the fly brain (Borst *et al.*, 2010). However, this has been primarily aimed at understanding visual processing in the optic lobes of the brain. Early work on blowflies showed that large neurons in the lobula plate of blowflies spiked in response to moving visuals, and this preparation was soon taken up by a number of laboratories to investigate whether a neural code existed for integrating motion information across the fly retina (Egelhaaf *et al.*, 2002). Some lobula neurons only fired when objects moved front to back, whereas others fired when objects moved up and down. An ongoing debate has ensued around the question of whether visual information is processed as a temporal code (where the timing of every spike counts) or as a rate code (where the amount of spikes is what matters) (Nemenman *et al.*, 2008). More recently, novel recording techniques have moved the research questions to *Drosophila*, where genetic tools such as labeling or silencing of neurons can be used to sharpen the debate (Borst, 2009). Surprisingly, there have been few studies to date asking whether visual responses in the optic lobes can be modulated by arousal states. One study in bees found that motion-responsive neurons in the lobula became quiescent during sleep (even in the presence of motion), and that these could be "awakened" by stimulating the contralateral eye (Kaiser and Steiner-Kaiser, 1983). A series of studies in crayfish found that responses to flickering lights were altered in the central brain of the arthropod during sleep (Mendoza-Angeles *et al.*, 2007, 2010; Ramon *et al.*, 2004). That the suppression of stimuli, as expected during sleep, might also be measurable in attention experiments may seem obvious, but no studies specifically designed experiments around that question in insects until quite recently.

As the best evidence for visual selective attention in insects came from *Drosophila* in the flight arena, it is perhaps not surprising that this paradigm should have provided the obvious platform for probing neural correlates of attention in flies. This was actually more difficult than expected, perhaps due to the small size of the fly brain. A recent study modified the original tethered fly design to specifically test for attention-like effects in the fly brain, correlated to behavioral choices made by flies (Tang and Juusola, 2011). Three small wires were inserted into the fly brain, one in either optic lobe and a reference wire in the dorsal central brain. Two kinds of signals were recorded from the fly brain, depending on the sampling frequency: a local field potential (LFP), which typically reveals 1–100 Hz effects resulting from the summed activity of groups of neurons, and spiking activity, sampled at much higher frequencies (up to 25 KHz), to reveal action potentials of neurons close to the electrode. Both signals reflect neuronal activity, although not necessarily simultaneously. In mammalian preparations, spiking activity of individual neurons is often registered by recording intracellularly or from multiple nearby electrodes at the same time (e.g., tetrodes). This ensures that the same neuron (or "unit") is identified through time, rather than a mixture of different spikes. In the current visual attention study, for technical reasons, spiking activity was most likely gathered from a mixture of nearby neurons.

Rather than using a circular drum around the fly, as in most previous flight arena experiments, the authors presented a continuously looping patterned strip to either eye. This way, visual responses to the left or to the right eye could be evoked separately during the brain recordings. Presenting competing motion stimuli to either eye also allowed the authors to ask whether torque behavior in the fly alternated left and right, much like for Wolf and Heisenberg's attention experiments using competing oscillating stimuli in the arena (Wolf and Heisenberg, 1980). A form of stimulus rivalry was indeed seen in this novel preparation, with flies alternating flight directions when competing gratings moved front to back for either eye (Fig. 3A). These sustained behavioral choices were markedly different than the classic, much smaller optomotor responses to continuous field rotations, suggesting flies were selecting one flow and ignoring the other for periods of time. The behavioral responses were clearly not just input driven as they were not a summation of either stimulus, and onset of choice behavior was variable, sometimes delayed for up to a second. Brain recordings during these experiments

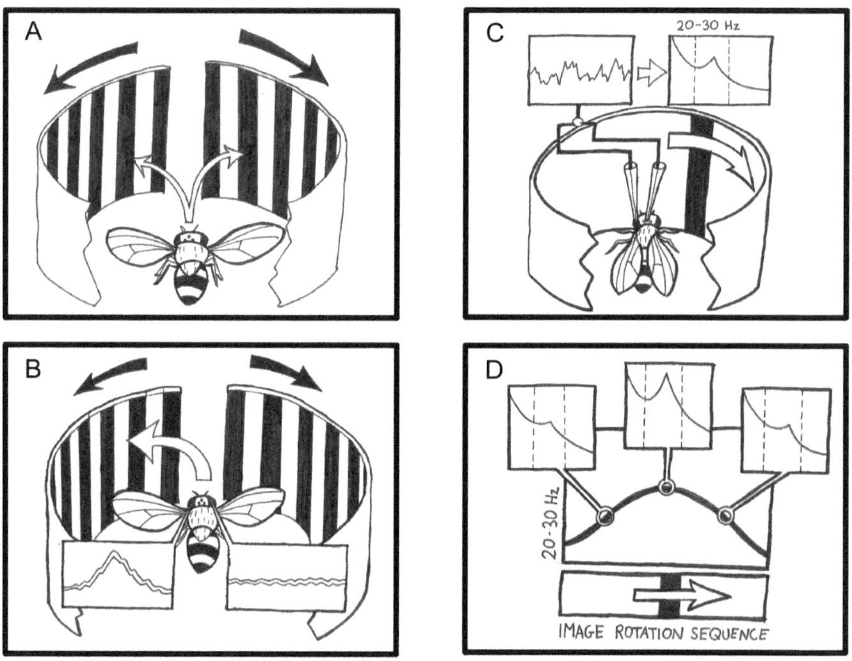

FIG. 3. Recording from the fly brain. (A) Competing moving gratings presented to either eye evoke alternating choice behavior (Tang and Juusola, 2011). (B) Recordings from either optic lobe reveal boosted brain activity (local field potential [LFP] amplitudes and spikes) on the behaviorally selected side (inset boxes) (Tang and Juusola, 2011). (C) Moving visual objects evoke 20–30 Hz LFP activity in the fly brain (van Swinderen and Greenspan, 2003). This is determined by a spectral analysis of the LFP signal (inset boxes). (D) The amplitude of the 20–30 Hz frequency band can be mapped onto the image rotation sequence, with the greatest amplitude occurring when the object sweeps in front of the fly.

revealed that the LFP was boosted on the side of the brain associated with eventual choice behavior (Fig. 3B). Indeed, the first neural responses (both LFP and spikes) started well before the behavioral response, at about 20 ms compared to an average of 300 ms for the behavior. The LFP was also anticipatory in nature, with a gradual shift in LFP amplitude between the left and right optic lobes predicting a subsequent shift in choice behavior (i.e., a left flight switching to a right flight). Finally, the boosted LFP on the side associated with behavioral choice occurred within a specific frequency range, between 20 and 50 Hz. A similar LFP effect was found in an earlier study: when flies initiate behavioral fixation (in closed loop) on a visual object, this is associated with a transient burst of 20–30 Hz activity (van Swinderen and Greenspan, 2003). In the more recent study, power in this range alternated between the optic lobes in correlation with the fly's behavior. Interestingly, boosted 20–50 Hz activity was even evident when the fly was not flying, but passively exposed to a moving grating only on one side. This raises the question of whether flies can attend to stimuli without actually performing a directed behavior, much like humans can attend to an object without necessarily walking toward it. The co-occurrence of 20–50 Hz LFP activity during passive exposure or active choice suggests that this may be the case.

The above work is consistent with the results from a series of earlier studies examining LFPs in tethered but nonflying *Drosophila* exposed to visual stimuli (van Swinderen, 2007a; van Swinderen and Brembs, 2010; van Swinderen and Greenspan, 2003; van Swinderen *et al.*, 2009). In these studies, LFPs were sampled at 300 Hz as a voltage differential between an electrode implanted in the left optic lobe and in the central brain. A recording preparation originally devised to measure neural correlates of sleep and wake in the fly brain (Nitz *et al.*, 2002) was adapted to visual experiments relevant to selective attention (van Swinderen and Greenspan, 2003). Visual salience could be evoked by a number of manipulations, such as heat, novelty, or odors, and these manipulations produced a transient increase in 20–30 Hz activity in the fly brain (Fig. 3C) which could be mapped onto the image position as it rotated around the fly (Fig. 3D). In general, the increase in salience did not last more than 20 s. The 20–30 Hz response could also be bound to another modality, such as odor. A repeating odor puff (a banana smell controlled by a valve driver) synchronized to the recurrence of a moving object (a vertical bar) specifically increased the salience of the visual, while the smell itself evoked responses in a different frequency range (70–80 Hz). Finally, this first study showed that classical conditioning (using heat as a US) could bias 20–30 Hz effects to either of two competing visual stimuli, and this selection/suppression effect could be detected even when the two objects were physically overlapping (by one object rotating around the fly twice as fast as the other).

The key to being able to assign attention-like effects in the fly brain to individual stimuli is to have them temporally tagged in some way. In human attention studies, competing visuals are often flickered at distinct frequencies, and the power or coherence of these separable frequencies can be detected by electro-encephalography (EEG) or magneto-encephalography (MEG) approaches (Vialatte *et al.*, 2010). Typically, attended flickering objects are represented in the brain activity by increased amplitude or coherence effects, and it is thus in principle possible to infer what flickering stimulus a human subject is paying attention to by measuring which tag has greater power. In the preceding attention experiments, virtual objects in an LED arena were rotated around the fly, 180° apart, with a period of 3 s (Fig. 4A). This way, each object swept in front of the fly at a different time (for a different 1.5 s epoch); therefore, each object was tagged by its specific timing when it was in front of the fly. Importantly, 20–30 Hz responses to salient objects (e.g., by making them novel) was increased when these swept in front of the fly while responses to the competing object were suppressed—even though it also swept in front of the fly (Fig. 4B). Such temporal tagging of either object is one way of associating brain activity with competing stimuli; another way would be to flicker them at distinct frequencies or move them at different speeds around the fly.

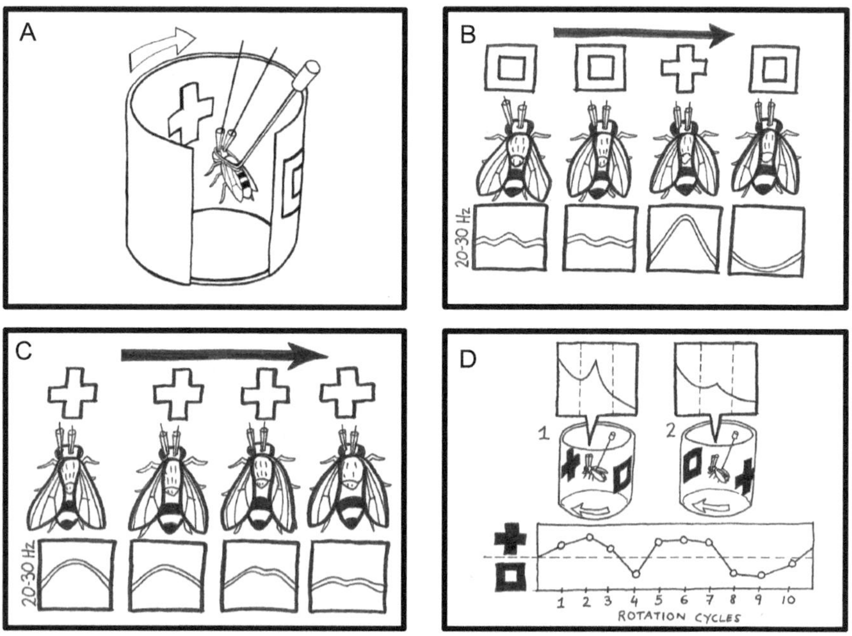

Fig. 4. Local field potential (LFP) recordings and visual attention (A) Two objects 180° apart rotate around the fly, once every 3 s, evoking 20–30 Hz responses at different times. (B) 20–30 Hz responsiveness is increased for visual novelty (the cross) and suppressed for the competing non-novel object (van Swinderen, 2007a; van Swinderen and Greenspan, 2003). (C) 20–30 Hz responsiveness to visual salience is abolished after three to four exposures of the stimulus (van Swinderen, 2007a, 2007b). (D) 20–30 Hz responsiveness for successive exposures of competing stimuli defines a time series, plotted as a 20–30 Hz ratio for successive image rotation cycles (van Swinderen, 2007b; van Swinderen and Brembs, 2010).

A recurrent argument in studies of attention, especially in smaller animals such as insects, is whether we are simply measuring habituation. Especially in the context of novelty conditioning, one could imagine that the transient responsiveness of the LFP to a novel object and subsequent attenuation might reflect desensitization effects in the eye. Behavioral observations, discussed earlier, show that flies can behave counter to expectations if habituation were the only mechanism involved here. For example, visual cuing can be separated from their targets by up to 2 s (Sareen *et al.*, 2011), visual learning can be position invariant (Tang *et al.*, 2004), or flies alternate their choice behavior to explore their visual environment, sometimes against optomotor expectations (Tang and Juusola, 2011). For electrophysiology, suppression effects argue against habituation, and novelty responses in the LFP have also been shown to be position invariant, arguing that this effect cannot be retinotopic (van Swinderen, 2007a). The argument could of course always be made that the LFP simply reflects habituation effect deeper in the brain, among neurons that code for the percepts (e.g., a cross or a square) rather than for the impacted photoreceptors. This argumentation, however, simply blurs the difference between attention and habituation; the dynamic properties of perceptual representations in the fly brain can be studied, regardless of what the phenomenon is called.

The study of visual learning and perception in *Drosophila* has provided valuable temporal information that is relevant to attention studies, although these data are often overlooked. We have seen in recent studies, discussed above, that fly choice behavior can be delayed by a second, following the presentation of competing visual stimuli (Tang and Juusola, 2011) and that a visual cue can remain in a fly's working memory for 2 s (Sareen *et al.*, 2011). In classical condition experiments, the performance index following training in the arena is usually represented as a single histogram describing choice bias over 2 min of tethered flight. Closer examination of this 2-min flight reveals that flies may require several seconds of exploration before "settling" on a decision, especially when nonelemental learning is involved, such as position invariance experiments (Tang *et al.*, 2004) or involving contradictory cues (Zhang *et al.*, 2007). These observations again argue against a habituation effect. The relevance of habit formation to visual learning has been studied specifically, with interesting implications for attention in flies. It appears that overtraining in the flight arena (by doubling the training time from 8 to 16 min) consolidates the motor component of learning (i.e., flying right is going to keep me from getting hot) while compromising the flexibility required to display fact learning (what object is not hot?) through alternate motor strategies (Brembs, 2009). On the other hand, short training sessions promoted the flexibility required for fact learning while preventing the rigidity of motor learning. These fascinating observations were possible by dissociating these different aspects of learning (motor from fact) in carefully designed flight arena experiments (Brembs, 2009). The implication for attention is that flexibility is required during early stages of fly learning, whereas extensive training overrides attention systems to promote defined motor responses in the learned context. These results have clear implications even beyond fly work: most learning studies involve some overtraining of sorts.

Timing effects are also evident in LFP activity in the fly brain. In one study, the design of novelty salience experiments involved presentation of two identical objects (the training phase), whereupon one of these objects would change (this being an LED arena) to a competing choice (van Swinderen, 2007a). Following this change, the LFP was transiently increased for the novel object and simultaneously suppressed for the non-novel object (independent of position on the visual field, since the objects were continuously rotating around the fly). It was found that 50 s of training produced a selection/suppression effect, but 25 s of training did not. As this was a recurrent stimulus scenario (where training and test phases alternated for 10 cycles), this suggested that 25 s of training was not long enough to make a previously seen object salient again, thereby defining a possible window of working memory for the fly (<25 s). Interestingly, this window length was comparable with the duration of salience effects following training, also around 20 s (van Swinderen and Greenspan, 2003).

The observation that brain LFP activity could be transiently increased for one visual tag and suppressed for a competing visual raised the question: what is a fly's attention span in this paradigm? Is it really 20 s? The answer most likely has less to do with absolute time than with the number of repeated presentations of an object. In the LED arena, virtual objects were made to rotate around the fly with a period of 3 s. This means that following a salience manipulation such as classical conditioning with heat, competing choices swept in front of the fly every 3 s. A cycle-by-cycle analysis (tied to the image rotation period) of LFP effects revealed that selection/suppression in the 20–30 Hz range were significant for three to four successive presentations of the visual objects (Fig. 4C). After that, individual sweeps could be biased for one or the other object, but this varied among flies meaning that on average the attention-like effects disappeared after four successive presentations of a salient object (van Swinderen, 2007a; van Swinderen and Brembs, 2010; van Swinderen *et al.*, 2009).

Salience can be set by the experimenter, using heat, novelty, odors, and the like, but presumably salience is also a dynamic property that depends on an animal's recent attention history. For example, something that has not been attended to for a while might become salient again, whereas an attended object loses salience after repeated exposures. The same preparation to test for salience effects in the LFP was applied in subsequent studies to measure alternation dynamics in 20–30 Hz selection/suppression directed to two competing objects presented for extended periods (van Swinderen, 2007b; van Swinderen and Brembs, 2010). The way this was done was by calculating a ratio of 20–30 Hz power for each image rotation cycle, where the numerator was summed 20–30 Hz power for one object (1.5 s of time when it was in front of the fly) and the denominator was summed 20–30 Hz power for the 1.5 s when the alternate object swept in front of the fly. This produced a time series of 20–30 Hz ratios (Fig. 4D) that could then be statistically compared with randomized data, using permutations of 20–30 Hz ratios from the same fly. Interestingly, this analysis revealed that even without experimenter-imposed salience, the LFP displayed a form of hysteresis where power would be biased for one or the other object for more cycles than expected by chance (van Swinderen, 2007b; van Swinderen and Brembs, 2010). This defined a possible "attention" span for flies, once again tied to image exposures rather than absolute time. In tethered flight, flies also alternate behavioral choices, and brain activity is correlated with these nonrandom alternations. Interestingly, changes in LFP activity in either optic lobe appeared to predict subsequent behavioral alternations (Tang and Juusola, 2011).

One of the implications of these and other visual attention studies in flies is that the animals have a flexible "spotlight" of attention that can cover a wide range of their visual field, even though the animals are fixed in space by a tether. We saw in the behavioral studies, discussed earlier, that cueing effects can occur 20° away from the visual target, and that the LVF is more sensitive to such cueing effects than the upper visual field (Sareen *et al.*, 2011). These and other experiments also showed that attention can be directed to one hemifield at a time, that is, the left eye versus the right eye (Tang and Juusola, 2011). The LFP recording experiments also provide some information regarding a possible attention spotlight in flies. First, the 20–30 Hz response appears to peak "in front" of the fly, meaning that LFP amplitude increases as the attended object sweeps by a ~90° wedge in front of the tethered animal. Importantly, when a competing nonsalient object sweeps by this same quadrant, 20–30 Hz responsiveness is suppressed, suggesting that the fly's window of attention is either elsewhere or shut down when the nonsalient competing object sweeps by. It is interesting to speculate why, on average, the LFP response is greatest directly in front of the fly in this particular preparation. One possibility is that flies in general move forward, so it would be logical for a default window of attention to be centered, on average, in front of the animal. However, this positional response is malleable. Attaching salience to a rotating object in the LED arena shifts the 20–30 Hz response by 0.5 s, bringing it closer to its recurrent point of origin (van Swinderen and Greenspan, 2003). Thus, a salient object rotating clockwise around the fly will evoke a 20–30 Hz response further to the left. This suggests that flies may be anticipating the objects appearance, and correspondingly shifting their window of attention to an earlier portion of the image rotation sequence.

It is especially striking that LFP effects in the fly brain display attention-like qualities even in animals that are not actively behaving in a goal-directed manner. This suggests that attention can be uncoupled from behavior, even in insects. However, such studies are also problematic because they lack a behavioral readout. Just as behavior without electrophysiology is never entirely convincing, the reverse is also true. Indeed, as for the study of sleep, behavioral tests remain the most convincing way of assessing whether attention exists in a simple animal. There have been very few studies combining electrophysiology and behavior in *Drosophila*, especially in the context of visual studies. One measured LFPs during fixation in the flight arena and another measured LFPs and spikes in a flight choice paradigm, both discussed above. The gold standard of electrophysiology is the patch-clamp recording, where information can be gained on how a single neuron's membrane potential is responding to a stimulus. Although this has been applied for a number of olfactory studies (Tanaka *et al.*, 2009) as well as a few visual studies (Joesch *et al.*, 2008; Schnell *et al.*, 2010), where the fly is often so restrained as to preclude any behavioral readout, only one study to date has succeeded in patching *Drosophila* brain neurons while animals are behaving, for example, in flight or walking. In this study (Maimon *et al.*, 2010), the authors show that the gain of visual interneurons (the well-studied VS cells, which respond to vertical motion) is doubled when animals are flying. Reduced membrane resistance in these neurons during flight suggests that synaptic drive from upstream motion-sensitive inputs is increased. This boosted visual response is consistent with the LFP results in the attention paradigm discussed above. Although the patch-clamp experiments were not designed as attention experiments *per se* (i.e., there was no "choice" in the flight behavior), it is worth speculating that the boosted intracellular activity reflects an attention-like process that gets activated once the fly starts flying; it seems logical that attention-like mechanisms need to be notched up in a flying animal, whereas a stationary animal may not require such active selection/suppression processes. Another study described a similar result, but employing instead optical imaging of visual neurons in walking flies (Seelig *et al.*, 2010). Together, these recent studies lend support to the idea that active behavior boosts attention processes in flies, although their viewpoint is that peripheral systems (e.g., neurons in the optic lobes) are directly controlled by behavioral state rather than by some central or brain-wide attention process.

IV. Related Behavioral Measures

Attention is a complex phenomenon, and to detect and measure attention in flies seems to therefore require complex devices. These have included, so far, tethered flight arenas (section II) and miniature brain recording paradigms (section III). The advantage of these sophisticated devices is that they can continuously track a signal (such as torques or LFP amplitudes), and in this way describe fine-grained changes that can be correlated with stimuli in attention experiments. This is in contrast, for example, to olfactory learning paradigms where performance is measured as a single distribution of flies between two tubes (Tully and Quinn, 1985). The lack of access to behavioral processes in most *Drosophila* population assays, which measure fly distributions, makes it difficult to study attention, which is most concerned with processes through time. That said, not every laboratory has easy access to the sophisticated tools and devices that have been used to study visual learning and attention. Indeed, there are to date only half a dozen laboratories in the world doing flight arena work, and even fewer recording from the fly brain during visual experiments. What other approaches can be used to study visual attention in flies? This problem is especially relevant for eventually applying molecular genetic strategies to the question. The sophisticated preparations that have unveiled attention in *Drosophila* are not amenable to genetic screens, partly because they are laborious single-animal assays, but also because many mutants do not fly readily in the arena, and are therefore unable to "report" their attentional states that way. Simpler and high-throughput paradigms need to be developed to further the study of attention in *Drosophila*, but it is imperative that these actually measure phenotypes relevant to attention.

Attention is as much about stimulus suppression as it is about selection (Bichot and Desimone, 2006). An animal must be able to ignore a competing cue, even though it may be sometimes stronger than the attended cue. The expectation for simple animals such as flies is that they will always respond "reflexively" to salient stimuli, such as a bright light or a moving grating. This is in fact not the case. There is increasing evidence that experience can modulate simple reflexes in flies, not by habituation but by learning or by an internal motivational state. One way of studying attention indirectly is by using assays that measure suppression of a simple reflex. Some of these assays can be easily scaled up for genetic screening purposes, and subsequent focused experiments using the sophisticated attention paradigms discussed above might then be used to confirm or further characterize attention-like phenotypes in these mutants. Two assays of this sort are the aversive phototaxic suppression (APS) paradigm and the visual maze (Fig. 5).

Flies are attracted to light and if given a choice will overwhelmingly prefer to walk down a lit corridor than a dark one. Such observations have guided some of the first genetic studies in this model, such as Seymour Benzer's countercurrent assays that identified the *photophobe* mutant (Ballinger and Benzer, 1988) and catalyzed similar designs for olfactory paradigms. Although phototaxis may appear to be a hard-wired reflex, it can be suppressed by training: associating a lit corridor with an aversive-tasting stimulus, such as quinine, will bias subsequent approaches (in retrials of the same fly) toward the alternate dark corridor, such that after 12–16 training sessions a wild-type will make the unnatural choice of choosing darkness more than half the time (Fig. 5A) (Le Bourg and Buecher, 2002; Seugnet *et al.*, 2009). To do this, flies must suppress their innate phototaxic tendency. It is hypothesized that this mechanism may be related to the suppression

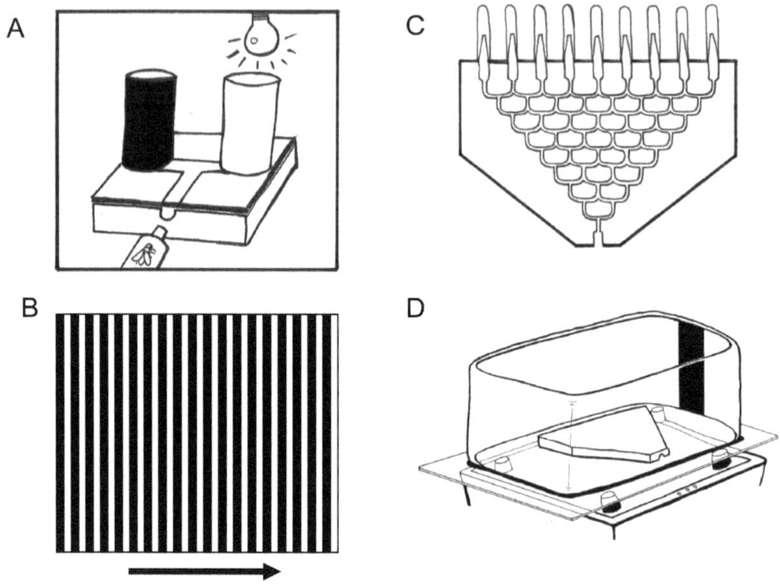

FIG. 5. Other behavioral paradigms 1. (A) The aversive phototaxis suppression (APS) assay measures suppression of a reflex to follow light, by associating a lit chamber with an aversive taste such as quinine (Le Bourg and Buecher, 2002). By 14–16 repeated trials, around half the flies will suppress phototaxis and enter the dark chamber instead (Seugnet *et al.*, 2009). (B) Flies display optomotor reflexes, by following the motion of a moving grating for example. (C) An eight-point choice maze placed over a moving grating will fractionate flies for visual behavior, yielding a score based on the distribution of flies among the nine collection tubes (van Swinderen and Flores, 2006). (D) Attention paradigms can be designed around the visual response, by for example providing visual distracters which suppress the optomotor reflex (van Swinderen, 2007a; van Swinderen and Flores, 2006).

involved in attention experiments, namely that central processes in the fly brain override input from the periphery to guide behavioral choices. Although the APS paradigm is quite laborious (single flies must be followed, collected, and re-run), it possible to imagine ways of multiplexing this paradigm to screen for visual suppression effects. Variations on this paradigm have also been applied to color choice behavior (Zhou *et al.*, 2010).

Flies have a tendency to follow motion (Fig. 5B). This reflex is termed an optomotor response (Gotz and Wenking, 1973; Gotz, 1980), and this behavior has been intensively studied in the flight arena as well as in tethered or freely walking paradigms. The expectation, as with phototaxis, is again that this is a hard-wired reflex. Although it is accurate to say that fly behavior is strongly dependant on the physical aspects of a motion stimulus (grating velocity, density, luminosity, and contrast have all been carefully studied with regard to their effect on the optomotor response (Borst *et al.*, 2010)), it is also true that individual flies do not always respond according to optomotor expectations on a trial-by-trial basis, and flies have been shown to "ignore" optomotor cues, as in the attention experiments described above. An indirect way of studying attention might then be to measure optomotor suppression behavior, much like phototaxic suppression. This would require an efficient way of first measuring visual behavior in flies, preferably via a population paradigm in freely walking flies.

Population paradigms aimed at segregating *Drosophila* populations for optomotor traits have been developed decades ago by Gotz, Heisenberg, and colleagues (Gotz, 1970; Bausenwein *et al.*, 1986). These involved rather bulky contraptions, featuring rotating drums and choice corridors, and an experiment could last several hours. Nevertheless, the first optomotor mutant, *optomotor blind*, was isolated using such a device (Pflugfelder and Heisenberg, 1995). A different approach was introduced in 2006 (van Swinderen and Flores, 2006), exploiting a simple design used originally in geotaxis assays for flies (Toma *et al.*, 2002). In this novel paradigm, flies navigated an eight-point choice maze while they were exposed to moving visuals displayed on a computer monitor (Fig. 5C). At each choice point, flies can make either a left or a right turn, and this decision is influenced by the moving visuals flowing beneath them on the monitor. Each experiment, comprising about 30 flies running the maze, is completed in about 2 min, and a total of 4–8 maze runs yields sufficient data to quantify visual responsiveness in a strain. Upon completing the maze, flies end in one of nine collection tubes, and a weighted average of the distribution of flies among the nine tubes provides a visual response score for the strain. For an experiment, visuals are run in either direction (left to right and right to left) to balance any possible odor trails resulting from previous experiments or unaccounted-for asymmetries in the visual environment of the maze. Typically, wild-type flies will produce a visual response of 0.7 ± 0.1 to moving green/black gratings. The maze paradigm has recently been improved to a modular design, and the application of infra-red fly counters has allowed for the system to be fully automated, from introducing the flies to statistical analyses, thus providing a highly efficient way of assaying visual behavior in *Drosophila* (Evans *et al.*, 2011).

Experiments in the maze revealed that visual responsiveness in wild-type flies could be quite fickle, with the geometry of objects outside the transparent maze having a significant influence on the visual response. For example, one anecdotal observation was that visual responsiveness could be abolished by a person standing next to the maze during an experiment. This suggested that flies navigating the maze were partitioning their perceptual resources between the moving grating beneath their feet and surrounding objects around the maze. Although this called for careful control of the surrounding maze environment, it also implied that attention-like processes might be interfering with the optomotor response. Following this serendipitous observation, the maze paradigm was redesigned to allow for visual distracters to be presented to one or either side of the maze

(Fig. 5D), initially as a static black cardboard bars (van Swinderen and Flores, 2006). The addition of these distracters showed that (1) flies could see through the maze and were attracted to these objects, and (2) the presence of a competing visual distracter could abolish the visual response to moving gratings in the maze. Refinement of the distracter design, by placing it on liquid crystal display (LCD) panels to the sides of the maze, allowed for quantification of the distracter effect in subsequent studies (van Swinderen *et al.*, 2009).

The ability to completely abolish optomotor responsiveness in wild-type flies by a simple static visual placed to the side of the maze suggested that the response to moving gratings was not as hard-wired as once thought, at least in wild-type flies. Indeed, responsiveness to the static distracter suggested that wild-type flies could suppress their reflex to follow a moving grating, if they were instead paying attention to a distant object. This implied that flies were constantly engaged in optomotor suppression to some level, and that strains defective in visual attention would not be able to suppress the response to the moving grating as well. This hypothesis was first tested by a selective breeding scheme, much like the original geotaxis breeding experiments that isolated "hi" and "low" geotaxis lines (Toma *et al.*, 2002). Selective breeding of flies with increased visual responsiveness to a moving grating for over 20 generations produced individuals that were insensitive to the presence of a static distracter outside the maze (van Swinderen and Flores, 2006). Electrophysiology experiments on these genetic variants revealed that these flies indeed were characterized by an attention defect, using a novelty salience paradigm previously applied to wild-type flies. The simple maze design was thus validated as a useful approach to screen for *Drosophila* attention mutants, by paradoxically looking for strains with *increased* visual responsiveness to the grating. Increased responsiveness would thus be suggestive of a *failed* ability to suppress a prepotent optomotor reflex, much like failed phototaxis suppression would also signify a breakdown in fly cognitive processes (Seugnet *et al.*, 2008, 2009). These two assays, the optomotor maze and the APS paradigm, both reveal potential attention phenotypes by a failure to suppress an innate reflex. The same rationale, measuring increased visual reflexes, has also been used to study attention defects in humans (Rubinstein and Abel, 2011; Williams *et al.*, 2006). Results from first genetic screens in *Drosophila* using these assays will be discussed in the next section.

There are other *Drosophila* walking assays that can inform about visual selective attention, although these are mostly single-animal paradigms that measure selection of objects rather than suppression of alternate choices. Recent developments combining flexible visual displays with fly tracking software have created interesting possibilities for designing attention experiments in walking flies (Branson *et al.*, 2009). Most of the original open-field walking assays were based on Buridan's paradigm, originally used by Karl Gotz and colleagues to describe the visual control of walking behavior in flies (Gotz, 1980) (Fig. 6). In this paradigm, flies'

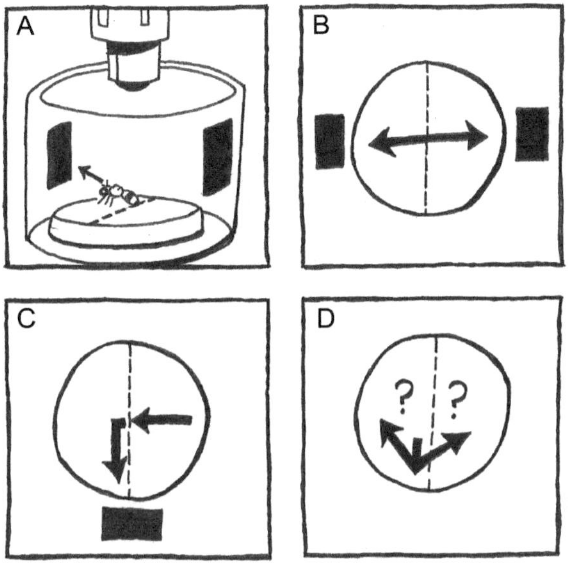

FIG. 6. Other behavioral paradigms 2. (A) In Buridan's paradigm, are placed in a chamber with competing visuals displayed on the wall, and walking choice behavior is tracked. (B) Flies will typically walk back and forth between the competing visuals (Gotz, 1980). (C) Changing the position of visual cues will re-orient fly locomotion. (D) Flies can still orient to temporarily invisible landmarks, based on their prior experience, thereby displaying a spatial memory (Neuser *et al.*, 2008).

wings are cut off and the following day they are placed in a small circular arena, usually surrounded by a moat of water (Fig. 6A). Visual objects are displayed on the walls of the arena, and the original observation by Gotz was that flies will continuously walk back and forth between two vertical bars positioned 180° apart on the arena (Fig. 6B, Buridan's ass, in Greek mythology, was similarly mesmerized by two choices). Numerous variations on this paradigm have revealed object preferences in flies, but it was not until the development of fly tracking and closed-loop feedback to LED arenas that experiments could be performed relating more specifically to visual selective attention. In a series of remarkable studies, Roland Strauss and colleagues exploited the fine temporal control of the LED system to change visual objects as the fly walked in the arena (Neuser *et al.*, 2008; Schuster *et al.*, 2002; Strauss and Pichler, 1998). For example, a visual target could disappear as the fly walked toward it, and a distracter could appear at another position on the drum (Fig. 6C). Using such a paradigm, it was shown that wild-type *Drosophila* have a spatial orientation memory for object positions in the arena, and will even walk toward *invisible* objects based upon their working memory of where the visual target used to be (Fig. 6D) (Neuser *et al.*, 2008). One of the hallmarks of top-down attention in humans is that it interacts with memory systems, and by demonstrating a memory component to visual fixation in flies (without the object actually being present) effectively separates the visual response from an immediate visual input. The implication for attention in flies is tremendous, as this could be analogous to the fly attending to a memory. To understand the relationship between attention and memory is an obvious next step in the field, and demonstrations of both phenomena happening concurrently are crucial to moving the field forward.

V. Neural Substrates of Visual Attention in *Drosophila*

There is comparatively much less known about the molecular and neural underpinnings of vision in *Drosophila* than there is for olfaction. Consequently, even less is known about the neural substrates of visual attention in the fly. Nevertheless, a synthesis from genetic work done in the multiple paradigms discussed earlier paints a preliminary picture of neurons and molecules likely to modulate attention in the fly.

Olfactory learning in *Drosophila* is centered on the mushroom bodies (Fig. 7A) (Heisenberg *et al.*, 1985), which are paired structures in the central brain with direct input from the antennal lobes, which process odors (Keene and Waddell, 2007). Each mushroom body comprises ~2500 neurons, called Kenyon cells, and these are organized into distinct lobes that have been associated with different temporal phases of odor learning (Krashes *et al.*, 2007). Early on, the mushroom bodies were found not to be required for simple visual learning in the flight arena (Wolf *et al.*, 1998), and this was perhaps not surprising as there are no direct inputs from the optic lobes to the mushroom bodies in the fly (Heisenberg, 2003). What was surprising was that almost every subsequent study on visual learning in *Drosophila* has implicated the mushroom bodies for more complex forms of visual learning. The first indication of this was for context generalization: ablating the mushroom bodies (chemically or genetically) rendered flies unable to show learning in a different context, such as a different background color illumination (Liu *et al.*, 1999). Then, in a series of publications from the Guo laboratory, the mushroom bodies were also found to be required for resolving contradictory visual cues (Tang and Guo, 2001; Zhang *et al.*, 2007) and establishing salience levels for visual stimuli (Xi *et al.*, 2008). These results confirmed that the mushroom bodies are not just involved in olfactory processing. This had already been suggested by studies finding the mushroom bodies to be involved in motor control. Ablating the mushroom bodies appears to increase locomotor activity, suggesting that these neurons are somehow involved in suppressing motor output (Martin *et al.*, 1998). One question raised at this point is whether the mushroom bodies form part of a

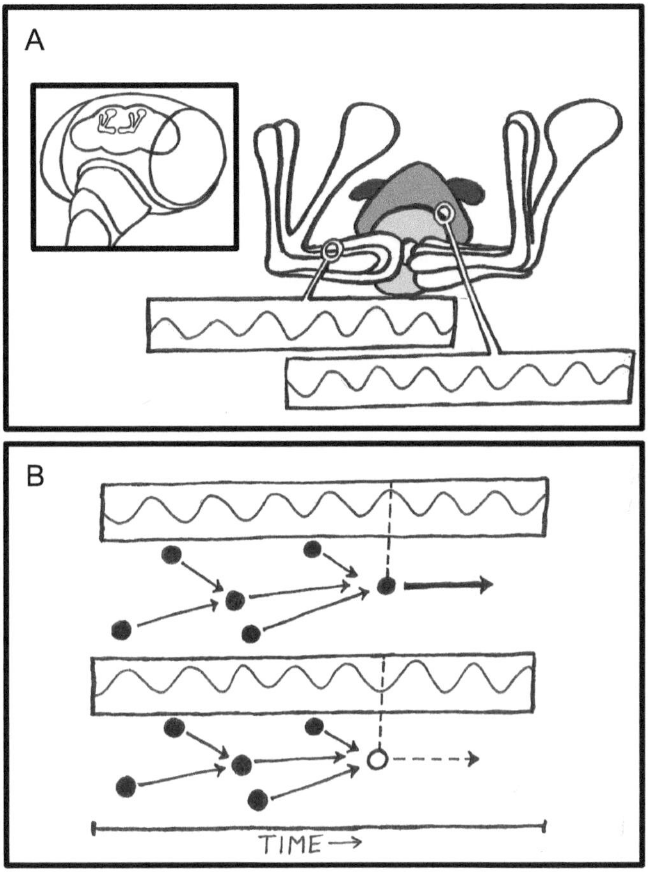

FIG. 7. A model of attention in the *Drosophila* brain. (A) The mushroom bodies (white) have been associated with olfactory memory and contextual visual learning. The central complex (gray), in particular the fan-shaped body (dark gray), has been associated with simple visual learning (Liu *et al.*, 2006). Both structures might be involved, together with neuromodulators such as dopamine, in generating patterns of oscillatory neural activity in the fly brain. Inset, a fly head with brain and mushroom bodies drawn. (B) Stimuli produce consequences (e.g., a behavior or a memory) via a polychromous network of precisely timed neuronal firing events (Izhikevich, 2006). Subthreshold oscillations could enable or suppress these polychronous groups depending on the phase of the background oscillation (black circle: firing neuron; open circle: nonfiring neuron).

broader suppression circuit because context generalization (for example) by definition implies suppression of nonpredictive cues, such as background color. Are these different forms of suppression related?

Recent work by Bjoern Brembs, using the flight arena, addressed this question in an indirect way. As discussed earlier, overtraining in the arena was found to promote motor learning at the expense of behavioral flexibility (Brembs, 2009). This habit formation paradigm was found to be mushroom body–dependent; without the mushroom bodies, flies behaved as if they were overtrained even with the short training paradigm. This suggested that the mushroom bodies were required to promote behavioral flexibility, thereby allowing flies to display their learning via different motor programs. Without the mushroom bodies, motor learning took over and flies relied on just one skill (e.g., flying only left). Such lost flexibility may be useful in some scenarios, such as for behavioral responses that need not change, but will of course be maladaptive for behaviors where attention is involved. A synthesis with the preceding results involving the mushroom bodies in motor suppression, nonelemental learning, and setting salience thresholds would propose that these neurons are providing a gating mechanism for visual behavior: they suppress motor programs in order to promote a level of behavioral flexibility tied to the outcome of a selective attention contest among competing percepts in the brain. With increased training and habit formation, this circuit is somehow bypassed. The relevance of the mushroom bodies to visual selective attention has since been supported by electrophysiological studies where these structures have been silenced during salience paradigms (van Swinderen *et al.*, 2009).

Dopamine is a neurotransmitter that has been tightly associated with aversive olfactory conditioning in flies, but it now appears that this molecule may play a role in attention as well. Dopamine was originally proposed as a carrier for negative valence cues, such as electric shocks, whereas another neurotransmitter, octopamine, signaled rewarding cues such as sugar (Schwaerzel *et al.*, 2003). Dense dopamimergic innervations of the mushroom bodies supported this role in olfactory learning and memory, and convincing models have been proposed whereby coincident dopamine and olfactory input changes mushroom body neuron physiology, and that this memory trace biases subsequent odor presentations to guide the behavioral choices made by flies (Keene and Waddell, 2007). Nevertheless, dopamine has been found to affect several behaviors in addition to olfactory learning. Broadly, dopamine modulates general arousal levels: decreased dopamine promotes sleep, whereas increased dopamine promotes locomotion (Andretic *et al.*, 2005). Specifically, dopamine seems to regulate arousal thresholds for a variety of stimuli, from mechanical shocks to visual flashes or odors. Compromising dopamine signaling acutely, either by increasing or decreasing it, seems to have a common consequence: arousal thresholds are decreased (see (van Swinderen and Andretic, 2011) for a review of dopaminergic effects on fly behavior). As selective attention is also concerned with dynamic setting of arousal thresholds (to competing stimuli), it is possible that there exists a connection between dopamine's role in arousal and a potential role in attention.

The possible overlap between systems affecting olfactory learning and visual learning and attention extends beyond the mushroom body neurons and dopamine. Some olfactory learning and memory mutants are also visual learning and attention mutants. This was first discovered by testing classical mutants such as *dunce*^{1}, *rutabaga*^{2080}, and *amnesiac*^{1} in Pavolvian conditioning paradigms associating colors with shaking in walking flies (Folkers, 1982). The suspicion that some of these mutations might be multimodal was confirmed by classical conditioning in the flight arena, for mutants that were capable of maintaining sustained flight (Gong *et al.*, 1998). Visual learning and memory deficits in *dunce*^{1} and *rutabaga*^{2080} raised the possibility that these mutations were affecting more general aspects of memory formation that impact both sensory modalities. Although both of these genes, which are involved in cyclic AMP signaling in neurons (Davis *et al.*, 1995), are strongly expressed in the mushroom bodies (Nighorn *et al.*, 1991), they are also expressed in other structures, such as in neurons of the central complex—set of structures positioned between the mushroom bodies (Fig. 7A). The role of the central complex in simple visual learning was confirmed through genetic rescue of the *rutabaga*^{2080} visual learning defect (Liu *et al.*, 2006). Expression of a wild-type copy of *rutabaga* in the fan-shaped body of the central complex restored simple visual learning (associating a shape with heat), and finer-grained rescue experiments revealed that different layers of the fan-shaped body could rescue distinct learning of distinct visual features. Subsequent *rutabaga* rescue experiments demonstrated that the ellipsoid body, another central complex structure, is also involved in simple visual learning, and that an interaction between the fan-shaped body and the ellipsoid body is likely in the control of visual learning in flies (Pan *et al.*, 2009). These two structures show up again as necessary for visual performance in Buridan's paradigm (Fig. 6), and the ellipsoid body specifically was also found to be crucial for spatial learning in walking flies (Neuser *et al.*, 2008).

There thus appears to be a separation between olfactory learning in the mushroom bodies and visual learning in the central complex. The fact that complex visual learning requires the mushroom bodies complicates this picture. One possibility is that the mushroom bodies are required for attention-like processes, as suggested earlier. Why would an "olfactory" neuropil be recruited for visual attention and nonelemental visual learning? One radical possibility is that a primary function for these neurons is stimulus gating, attention, and context generalization, but that simple odor learning assays were not designed to test for this, thereby yielding learning mutants that were really attention mutants. More likely is the possibility that the mushroom bodies are multimodal processing centers. Indeed, olfaction in insects may have predated mushroom body evolution (Strausfeld *et al.*, 2009). Although the mushroom bodies are clearly involved with olfactory discrimination and memory in flies, the neuronal architecture and connectivity of these structures may have been exploited for brain-wide attention processes as well. For example, olfactory input to the mushroom bodies has been shown in locusts to generate a 20–30 Hz oscillation, which can then be used as a timer of sorts to gate the occurrence of spiking activity in these neurons (Laurent, 2002). The 20–30 Hz oscillation is propagated by successive feed-forward and feed-back systems to target neurons beyond the mushroom bodies (Cassenaer and Laurent, 2007), thus setting up an oscillation in the brain that could potentially be used to gate other modalities in addition to olfaction. The fact that such a brain-wide currency (e.g., a 20–30 Hz oscillation) was generated through olfactory input may be irrelevant; in other creatures such an oscillation may be generated through other modalities. Thus, one hypothesis for the relevance of the mushroom bodies (and perhaps dopamine) for visual attention is that these neurons are a necessary component for generating a brain-wide oscillation that gets used for dynamically selecting and suppressing stimuli from the environment (Fig. 7B). In support of a multimodal role for these neurons traditionally associated with olfaction, transiently silencing mushroom body neurons attenuates the 20–30 Hz response in the fly brain associated with visual salience (van Swinderen *et al.*, 2009), as does transiently silencing dopamine (Andretic *et al.*, 2005).

The possibility of screening for visual attention mutants was introduced with the optomotor maze (Fig. 5), discussed earlier. It was suggested that attention-defective mutants might display *increased* visual responsiveness in the maze because they might be unable to suppress an optomotor reflex. As discussed above, silencing either the mushroom bodies or dopaminergic neurons increased visual responsiveness in this paradigm. A similar effect was seen in a screen of existing *Drosophila* mutants, with *dunce*^{1} and *rutabaga*^{2080} displaying stronger responsiveness to a moving grating than any other mutant tested (van Swinderen, 2007a). Predictably, these learning mutants were also unresponsive to stationary distracters. Brain recordings showed that LFP correlates of visual attention were also compromised in these mutants: there was no detectable 20–30 Hz selection/suppression effect following novelty salience (van Swinderen, 2007a). *Dunce*^{1} visual defects could be rescued by expressing the wild-type protein in mushroom body and central complex neurons (van Swinderen *et al.*, 2009), and it appears that *dunce* expression is required during the pupal stage of development; rescue only at the larval stage or only at the adult stage was not possible (van Swinderen, 2007a; van Swinderen *et al.*, 2009). This suggests that the *dunce* gene (phosphodiesterase II) is required during development to "wire" a circuit that is subsequently necessary for attention-like behavior in adult flies. This result contrasts with the role of these molecules in olfactory learning, which can be restricted to expression in adult neurons (McGuire *et al.*, 2003). Consequently, the role of *dunce* or *rutabaga* in attention processes is probably not direct: many molecules can disrupt wiring during development, and one prediction would be that many of these might produce attention-like defects (and consequently increased visual responsiveness in adult flies), for different reasons.

The connection between visual attention and memory mutants is further exemplified by the *radish*^{1} mutation. *radish*^{1}, originally isolated as a mutant for anesthesia resistant memory (ARM) (Folkers *et al.*, 1993), is the sole member of a memory consolidation pathway that is proposed to be parallel to the *dunce* or *rutabaga* pathway (Isabel *et al.*, 2004). In olfactory memory assays, *radish*^{1} flies are not able to remember after they have been briefly anesthetized (by cooling), whereas wild-type flies are able to carry a learned response through this transient manipulation. The *radish* gene, with no close homology to human genes (Folkers *et al.*, 2006), is expressed strongly in the mushroom bodies, as are *dunce* and *rutabaga*. The connection between *radish*^{1} and visual attention was uncovered using the same optomotor screening device that showed increased responsiveness in *dunce*^{1} and *rutabaga*^{2080}, except that *radish*^{1} was found to be less responsive to moving gratings and more distractible than wild (van Swinderen and Brembs, 2010). Brain recordings and behavioral experiments in the flight arena confirmed that *radish*^{1} is also an attention mutant, but different than *dunce*^{1}: *radish*^{1} flies displayed a 1–2 Hz behavioral "jitter" when presented with competing visuals in the arena. Furthermore, 20–30 Hz responsiveness to novelty (in brain recordings) lasted only one exposure, compared to 3 or 4 in wild type (van Swinderen and Brembs, 2010). In other words, *radish*^{1} was found to have a short attention span, possibly caused by an inherent 1–2 Hz oscillation that may periodically reset visual perception in the animal.

It is by now clear that learning and memory mutants originally isolated in olfactory paradigms are afflicted with other behavioral or brain defects when one looks more closely. Indeed, an optomotor screen of a panel of long-term memory mutants revealed yet another batch of candidates for visual attention studies (van Swinderen *et al.*, 2009). One open-ended question at this point is whether these attention-like effects cause the memory defects in any way. It is likely that attention and memory are intertwined and mutually causal: attention promotes memory formation, but memory also guides attention (Chun and Turk-Browne, 2007). Attention deficits are probably an important factor in eventual defective memory formation, consolidation, or retrieval. It is perhaps not so surprising that the same genes and neurons that have been central to understanding (olfactory) memory formation are being found to be relevant for attention. However, there is an inherent bias in this position: all of the visual learning or attention mutants were found by testing existing olfactory learning and memory mutants, many of which were uncovered in forward-genetic screens. To be truly convincing, visual attention and visual learning screens need to uncover their own mutants from scratch, as was done for olfactory paradigms. Fortunately, efficient visual paradigms now exist to do this (Evans *et al.*, 2011).

VI. Conclusion and Future Directions

Attention has traditionally been compared to a "spotlight" or a "zoom lens" (Eriksen and St James, 1986; LaBerge, 1983). To some extent, this is what attention feels like for us, especially as we consciously focus on a visual target. This anthropocentric view may have distracted us from a broader definition of attention that may apply more readily to other animals. Attention is also concerned with suppressing incoming sensory information from multiple modalities, and this capacity may have evolved very early in animal evolution to provide experience-dependent flexibility in motile creatures. Rather than being simple stimulus-driven systems, animals became able to ignore the barrage of sensory information around them, to only process immediately relevant stimuli, and thus to learn. The emergence of experience-dependent stimulus-suppression mechanisms may have coincided with evolution of simple brains as well as with sleep—another stimulus-suppression mechanism. Whether stimulus selection and suppression involve different mechanisms is unclear, although some combined form of local gain control coupled to suppression of competing stimuli is likely in mammals (Moran and Desimone, 1985). The last 30 years of visual studies in *Drosophila* have provided good evidence for attention in insects. Future research will determine whether fly brains attend the same way we do.

There are two broad approaches to studying attention-like processes in *Drosophila*. One is to devise very sophisticated behavioral or brain recording devices to detect the difficult measures relevant to attention: stimulus selection/suppression, context, cueing, distraction. This has been the traditional approach in the field, namely to treat the fly as much as possible like a "mini-human." This approach has clearly been successful and has been perhaps the best way to provide to level of confidence among the scientific community at large that simple animals such as flies may have a selective attention. Unfortunately, because the methods used to investigate attention in flies convincingly can be so difficult, they are less amenable to exploiting *Drosophila* in the way that it should: by high-throughput behavioral screens coupled to molecular genetics. It is by homing in on key behavioral measures that mysteries such as circadian rhythms or learning and memory were unraveled in the *Drosophila* model (Vosshall, 2007). To expect to do the same for attention using the sophisticated devices described in the first two sections of this review is perhaps unrealistic. Instead, it may be more productive to identify key measures that reflect attention-like processes in the fly. One of these is clearly the suppression of reflexes, and efficient screens can be designed around this problem, only then followed by the more sophisticated approaches in select strains of interest. This solution is comparable with the approach taken by *Drosophila* sleep researchers. To study sleep requires satisfying multiple criteria, from measuring arousal thresholds to homeostasis or brain recordings or conducting some pharmacology (Shaw and Franken, 2003). However, the *Drosophila* sleep community convinced themselves that one measure—5 min of continued quiescence (Shaw *et al.*, 2000)—was sufficient to proceed with sleep screens, before going back to check the other criteria in strains of interest. This insight has been crucial to moving the sleep field forward, even beyond *Drosophila*. Such a strategy for visual attention in *Drosophila* is likely to produce a similar outcome.

References

Andretic, R., van Swinderen, B., and Greenspan, R.J. (2005). Dopaminergic modulation of arousal in Drosophila. *Curr. Biol.* **15**, 1165–1175.

Ballinger, D.G., and Benzer, S. (1988). Photophobe (Ppb), a Drosophila mutant with a reversed sign of phototaxis; the mutation shows an allele-specific interaction with sevenless. *Proc. Natl. Acad. Sci. U S A* **85**, 3960–3964.

Bausenwein, B., Wolf, R., and Heisenberg, M. (1986). Genetic dissection of optomotor behavior in Drosophila melanogaster. Studies on wild-type and the mutant optomotor-blindH31. *J. Neurogenet.* **3**, 87–109.

Bichot, N.P., and Desimone, R. (2006). Finding a face in the crowd: parallel and serial neural mechanisms of visual selection. *Prog. Brain Res.* **155**, 147–156.

Borst, A. (2009). Drosophila's view on insect vision. *Curr. Biol.* **19**, R36–R47.

Borst, A., Haag, J., and Reiff, D.F. (2010). Fly motion vision. *Annu. Rev. Neurosci.* **33**, 49–70.

Branson, K., Robie, A.A., Bender, J., Perona, P., and Dickinson, M.H. (2009). High-throughput ethomics in large groups of Drosophila. *Nat. Methods* **6**, 451–457.

Brembs, B. (2009). Mushroom bodies regulate habit formation in Drosophila. *Curr. Biol.* **19**, 1351–1355.

Brembs, B., and Heisenberg, M. (2001). Conditioning with compound stimuli in Drosophila melanogaster in the flight simulator. *J. Exp. Biol.* **204**, 2849–2859.

Brembs, B., and Hempel de Ibarra, N. (2006). Different parameters support generalization and discrimination learning in Drosophila at the flight simulator. *Learn Mem.* **13**, 629–637.

Brembs, B., and Wiener, J. (2006). Context and occasion setting in Drosophila visual learning. *Learn Mem.* **13**, 618–628.

Buzsaki G,,Buzsaki G. (2006). *Rhythms of the Brain.* Oxford University Press, New York.

Cassenaer, S., and Laurent, G. (2007). Hebbian STDP in mushroom bodies facilitates the synchronous flow of olfactory information in locusts. *Nature* **448**, 709–713.

Chun, M.M., and Turk-Browne, N.B. (2007). Interactions between attention and memory. *Curr. Opin. Neurobiol.* **17**, 177–184.

Collett, T. (1980). Angular tracking and the optomotor response: an analysis of visual reflex interaction in the hoverfly. *J. Comp. Physiol.* **140**, 145–158.

Davis, R.L., Cherry, J., Dauwalder, B., Han, P.L., and Skoulakis, E. (1995). The cyclic AMP system and Drosophila learning. *Mol. Cell Biochem.* **149–150**, 271–278.

Dickinson, M.H., Lehmann, F.O., and Gotz, K.G. (1993). The active control of wing rotation by Drosophila. *J. Exp. Biol.* **182**, 173–189.

Dill, M., Wolf, R., and Heisenberg, M. (1993). Visual pattern recognition in Drosophila involves retinotopic matching. *Nature* **365**, 751–753.

Egelhaaf, M., Kern, R., Krapp, H.G., Kretzberg, J., Kurtz, R., and Warzecha, A.K. (2002). Neural encoding of behaviourally relevant visual-motion information in the fly. *Trends Neurosci.* **25**, 96–102.

Eriksen, C.W., and St James, J.D. (1986). Visual attention within and around the field of focal attention: a zoom lens model. *Percept. Psychophys.* **40**, 225–240.

Evans, O., Paulk, A.C., and van Swinderen, B. (2011). An automated paradigm for Drosophila visual psychophysics. *PLoS One* **6**(6); e21619.

Folkers, E. (1982). Visual learning and memory of Drosophila melanogaster Wild Type CS and the mutants Dunce, Amnesiac, Turnip, and Rutabaga. *J. Insect Physiol.* **28**, 535–539.

Folkers, E., Drain, P., and Quinn, W.G. (1993). Radish, a Drosophila mutant deficient in consolidated memory. *Proc. Natl. Acad. Sci. U S A* **90**, 8123–8127.

Folkers, E., Waddell, S., and Quinn, W.G. (2006). The Drosophila radish gene encodes a protein required for anesthesia-resistant memory. *Proc. Natl. Acad. Sci. U S A* **103**, 17496–17500.

Fries, P., Reynolds, J.H., Rorie, A.E., and Desimone, R. (2001). Modulation of oscillatory neuronal synchronization by selective visual attention. *Science* **291**, 1560–1563.

Gong, Z., Xia, S., Liu, L., Feng, C., and Guo, A. (1998). Operant visual learning and memory in Drosophila mutants dunce, amnesiac and radish. *J. Insect Physiol.* **44**, 1149–1158.

Gotz, K.G. (1970). Fractionation of Drosophila populations according to optomotor traits. *J. Exp. Biol.* **52**, 419–436.

Gotz, K.G. (1980). Visual guidance in Drosophila. *Basic Life Sci.* **16**, 391–407.

Gotz, K.G., and Wenking, H. (1973). Visual control of locomotion in the walking fruitfly Drosophila. *J. Comp. Physiol.* **85**, 235–266.

Heisenberg, M. (2003). Mushroom body memoir: from maps to models. *Nat. Rev. Neurosci.* **4**, 266–275.

Heisenberg, M., Borst, A., Wagner, S., and Byers, D. (1985). Drosophila mushroom body mutants are deficient in olfactory learning. *J. Neurogenet.* **2**, 1–30.

Heisenberg, M., and Wolf, R. (1984). *Vision in Drosophila: Genetics of Microbehavior.* Springer-Verlag, Berlin, New York.

Heisenberg, M., Wolf, R., and Brembs, B. (2001). Flexibility in a single behavioral variable of Drosophila. *Learn Mem.* **8**, 1–10.

Isabel, G., Pascual, A., and Preat, T. (2004). Exclusive consolidated memory phases in Drosophila. *Science* **304**, 1024–1027.

Itti, L., and Koch, C. (2000). A saliency-based search mechanism for overt and covert shifts of visual attention. *Vision Res.* **40**, 1489–1506.

Izhikevich, E.M. (2006). Polychronization: computation with spikes. *Neural Comput.* **18**, 245–282.

James, W. (1890). *The Principles of Psychology.* Henry Holt and Co., New York.

Joesch, M., Plett, J., Borst, A., and Reiff, D.F. (2008). Response properties of motion-sensitive visual interneurons in the lobula plate of Drosophila melanogaster. *Curr. Biol.* **18**, 368–374.

Kaiser, W., and Steiner-Kaiser, J. (1983). Neuronal correlates of sleep, wakefulness and arousal in a diurnal insect. *Nature* **301**, 707–709.

Keene, A.C., and Waddell, S. (2007). Drosophila olfactory memory: single genes to complex neural circuits. *Nat. Rev. Neurosci.* **8**, 341–354.

Krashes, M.J., Keene, A.C., Leung, B., Armstrong, J.D., and Waddell, S. (2007). Sequential use of mushroom body neuron subsets during drosophila odor memory processing. *Neuron* **53**, 103–115.

LaBerge, D. (1983). Spatial extent of attention to letters and words. *J. Exp. Psychol. Hum. Percept. Perform.* **9**, 371–379.

Laurent, G. (2002). Olfactory network dynamics and the coding of multidimensional signals. *Nat. Rev. Neurosci.* **3**, 884–895.

Le Bourg, E., and Buecher, C. (2002). Learned suppression of photopositive tendencies in Drosophila melanogaster. *Anim. Learn Behav.* **30**, 330–341.

Liu, G., Seiler, H., Wen, A., Zars, T., Ito, K., Wolf, R., Heisenberg, M., and Liu, L. (2006). Distinct memory traces for two visual features in the Drosophila brain. *Nature* **439**, 551–556.

Liu, L., Wolf, R., Ernst, R., and Heisenberg, M. (1999). Context generalization in Drosophila visual learning requires the mushroom bodies. *Nature* **400**, 753–756.

Maimon, G., Straw, A.D., and Dickinson, M.H. (2008). A simple vision-based algorithm for decision making in flying Drosophila. *Curr. Biol.* **18**, 464–470.

Maimon, G., Straw, A.D., and Dickinson, M.H. (2010). Active flight increases the gain of visual motion processing in Drosophila. *Nat. Neurosci.* **13**, 393–399.

Martin, J.R., Ernst, R., and Heisenberg, M. (1998). Mushroom bodies suppress locomotor activity in Drosophila melanogaster. *Learn Mem.* **5**, 179–191.

Maunsell, J.H. (2004). Neuronal representations of cognitive state: reward or attention? *Trends Cogn. Sci.* **8**, 261–265.

Maunsell, J.H., and Treue, S. (2006). Feature-based attention in visual cortex. *Trends Neurosci.* **29**, 317–322.

McGuire, S.E., Le, P.T., Osborn, A.J., Matsumoto, K., and Davis, R.L. (2003). Spatiotemporal rescue of memory dysfunction in Drosophila. *Science* **302**, 1765–1768.

Mendoza-Angeles, K., Hernandez-Falcon, J., and Ramon, F. (2010). Slow waves during sleep in crayfish. Origin and spread. *J. Exp. Biol.* **213**, 2154–2164.

Mendoza-Angeles, K., Cabrera, A., Hernandez-Falcon, J., and Ramon, F. (2007). Slow waves during sleep in crayfish: a time-frequency analysis. *J. Neurosci. Methods* **162**, 264–271.

Moran, J., and Desimone, R. (1985). Selective attention gates visual processing in the extrastriate cortex. *Science* **229**, 782–784.

Nemenman, I., Lewen, G.D., Bialek, W., and de Ruyter van Steveninck, R.R. (2008). Neural coding of natural stimuli: information at sub-millisecond resolution. *PLoS Comput. Biol.* **4**, e1000025.

Neuser, K., Triphan, T., Mronz, M., Poeck, B., and Strauss, R. (2008). Analysis of a spatial orientation memory in Drosophila. *Nature* **453**, 1244–1247.

Nighorn, A., Healy, M.J., and Davis, R.L. (1991). The cyclic AMP phosphodiesterase encoded by the Drosophila dunce gene is concentrated in the mushroom body neuropil. *Neuron* **6**, 455–467.

Nitz, D.A., van Swinderen, B., Tononi, G., and Greenspan, R.J. (2002). Electrophysiological correlates of rest and activity in Drosophila melanogaster. *Curr. Biol.* **12**, 1934–1940.

Pan, Y., Zhou, Y., Guo, C., Gong, H., Gong, Z., and Liu, L. (2009). Differential roles of the fan-shaped body and the ellipsoid body in Drosophila visual pattern memory. *Learn Mem.* **16**, 289–295.

Pflugfelder, G.O., and Heisenberg, M. (1995). Optomotor-blind of Drosophila melanogaster: a neurogenetic approach to optic lobe development and optomotor behaviour. *Comp. Biochem. Physiol. A Physiol.* **110**, 185–202.

Poggio, T., and Reichardt, W. (1976). Visual control of orientation behaviour in the fly. Part II. Towards the underlying neural interactions. *Q Rev. Biophys.* **9**, 377–438.

Posner, M.I., Snyder, C.R., and Davidson, B.J. (1980). Attention and the detection of signals. *J. Exp. Psychol.* **109**, 160–174.

Ramon, F., Hernandez-Falcon, J., Nguyen, B., and Bullock, T.H. (2004). Slow wave sleep in crayfish. *Proc. Natl. Acad. Sci. U S A* **101**, 11857–11861.

Reichardt W,,Reichardt W. (1969). *Movement Perception in Insects*. Academic Press, New York.

Robbins, T.W. (2002). The 5-choice serial reaction time task: behavioural pharmacology and functional neurochemistry. *Psychopharmacology (Berl)* **163**, 362–380.

Rubinstein, N.J., and Abel, L.A. (2011). Optokinetic nystagmus suppression as an index of the allocation of visual attention. *Invest Ophthalmol. Vis. Sci.* **52**, 462–467.

Sareen, P., Wolf, R., and Heisenberg, M. (2011). Attracting the attention of a fly. *Proc. Nat. Acad. Sci, U S A* **108**, 7230–7235.

Schnell, B., Joesch, M., Forstner, F., Raghu, S.V., Otsuna, H., Ito, K., Borst, A., and Reiff DF,,Reiff DF. (2010). Processing of horizontal optic flow in three visual interneurons of the Drosophila brain. *J. Neurophysiol.* **103**, 1646–1657.

Schuster, S., Strauss, R., and Gotz, K.G. (2002). Virtual-reality techniques resolve the visual cues used by fruit flies to evaluate object distances. *Curr. Biol.* **12**, 1591–1594.

Schwaerzel, M., Monastirioti, M., Scholz, H., Friggi-Grelin, F., Birman, S., and Heisenberg, M. (2003). Dopamine and octopamine differentiate between aversive and appetitive olfactory memories in Drosophila. *J. Neurosci.* **23**, 10495–10502.

Seelig, J.D., Chiappe, M.E., Lott, G.K., Dutta, A., Osborne, J.E., Reiser, M.B., and Jayaraman, V. (2010). Two-photon calcium imaging from head-fixed Drosophila during optomotor walking behavior. *Nat. Methods* **7**, 535–540.

Seugnet, L., Suzuki, Y., Stidd, R., and Shaw, P.J. (2009). Aversive Phototaxic Suppression: evaluation of a short-term memory assay in Drosophila melanogaster. *Genes Brain Behav.*

Seugnet, L., Suzuki, Y., Vine, L., Gottschalk, L., and Shaw, P.J. (2008). D1 receptor activation in the mushroom bodies rescues sleep-loss-induced learning impairments in Drosophila. *Curr. Biol.* **18**, 1110–1117.

Shaw, P.J., and Franken, P. (2003). Perchance to dream: solving the mystery of sleep through genetic analysis. *J. Neurobiol.* **54**, 179–202.

Shaw, P.J., Cirelli, C., Greenspan, R.J., and Tononi, G. (2000). Correlates of sleep and waking in Drosophila melanogaster. *Science* **287**, 1834–1837.

Strausfeld, N.J., Sinakevitch, I., Brown, S.M., and Farris, S.M. (2009). Ground plan of the insect mushroom body: functional and evolutionary implications. *J. Comp. Neurol.* **513**, 265–291.

Strauss, R., and Pichler, J. (1998). Persistence of orientation toward a temporarily invisible landmark in Drosophila melanogaster. *J. Comp. Physiol. [A]* **182**, 411–423.

Tammero, L.F., Frye, M.A., and Dickinson, M.H. (2004). Spatial organization of visuomotor reflexes in Drosophila. *J. Exp. Biol.* **207**, 113–122.

Tanaka, N.K., Ito, K., and Stopfer, M. (2009). Odor-evoked neural oscillations in Drosophila are mediated by widely branching interneurons. *J. Neurosci.* **29**, 8595–8603.

Tang, S., and Guo, A. (2001). Choice behavior of Drosophila facing contradictory visual cues. *Science* **294**, 1543–1547.

Tang, S., and Juusola, M. (2011). Intrinsic activity in the fly brain gates visual information during behavioral choices. *PLoS One* **5**, e14455.

Tang, S., Wolf, R., Xu, S., and Heisenberg, M. (2004). Visual pattern recognition in Drosophila is invariant for retinal position. *Science* **305**, 1020–1022.

Toma, D.P., White, K.P., Hirsch, J., and Greenspan, R.J. (2002). Identification of genes involved in Drosophila melanogaster geotaxis, a complex behavioral trait. *Nat. Genet.* **31**, 349–353.

Treisman, A.M., and Gelade, G. (1980). A feature-integration theory of attention. *Cogn. Psychol.* **12**, 97–136.

Treue, S. (2003). Visual attention: the where, what, how and why of saliency. *Curr. Opin. Neurobiol.* **13**, 428–432.

Tully, T., and Quinn, W.G. (1985). Classical conditioning and retention in normal and mutant Drosophila melanogaster. *J. Comp. Physiol. [A]* **157**, 263–277.

van Swinderen, B. (2005). The remote roots of consciousness in fruit-fly selective attention? *Bioessays* **27**, 321–330.

van Swinderen, B. (2007a). Attention-like processes in Drosophila require short-term memory genes. *Science* **315**, 1590–1593.

van Swinderen, B. (2007b). The attention span of a fly. *Fly* **1**, 187–189.

van Swinderen, B., and Andretic, R. (2011). Dopamine in Drosophila: setting arousal thresholds in a miniature brain. *Proc. Biol. Sci.*

van Swinderen, B., and Brembs, B. (2010). Attention-like deficit and hyperactivity in a Drosophila memory mutant. *J. Neurosci.* **30**, 1003–1014.

van Swinderen, B., and Flores, K.A. (2006). Attention-like processes underlying optomotor performance in a Drosophila choice maze. *J. Neurobiol.*

van Swinderen, B., and Greenspan, R.J. (2003). Salience modulates 20–30 Hz brain activity in Drosophila. *Nat. Neurosci.* **6**, 579–586.

van Swinderen, B., McCartney, A., Kauffman, S., Flores, K., Agrawal, K., Wagner, J., and Paulk, A. (2009). Shared visual attention and memory systems in the Drosophila brain. *PLoS One* **4**, e5989.

Vialatte, F.B., Maurice, M., Dauwels, J., and Cichocki, A. (2010). Steady-state visually evoked potentials: focus on essential paradigms and future perspectives. *Prog. Neurobiol.* **90**, 418–438.

Vosshall, L.B. (2007). Into the mind of a fly. *Nature* **450**, 193–197.

Williams, I.M., Mulhall, L., Mattingley, J., Lueck, C., and Abel, L. (2006). Optokinetic nystagmus as an assessment of visual attention to divided stimuli. *J. Clin. Neurosci.* **13**, 828–833.

Wolfe, J.M., Alvarez, G.A., and Horowitz, T.S. (2000). Attention is fast but volition is slow. *Nature* **406**, 691.

Wolf, R., and Heisenberg, M. (1980). On the fine structure of yaw torque in visual flight orientation of Drosophila melanogaster. *J. Comp. Physiol.* **140**, 69–80.

Wolf, R., and Heisenberg, M. (1991). Basic organization of operant behavior as revealed in Drosophila flight orientation. *J. Comp. Physiol. [A]* **169**, 699–705.

Wolf, R., and Heisenberg, M. (1997). Visual space from visual motion: turn integration in tethered flying Drosophila. *Learn Mem.* **4**, 318–327.

Wolf, R., Wittig, T., Liu, L., Wustmann, G., Eyding, D., and Heisenberg, M. (1998). Drosophila mushroom bodies are dispensable for visual, tactile, and motor learning. *Learn Mem.* **5**, 166–178.

Xi, W., Peng, Y., Guo, J., Ye, Y., Zhang, K., Yu, F., and Guo, A. (2008). Mushroom bodies modulate salience-based selective fixation behavior in Drosophila. *Eur. J. Neurosci.* **27**, 1441–1451.

Zhang, K., Guo, J.Z., Peng, Y., Xi, W., and Guo, A. (2007). Dopamine-mushroom body circuit regulates saliency-based decision-making in Drosophila. *Science* **316**, 1901–1904.

Zhou, M., Lei, Z., Li, H., Yi, W., Zhang, Z., and Guo, A. (2010). NMDA receptors-dependent plasticity in the phototaxis preference behavior induced by visual deprivation in young and adult flies. *Genes Brain Behav.* **9**, 325–334.

THE ROLES OF *FRUITLESS* AND *DOUBLESEX* IN THE CONTROL OF MALE COURTSHIP

Brigitte Dauwalder

University of Houston, Department of Biology and Biochemistry, 369 SR2, Houston, TX 77204-5001, USA

Male courtship in *Drosophila melanogaster* is a robust innate behavior that is shaped by sensory input and experience. It is regulated by the general sex-determination pathway through the sex-specific forms of *fruitless* and *doublesex*. Recent findings have shown that both *fruitless* and *doublesex* are required for courtship. This chapter reviews the role of these proteins and the neurons that express them in the regulation of courtship behavior. In particular it discusses how *doublesex* and *fruitless* contribute to the generation of sexually dimorphic neurons, the role of cell death, and the emerging information about circuits that underlie the behavior.

I. Introduction

Courtship in *Drosophila melanogaster* consists of an innate set of behavioral steps that are shaped by sensory input and prior experience. The behavior is characterized by a series of steps that are performed sequentially and can relatively easily be observed and quantified in the laboratory. It is therefore well suited for genetic analysis and to address fundamental questions about the molecular control of

INTERNATIONAL REVIEW OF NEUROBIOLOGY, VOL. 99
DOI: 10.1016/B978-0-12-387003-2.00004-5

complex behaviors. The mating behavior of *D. melanogaster* males has been described by Bastock and Manning (1955) and Spieth (1974) and reviewed in detail (Greenspan, 1995; Hall, 1994; Villella and Hall, 2008). This chapter will summarize the behavior and its control by the general sex-determination pathway, specifically the roles of *fruitless* and *doublesex*. It will then review the latest findings on the roles of some of the *fruitless* and *doublesex* expressing neurons and how sexual dimorphism is generated by cell death. Finally, the search for downstream targets is discussed, as well as the role of non-neuronal input through diffusible factors from the fat body.

II. The Behavior

A. Male–Female Courtship

A protocol on how to set up the behavioral assay has recently been published (Ejima and Griffith, 2007). In short, the following steps can be observed (Fig. 1):

1. *Orientation toward the female.* The time to first orientation toward the female is measured as latency. The time it takes the male to orient toward the female is highly dependent on a variety of sensory stimuli. Vision and pheromone detection, but also auditory cues play an important role (Ejima and Griffith, 2008; Griffith and Ejima, 2009b). Performing courtship assays in red light will increase latency because visual stimuli are absent. Assays performed in red light are therefore a sensitive way to examine the role and perception of pheromones. In bright light, where visual input plays a significant role, *white* mutant males, which have impaired vision, will show a long latency compared to w^+ males. This can be an issue with transgenic strains whose marker w^+ expression is very weak. Latency also depends on the size of the courtship chamber.
2. *Tapping* of the female's abdomen with the forelegs. Male forelegs have been shown to harbor gustatory cells. Three gustatory receptors have been identified with a role in courtship (Gr32a, Gr33a and Gr68). The behavior of mutants suggests that these receptors are involved in inhibiting male–male courtship, in the perception of inhibitory pheromones and the control of correct courtship song (Koganezawa *et al.*, 2010; Miyamoto and Amrein, 2008; Moon *et al.*, 2009).

orientation →	tapping →	courtship song →	attempted copulation →	copulation
vision olfaction pheromones gustatory input auditory input	pheromones gustatory input	abdominal ganglia brain neurons		female receptive?

FIG. 1. Courtship steps displayed by *Drosophila melanogaster* males.

3. "*Singing*" of the courtship song by extension and vibration of the wing that is closest to the female. This song is male- and species-specific. This unilateral wing vibration produces a species-specific courtship song that consists of sine song (a humming sound) and a pulse song with rhythmic elements, which stimulate the female to mate (Von Schilcher, 1976). The pulse song contains species-specific information (Kyriacou and Hall, 1986; Kyriacou *et al.*, 1990). The song can be recorded and, when played to a virgin receptive female, stimulates her willingness to mate.
4. *Licking* of the female's genitalia.
5. *Attempted copulation:* Curling of the male's abdomen in attempts to copulate.
6. *Copulation.* Whether or not copulation occurs is mostly dictated by the female. If she is too young or has previously mated, and is therefore unreceptive, she will run away, kick the male with her legs and protrude her ovipositor, and copulation will generally not occur. If she is mature and ready to copulate, she will slow down, open her anal plates and copulate. Copulation takes about 20 min.

In general, although the different steps of courtship have been shown to be regulated by different regions of the central nervous system (CNS) by mosaic analysis, they are stereotypically displayed in this order. Spieth (1974) has suggested that each step may be involved in elevating arousal levels, with thresholds that trigger later steps.

Courtship assays are generally performed in plexigas chambers or small tubes. Overall courtship is measured as courtship index (CI), the fraction of time a male performs any of the courtship steps during the observation period. The "wing-expansion index" WEI measures the fraction of time a wing is extended toward the female. The CI measured for wild-type flies is dependent of the size of the chamber and, as mentioned before, whether or not the assay is performed in white or red light.

B. Male–Male Courtship or How to Discriminate Between Males and Females

Mature wild-type males generally do not court other males. This is mainly due to the inhibitory effect of male-specific pheromones on the cuticle of adult males, and the rejection behavior of courted males. Young males elicit courtship from mature males because the full repertoire of sex-specific pheromones is established during the first 2 days after eclosion. Young males also court males and females as the full behavioral competence may only be established after eclosion (McRobert and Tompkins, 1983). A number of mutations cause male–male courtship, often without affecting male–female courtship, indicating a loss of the ability to discriminate males and females. This may be an indication that multiple mechanisms are in place in wild-type males to suppress male–male courtship. In this context it is interesting that the silencing of mushroom-body neurons, a structure not basically required for courtship (Kido and Ito, 2002), causes male–male courtship, suggesting an inhibitory function of these neurons (Kitamoto, 2002).

The discrimination between males and females is in large part mediated by sex-specific pheromones (Ferveur, 2005). As most of them have low volatility, recognition is probably mostly gustatory. Gustatory sensillae are present on the maxillary palp, the proboscis, and the forelegs, which all make contact with the female during courtship. Gustatory neurons of the legs project to thoracic ganglia and the suboesophageal ganglion (SOG) (Possidente and Murphey, 1989), olfactory neurons to the glomeruli of the antennal lobes. Interestingly, sex-specific branching of neurons has been observed in several of these structures that is dependent on the sex-specific regulators *fruitless* and *doublesex* (see Section III.C). Moreover, although specific receptors and pheromones are playing an important role, intriguingly, the specificity of some of the responses may lie in the wiring and projections of specific neurons (see Section III.C). Besides pheromones, many sensory inputs contribute to efficient mate recognition and courtship. Vision, smell, taste, and auditory stimuli all contribute to the behavior (Griffith and Ejima, 2009b). In the absence of any one of these stimuli, courtship still proceeds, even though at lowered efficiency. Pheromones also play an important role in species recognition (Ferveur, 1997). In this context, an important question still largely unclear is how sender and receiver of the signal co-evolve. A recent study has suggested that the *desat1* gene involved in the biosynthesis of pheromones plays a role: a mutation in *desat1* simultaneously altered both sex pheromone emission and perception in *D. melanogaster* (Bousquet *et al.*, 2009).

C. The Role of the Female

Although the female seems to have a mainly passive role in courtship to the observer, she is the one that ultimately decides acceptance by slowing down and allowing copulation. Besides sensory input from the courting male (his courtship song, for example), major factors that influence female receptivity are her age (very young females are not sexually mature), and whether or not she has previously mated. Male accessory gland proteins that are transferred during copulation cause a profound change in the female's behavior and lead her to reject other males. These postmating behaviors of *D. melanogaster* females are well-studied and described (Ferveur, 2010; Kubli, 1992; Wolfner, 2009).

D. Males Remember: Courtship Conditioning

Although male courtship behavior is a fairly stereotyped innate behavior, it is shaped by sensory input as described above. It is also influenced by the previous mating experience of the male. If a male encounters a female that rejects him because she is too young or already mated, the rejected male will form a lasting memory of the rejection. It has been shown that mated females carry cisVA on their cuticle that was transferred from the male during mating. cisVA is repulsive to males and its perception is associated with the rejection (Ejima *et al.*, 2007). If previously rejected males subsequently encounter a virgin female that is ready to mate, they show an increased latency to court. The memory decays over time. This associative learning ("courtship conditioning") has been shown to be dependent on the same genes and brain structures that are generally required for associative aversive learning in *Drosophila*. The courtship conditioning assay is therefore often used as a paradigm to study learning and memory (reviewed in Ejima and Griffith (2011); Griffith and Ejima (2009a)).

III. The Genes, the Neurons

A. Fruitless and Doublesex

The master regulators of the somatic sex-determination pathway also regulate sexual behavior (Belote and Baker, 1987; McRobert and Tompkins, 1985; Taylor *et al.*, 1994). Sex in *Drosophila* is determined cell autonomously by a cascade of sex-specific alternative splicing (Baker, 1989) (Fig. 2). The primary signal lies in the ratio of X chromosomes to autosomes. The XX:AA = 1 ratio in females leads to

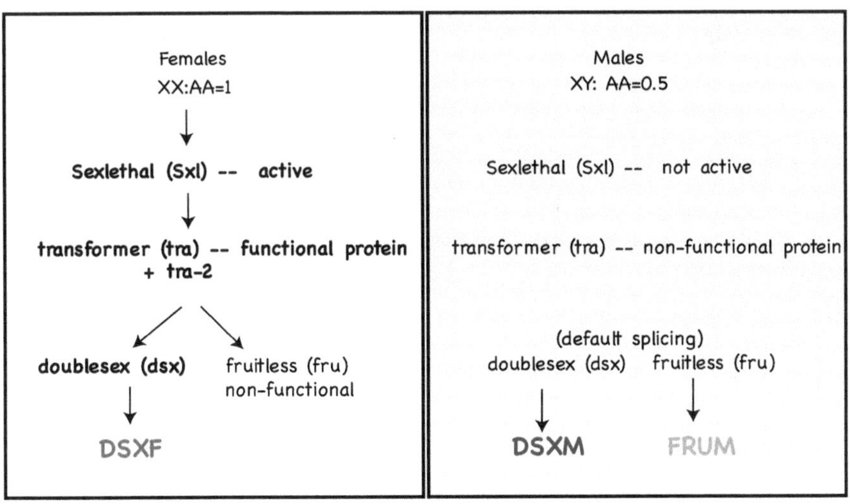

FIG. 2. Regulators of the *Drosophila melanogaster* sex-determination pathway.

the activation of *Sexlethal (Sxl)*, a splicing factor. *Sxl* autoregulates itself to produce an active protein that regulates splicing of the *transformer* (*tra*) pre-mRNA. Only *tra* RNA that is spliced in the female mode can make a functional protein (TRAF) because the default male splice form contains an early stop codon. TRAF is also a splicing factor that, together with TRA-2, regulates the sex-specific alternative splicing of the two downstream target pre-mRNAs *doublesex (dsx)* and *fruitless (fru)*. This leads to the production of the female-specific DSXF protein. The female-specific splicing of *fru* results in a nonfunctional FRU protein. In males, default splicing in the absence of TRA leads to the production of the male-specific DSXM protein. The default splicing of *fru* pre-mRNA in males leads to the generation of a functional male-specific FRUM protein. In summary, males end up with DSXM and FRUM, and females with DSXF (and no FRUM). In genotypically normal XX females that carry a mutation in *tra* or *tra-2*, *dsx* and *fru* are spliced in the male mode. These animals look like males and court like males, confirming that male courtship behavior is regulated by the sex-determination pathway. When *dsx* and *fru* mutants were examined for their courtship behavior, *dsx* mutants showed a reduction in courtship, but still courted (Villella and Hall, 1996). Furthermore, induction of DSXM in females did not initiate courtship. In contrast, strong *fruitless* mutant males no longer courted. Weaker *fru* mutants showed strongly reduced courtship toward females and they courted males, and these mutants became "famous" for the male courtship chains they formed (a male courting a male, followed by a courting male, followed by another courting male, and so on) (Gailey and Hall, 1989; Ito *et al.*, 1996; Ryner *et al.*, 1996). Based on these mutant findings, *fruitless* was long thought to be the only regulator of male courtship behavior. Fruitless is expressed in about 2000 neurons in the male nervous system as determined by antibodies (Lee *et al.*, 2000). This expression pattern was reproduced in transgenic flies in which Gal4 sequences had been introduced into the *fru* locus by homologous recombination such that expression of Gal4 was controlled by the *fru* P1 promoter (Dornan *et al.*, 2005; Manoli *et al.*, 2005; Stockinger *et al.*, 2005). The fru P1 promoter directs expression of the *fru* transcripts that are subject to sex-specific alternative splicing. In the *fru-Gal4* construct P1-controlled expression of Gal4 leads to the presence of Gal4 in the cells that normally express *fru* in both sexes. This is because the male-specific presence of FRUM is due to alternative splicing of the *fru* pre-mRNA and translational control, and not regulated at the level of transcription. The availability of Gal4 lines that express Gal4 specifically in *fru* neurons in both sexes has opened the door to extensive manipulation of these neurons by expression of a variety of UAS-transgenes. In addition, these constructs have made it possible to address an important question: Are the neurons that express FRUM in males male-specific neurons that are absent in females (in which case one would not expect to see Gal4 expression in females), or are the neurons present in females, but just not expressing FRUM? When the *fru-Gal4* flies were crossed to *UAS-reporters* and expression visualized, the answer was that the neurons are generally present in both sexes (Dornan *et al.*, 2005; Manoli *et al.*, 2005; Stockinger *et al.*, 2005). This indicates that the neurons are present in both sexes, but probably work differently because they express FRUM in males, but not in females. Indeed at first glance it appeared that there were no sex-specific differences. However, as the analyses became more detailed since these initial experiments, a growing number of instances have been found in which the cell number in *fru* clusters differs between males and females, or where small clusters are only present in one sex. Most importantly, in a number of cases it has been found that the projection pattern of these neurons differs between males and females. This is likely to be of significant biological importance, influencing the way these neurons work and connect. Interestingly, in many of these instances it has also been shown that *doublesex* plays an important role in the generation of these differences (see below).

B. *dsx* and *fru* Are Both Required for Male Courtship Behavior

Another important transgene was used to address the question whether FRUM is sufficient to induce male courtship behavior. In this transgenic line the *fru* gene had been manipulated to always express the male FRUM protein, also in females (Demir and Dickson, 2005; Manoli *et al.*, 2005). Indeed, females that express FRUM from this transgene are capable of performing male courtship, even though at lower rate and with some steps missing. These experiments clearly demonstrated the very important role FRUM has in setting up the competence of the nervous system to display courtship. But what could be the factor(s) that are missing in the FRUM females in order to obtain complete male courtship behavior? In the past few years it has become increasingly clear that while *fruitless* has a very important role in conferring the ability to court in the nervous system, *doublesex* is also required for full and normal male courtship. Notably, the courtship "song" of FRUM females was not a real courtship song: It lacked "sine" song and showed aberrant features in the "pulse" song (Rideout *et al.*, 2007). As it turns out, both FRUM and DSXM are required and females that express both FRUM and DSXM (as is the case in *tra* mutant females) have normal male courtship song (Rideout *et al.*, 2007; von Schilcher and Hall, 1979). It has previously been shown that the male differentiation of the ventral thoracic ganglia is important for courtship song (von Schilcher and Hall, 1979). It is intriguing that a male-specific cluster of neurons is present in the thoracic ganglia that expresses both FRUM and DSXM (Lee *et al.*, 2002; Rideout *et al.*, 2007; Sanders and Arbeitman, 2008). The sexual dimorphism in cell number is dependent on both *fru* and *dsx* (Lee *et al.*, 2002; Rideout *et al.*, 2007; Sanders and Arbeitman, 2008) with a major role for *dsx*. In fact, it has been shown that the female-specific DSXF activates cell death genes that lead to the absence of these cells in females (Sanders and Arbeitman, 2008).

These sexually dimorphic neurons are good candidates for mediating courtship song. However, as they do not make direct contact with the flight muscles, their exact function and involvement in courtship song is not clear at present. There are other neurons in the thoracic ganglia that are sexually dimorphic: A group of serotonergic neurons that differ in their cell number in males and females in a DSXM- and FRUM-dependent way (Billeter *et al.*, 2006; Taylor and Truman, 1992), as well as a group of abdominal neuroblasts for which DSXM prolongs division in males (Billeter *et al.*, 2006; Taylor and Truman, 1992).

There is another important sexual dimorphism in the abdominal ganglia: Sexually dimorphic axons have been identified in the prothoracic ganglion that cross the midline in males, but not in females. These are the projections of foreleg gustatory receptor neurons (Boll and Noll, 2002; Possidente and Murphey, 1989). The differential midline crossing of their axons has been found to be dependent on both *fru* and *dsx*, with a major role for *fru* (Mellert *et al.*, 2009). This is one of the rare cases where a downstream effector has been identified that is involved in mediating the differential effects of *fru* and *dsx*: It has been found that the midline crossing of the gustatory receptor axons in males requires the presence of FRUM to repress *roundabout* (*robo*) signaling. *robo* receptors have a wide role in regulating axonal pathfinding. For the gustatory neurons, reduced *robo* signaling is required to allow midline crossing, and FruM appears to be involved in the reduction of an inhibitory signal that keeps the axons from crossing in females (Mellert *et al.*, 2009).

These findings are in keeping with previous mosaic mapping experiments that examined regions of the nervous system that need to be male in order for male courtship to occur. Mosaic mapping using different techniques has revealed anatomical foci in the ventral ganglia as well as the brain (Ferveur *et al.*, 1995; Hall, 1979; Hotta and Benzer, 1972). The above examples illustrate roles of the ventral ganglia in sensory perception, but also in motor output function, with courtship song as a prominent example.

C. Sexual Dimorphism in *fru* and *dsx* Expressing Neurons in the Brain

The crucial function of male brain regions in the regulation of the behavior that was found in the mosaic studies has since been confirmed many times. The constant refinement of these maps and the identification of neuronal cell groups with potential specific roles are a vibrant and exciting part of current research into the control of courtship behaviors. New tools and techniques are continuing to allow more detailed insight. Not surprisingly, *fru*, but also *dsx* expressing cells are found in the brain regions that are required for courtship. A number of brain clusters have been identified by now that are sexually dimorphic. The first one to be discovered was a FRUM expressing cluster of interneurons in a region just above the antennal lobe called mAL. Males and females differ in the number of neurons and their projection patterns. It was found that the higher number of cells in males is caused by FRUM, which inhibits cell death of the neurons at the pupal stage and determines their projection pattern (Kimura *et al.*, 2005). Projections of these neurons are seen in the lateral protocerebrum and the region of the SOG and the use of a presynaptic marker suggests that the SOG is the input region, and the lateral protocerebrum the output site. The SOG receives input from gustatory neurons and it is an intriguing hypothesis that sex-specific gustatory neurons may make contact with mAL neurons. In fact, a foreleg gustatory neuron that expresses receptor Gr32a projects toward mLA extensions in the SOG (Koganezawa *et al.*, 2010). Males with inactivated Gr32a neurons or males in which the projections from this neuron have been interrupted fail to show the unilateral wing extension that is typical for *D. melanogaster* courtship and often extend both wings instead. Their altered courtship song is likely the cause for their lowered mating success. These findings suggest a pathway of information flow: from gustatory input to higher order integration centers in the brain to output neurons and abdominal neurons that mediate the wing extensions.

In addition to gustatory input, olfactory input plays a role in courtship as well. Olfactory receptor neurons project to the antennal glomeruli and receptor neurons and a subset of antennal glomeruli have been identified that express fru or are contacted by fru neurons (Datta *et al.*, 2008; Kimura *et al.*, 2005; Stockinger *et al.*, 2005). Changing the sex of antennal glomeruli that are contacted by *fru*-positive neurons induces male–male courtship, and silencing of these neurons leads to a drastic reduction in courtship also toward females (Stockinger *et al.*, 2005).

Another sexually dimorphic neuronal brain cluster that expresses *fru* is P1. P1 neurons are only present in males. They express both FRUM and DSXM and several studies have now shown their importance in the control of the first steps of courtship. P1 neurons are located in the posterior part of the brain, close to the mushroom bodies, an area that has been implicated in the initiation of courtship in the mosaic studies mentioned earlier. It has been found that FRUM is required to define the normal branching pattern of these neurons. In females, the P1 cluster is absent due to the presence of DSXF that triggers cell death of these neurons. When FRUM and DSXM are expressed in these cells in otherwise normal females by the induction of *tra* mutant clones (tra mutant cells assume a male fate and express both FRUM and DSXM), P1 neurons are present and the females will initiate courtship behavior, suggesting a role for these neurons in the initiation of courtship (Kimura *et al.*, 2008). Two recent studies have further corroborated the importance of this cluster in the initial steps of courtship. In one of the studies, activation of P1 neurons triggered tapping and wing extension, two early steps of courtship. A transgenic strain that expressed the heat-activatable channel dTRPA1 in *fru* neurons was used to create mosaic males in which dTRPA1 was present only in subsets of *fru* neurons. When the temperature was raised, some of the mosaic males initiated tapping and wing extension in the absence of a female, with characteristics very similar to males encountering a female. These behaviors were highly correlated with expression of dTRPA1 in P1 neurons as well as a second cluster nearby, P2b, consisting of descending interneurons (Kohatsu *et al.*, 2011). Furthermore, touching of the foreleg of a tethered male with a female abdomen triggered neuronal activity in P1 neurons, suggesting a pathway from sensory input to integration in P1 neurons to potential output through P2b descending neurons.

TRPA1 induction was also used in the screening of driver lines that were expressed in subsets of *fru* neurons (von Philipsborn *et al.*, 2011). The males initiated wing extensions when P1 neurons and pIP10 neurons in the brain were activated by raised temperatures. The pIP10 neuron is a descending neuron with axons terminating in the mesothoracic ganglion with likely input from P1. It would therefore be well positioned to mediate the command for courtship song. Although activation of P1 and pIP10 neurons led to wing extension, it did not lead to structured courtship song. Three other types of neurons in the mesothorax were identified that appear to provide the typical song features of the courtship song. These results were similar to earlier experiments in which headless flies displayed wing extensions when abdominal *fru* neurons (in a pretty broad pattern) were activated through the activation of a light-activatable channel (Clyne and Miesenbock, 2008).

In contrast to *fru*, alternative splicing of *dsx* produces two different functional proteins in males and females (DSXM and DSXF). DSXM and DSXF are expressed in a sexually dimorphic pattern in the CNS as well (in addition to their expression in non-neuronal tissues) (Lee *et al.*, 2002; Sanders and Arbeitman, 2008). A central role for *dsx* expressing neurons in behavior was demonstrated recently by using *dsx-Gal4* transgenic flies in which Gal4 was placed into the *dsx* locus. The silencing of *dsx-Gal4* expressing neurons reduced male courtship drastically and abolished courtship song (Rideout *et al.*, 2010). Sanders and Arbeitman (2008) found that the sex-specific number of neurons that express *dsx* is determined by cell death that is regulated by the amount of DSX and by the presence of the specific male or female DSX isoform. For example, as a consequence of DSXF expression in females, cell death of some of the neurons occurs during metamorphosis (it is blocked in a mutant in which several cell death genes are deleted). Sex-specific cell death, or more specifically, cell death that occurs in females due the presence of DSXF, thus emerges as a recurrent and important mechanism that creates sexually dimorphic numbers of neurons (reviewed in Kimura (2011). Although the difference in cell numbers is fairly modest, the observed differences in projections from these neurons are pronounced and could significantly influence the potential interactions of these neurons with other neurons. Not surprisingly, the function of DSXF in regulating cell death does not depend on *fru* (Sanders and Arbeitman, 2008), as no Fru protein is expressed in females normally.

In summary, there are a number of possible mechanisms that can underlie the sexually dimporphic function of *fru* and *dsx*-expressing neurons (diagrammed in Fig. 3). In the male neurons that express FRUM and, in many instances, DSXM, it

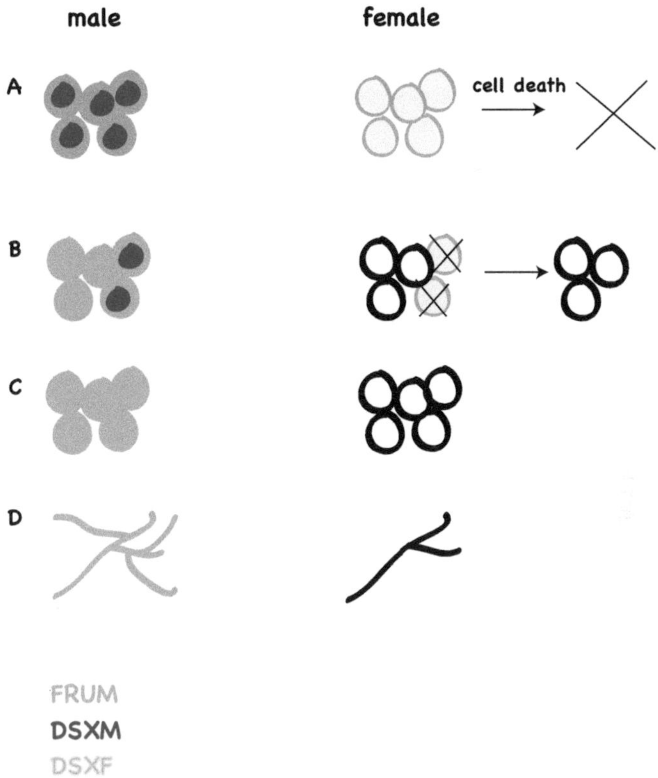

FIG. 3. Mechanisms that create sexual dimorphism in fru and dsx expressing neurons. (A) Absence of a particular cluster in females due to DSXF-mediated cell death in females. A cell cluster is shown in which all cells in the cluster express FRUM (green) and DSXM (blue) in males, and DSXF (pink) in females. DSXF-expressing cells undergo cell death. As a result, the cell cluster is absent in females, whereas it is present in males (no cell death occurs in the male cells that express FRUM and DSXM). (B) Reduced cell number in a particular cluster due to DSXF-mediated cell death of some of the neurons. Only some of the cells in the cluster express DSX (DSXM in males and DSXF in females). In females, the DSXF expressing cells undergo cell death. This leads to a different number of neurons in males and females. The remaining cells in the female are nevertheless qualitatively different from their male counterparts because they do not express FRUM. (C) Equal number of cells in a particular cluster. Although the cell number in the cluster is the same in males and females, neurons in the male express FRUM, the corresponding neurons in the female do not. (D) FRUM expressing neurons have different neuronal extensions than the corresponding female axons. Due to the presence of FRUM, projection patterns are different in males. (For interpretation of the references to color in this figure legend, the reader is referred to the web version of this book.)

is likely that these two regulators determine at least some of their function and projections. The mechanisms of this functional determination are unknown, as are the genes that mediate it. Also, what makes *fru* expressing clusters different from each other? The number of neurons that express both *fru* and *dsx* is limited

(Rideout *et al.*, 2007; Sanders and Arbeitman, 2008). Therefore, sexual dimorphism by the observed mechanism of DSX-mediated cell death is likely not going to account for all sexually dimorphic cell groups. In addition, it is worth noting that there are a number of *fru* neurons that do not show differences in cell number, or even projections, between males and females. *fru* neurons that are present in both sexes are nevertheless molecularly dimorphic: They express FRUM or DSXM in males, whereas their female counterparts do not express FRUM or DSXM (some will express DSXF). It will be of particular interest to define the function of these neurons in females.

An attractive hypothesis is that *fru* neurons and *dsx* neurons form behavioral circuits. What constitutes such circuits and how these circuits interact in the behaving adult animal is an important next question to be addressed. Experiments have recently identified a likely *fru* circuit that mediates the recognition and response to cisVA. cisVA, a volatile pheromone, is attractive to females and inhibitory to males (Ejima *et al.*, 2007; Kurtovic *et al.*, 2007; Vosshall, 2008). It is perceived equally in both male and females by Or67d olfactory receptor neurons, evoking the same electrical response, indicating that perception by the receptor is not sex-specific. Or67d receptor neurons project to the antennal glomerulus DA1, one of several glomeruli that are contacted by *fru*-positive neurons. (The antennal lobe glomeruli are the first-order processing center for incoming olfactory receptor neurons.) DA1 is contacted by projection neurons that project to the lateral horn of the protocerebrum. These neurons differ in their projection patterns and branching in the two sexes, a difference that is dependent on *fru* (Datta *et al.*, 2008; Kimura *et al.*, 2008; Kimura *et al.*, 2005). Using photoactivated GFP to visualize the proximity of neurons, and calcium-imaging and electrophysiology to monitor activity, Ruta *et al.* have shown that the projection neurons that contact DA1 are part of a circuit that responds to cisVA. In the lateral horn they contact 4 sexually dimorphic clusters of presumptive third-order neurons (Ruta *et al.*, 2010). These neurons in turn make contact with neurons that extend processes into the ventral ganglia that may then be involved in the behavioral response to cisVA. Similarly, as described above, significant insight has been gained into the possible circuits that involve the P1 cluster and its control of initiation of courtship and its first steps such as tapping and wing extensions (Kohatsu *et al.*, 2011; von Philipsborn *et al.*, 2011).

Extensive mosaic mapping of *fru* neurons in male and female brains has recently been carried out and a detailed map of these neurons and their potential connections established by (Cachero *et al.*, 2010; Yu *et al.*, 2010). Both groups have mapped the identified projections onto a standard brain. These maps will facilitate the mapping of functional circuits in the future. The studies have confirmed a fair number of sexual dimorphism between males and females. In addition to changes in cell numbers as outlined above, in many instances second and third-order neurons have different projection patterns in males and females.

D. What Are the Downstream Effectors of *fru* and *dsx*?

Experiments with *tra* and *tra-2* temperature-sensitive mutants have indicated that the critical period to establish the basic competence for male courtship behavior is during late pupal stages (Arthur *et al.*, 1998; Belote and Baker, 1987). The temperature-shift experiments examined whether courtship was being displayed but did not quantify it or look at individual components. It is not known to what degree *fru* and/or *dsx* are also required for the optimal adult function of these neurons. The experiments mentioned in the section above did not use conditional mutants that would allow an assessment whether *fru* and *dsx* are required mainly to set up neuronal cell numbers and connections, or whether they are still required for the optimal sex-specific functioning of the neurons in the adult male.

An important next level of analysis will be to address the molecular consequences of *fru* or/and *dsx* expression in these neurons. What are the downstream targets? Do subsets express different cell adhesion proteins or cell migration guidance proteins, signaling pathways, and neurotransmitters? What differentiates the circuits: the incoming stimuli, the way this incoming information is processed, the connections the neurons make with others, and integration across a network of neurons? What are the output pathways, and which molecular and electrophysiological determinants define the specific behavioral response?

A number of potential downstream genes regulated by both *fru* and *dsx* have been identified in several screens, and their roles verified in a few cases (Arbeitman *et al.*, 2004; Dauwalder *et al.*, 2002; Fujii and Amrein, 2002; Goldman and Arbeitman, 2007). Interestingly, although both *dsx* and *fru* are transcription factors, except for the female yolk protein (yp) genes that are directly regulated by binding of DSX (An and Wensink, 1995; Coschigano and Wensink, 1993), no other genes are known to date that are directly bound and regulated by these factors, leaving open the question of direct or indirect regulation. Therefore, we do not know how these target genes are regulated. Based on the aforementioned *yp* studies, binding sites for DSXF and DSXM in the *yp* genes are well described. *yp* protein genes contain DSX binding sites in their promoter regions that can be bound by both DSXF and DSXM in a competing manner. This led to the model of DSXF acting as an activator, and DSXM as a repressor. We do not know whether the sequences that mediate DSX control are the same in all cell types (in neurons, for example). They may vary depending on tissue-specific co-regulators. This is a particularly interesting question in light of the crucial role DSXF plays in controlling sex-specific cell death. Are cell death genes directly controlled by DSXF? Do they have DSX-binding sites in their promoters? In some of the instances where DSXF-mediated cell death has been examined, it has been observed that the presence of DSXM affected the outcome of cell death (Sanders and Arbeitman, 2008). This suggests the presence of regulatory elements with characteristics similar to those in the *yp* genes. We know even less about how FRU regulates gene expression. No binding consensus sequences for FRU have been described. In the neurons that express both *fru* and *dsx*, some target genes may be activated by both transcription factors, as has been suggested for some targets. For example, in the case of the male-preferentially expressed *takeout* gene, it has been shown both *dsx* and *fru* are required for wild-type levels of the protein: transcript levels are reduced in *dsx* mutants, but they are also reduced in *fru* mutants (Dauwalder *et al.*, 2002). In addition, it was shown that expression of both DSXM and FRUM is required to express the gene at male levels in female fat body (Dauwalder, 2008). Interestingly, a number of identified *fru* targets that were identified in expression screens are expressed in the fat body, a secretory tissue that plays a role in courtship regulation (see below), but *fru* expression has so far not been observed in that tissue. It is therefore possible that *fru* regulation may occur indirectly, perhaps via diffusible factor(s) or/and as a consequence of neuronal activity in *fru* expressing neurons. As described earlier, a number of neurons express and require both *fru* and *dsx* and it is possible that both pathways may converge on some of the target genes directly or indirectly.

E. One Source of Likely Input Signals

1. *The Role of the Fat Body*

Although a major input into *fru* and *dsx* neurons that regulate courtship comes from sensory neurons, there is another source of input that is non-neuronal. The fat body, a major secretory tissue, has been shown to significantly contribute to courtship (Lazareva *et al.*, 2007). As mentioned earlier, females that express FRUM court other females, even though with reduced CI. However, when the fat body of these FRUM females is masculinized in addition, they have a near-normal CI. Similarly, when the fat body of normal males was feminized, the courtship of these males dropped significantly (Lazareva *et al.*, 2007). These results suggest that male-specific proteins from the fat body "talk to" the brain and contribute to the regulation of courtship behavior. Screens that examined downstream targets of *fru* and *dsx* have identified a number of genes that are expressed in the fat body. Intriguingly, fat body transcripts were also identified in experiments that examined genes whose expression changes in response to mating (Ellis and Carney, 2010a, 2010b). The biological function of most of them is not known yet. The best characterized ones to date are *fit*, *sx1*, *sx2*, and *takeout* (Dauwalder *et al.*, 2002; Fujii and Amrein, 2002). *takeout* is preferentially expressed in male head fat body and *takeout* mutant males have courtship defects. The Takeout protein is secreted into the hemolymph and has been shown to act as a secreted protein in courtship (Lazareva *et al.*, 2007). This is a likely mode of action also for other sex-specific genes from the fat body that interact with the nervous system to control courtship. Feminization of the fat body affects courtship significantly, beyond the effects seen in *takeout* mutants. This implies a significant role of secreted circulating molecules in the regulation of neuronal activity. The fat body secretes proteins into the hemolymph and an important (and unanswered) question is how hemolymph proteins interact with the CNS. Takeout has also been shown to be involved in feeding behavior and the regulation of longevity (Bauer *et al.*, 2010; Galikova and Flatt, 2010; Ringo *et al.*, 1992; Sarov-Blat *et al.*, 2000; Wilson *et al.*, 2003). As these are other known important functions of the fat body, courtship genes expressed in that tissue may also serve as a link between reproduction and metabolic processes. The takeout protein has characteristics of a soluble carrier protein of lipophilic molecules, such as odorant-binding proteins. It is most similar in sequence to juvenile hormone-binding proteins from other insects. It is not known whether Takeout binds JH or whether JH has a role in the control of male courtship behavior. Interestingly, *apterous* and *Met* mutants, in which JH levels are altered, show courtship defects (Ringo *et al.*, 1992; Wilson *et al.*, 2003). It will be interesting to see whether there is a role for this hormone in adult behavior. Recently a function for ecdysone, another major insect developmental hormone, has been identified in the control of sex-specific, *fru* expressing neurons. Mutants in the EcR-A receptor displayed male-male courtship and were found to have a size reduction in two antennal lobe glomeruli that express *fru* (Dalton *et al.*, 2009).

In summary, courtship behavior offers an excellent model to study how complex behavior is regulated. The emerging picture is one of sex-specific neurons that are required to establish the competence for courtship behavior, and which respond to sensory and environmental input as well as diffusible, hormone-like factors. The molecular basis of how the relevant neuronal circuits are established and how they function continues to be an exiting area for future studies.

Acknowledgment

Work in the author's laboratory was supported by grant IOS-0919697 from the National Science Foundation.

References

An, W., and Wensink, P.C. (1995). Integrating sex- and tissue-specific regulation within a single Drosophila enhancer. *Genes Dev.* **9**, 256–266.

Arbeitman, M.N., Fleming, A.A., Siegal, M.L., Null, B.H., and Baker, B.S. (2004). A genomic analysis of Drosophila somatic sexual differentiation and its regulation. *Development* **131**, 2007–2021.

Arthur Jr., B.I., Jallon, J.M., Caflisch, B., Choffat, Y., and Nothiger, R. (1998). Sexual behaviour in Drosophila is irreversibly programmed during a critical period. *Curr. Biol.* **8**, 1187–1190.

Baker, B.S. (1989). Sex in flies: the splice of life. *Nature* **340**, 521–524.

Bastock, M., and Manning, A. (1955). The courtship of *Drosophila melanogaster. Behaviour* **8**, 85–111.

Bauer, J., Antosh, M., Chang, C., Schorl, C., Kolli, S., Neretti, N., and Helfand, S.L. (2010). Comparative transcriptional profiling identifies takeout as a gene that regulates life span. *Aging (Albany, NY)* **2**, 298–310.

Belote, J.M., and Baker, B.S. (1987). Sexual behavior: its genetic control during development and adulthood in Drosophila melanogaster. *Proc. Natl. Acad. Sci. U S A* **84**, 8026–8030.

Billeter, J.C., Villella, A., Allendorfer, J.B., Dornan, A.J., Richardson, M., Gailey, D.A., and Goodwin, S.F. (2006). Isoform-specific control of male neuronal differentiation and behavior in Drosophila by the fruitless gene. *Curr. Biol.* **16**, 1063–1076.

Boll, W., and Noll, M. (2002). The Drosophila Pox neuro gene: control of male courtship behavior and fertility as revealed by a complete dissection of all enhancers. *Development* **129**, 5667–5681.

Bousquet, F., Houot, B., Chauvel, I., Dupas, S., and Ferveur, J.F. (2009). desat1 and the evolution of pheromonal communication in Drosophila. *Ann. N. Y. Acad. Sci.* **1170**, 502–505.

Cachero, S., Ostrovsky, A.D., Yu, J.Y., Dickson, B.J., and Jefferis, G.S. (2010). Sexual dimorphism in the fly brain. *Curr. Biol.* **20**, 1589–1601.

Clyne, J.D., and Miesenbock, G. (2008). Sex-specific control and tuning of the pattern generator for courtship song in Drosophila. *Cell* **133**, 354–363.

Coschigano, K.T., and Wensink, P.C. (1993). Sex-specific transcriptional regulation by the male and female doublesex proteins of Drosophila. *Genes Dev.* **7**, 42–54.

Dalton, J.E., Lebo, M.S., Sanders, L.E., Sun, F., and Arbeitman, M.N. (2009). Ecdysone receptor acts in fruitless-expressing neurons to mediate drosophila courtship behaviors. *Curr. Biol.* **19**, 1447–1452.

Datta, S.R., Vasconcelos, M.L., Ruta, V., Luo, S., Wong, A., Demir, E., Flores, J., Balonze, K., Dickson, B.J., and Axel, R. (2008). The Drosophila pheromone cVA activates a sexually dimorphic neural circuit. *Nature* **452**, 473–477.

Dauwalder, B. (2008). Systems behavior: of male courtship, the nervous system and beyond in Drosophila. *Curr. Genom.* **9**, 517–524.

Dauwalder, B., Tsujimoto, S., Moss, J., and Mattox, W. (2002). The Drosophila takeout gene is regulated by the somatic sex-determination pathway and affects male courtship behavior. *Genes Dev.* **16**, 2879–2892.

Demir, E., and Dickson, B.J. (2005). fruitless splicing specifies male courtship behavior in Drosophila. *Cell* **121**, 785–794.

Dornan, A.J., Gailey, D.A., and Goodwin, S.F. (2005). GAL4 enhancer trap targeting of the Drosophila sex determination gene fruitless. *Genesis* **42**, 236–246.

Ejima, A., and Griffith, L.C. (2007). Measurement of courtship behavior in Drosophila melanogaster. *Cold Spring Harb. Protoc.* doi: 10.1101/pdb.prot4847

Ejima, A., and Griffith, L.C. (2008). Courtship initiation is stimulated by acoustic signals in Drosophila melanogaster. *PLoS One* **3**, e3246.

Ejima, A., and Griffith, L.C. (2011). Assay for courtship suppression in Drosophila. *Cold Spring Harb. Protoc.* doi: 10.1101/pdb.prot5575

Ejima, A., Smith, B.P., Lucas, C., van der Goes van Naters, W., Miller, C.J., Carlson, J.R., Levine, J.D., and Griffith, L.C. (2007). Generalization of courtship learning in Drosophila is mediated by cis-vaccenyl acetate. *Curr. Biol.* **17**, 599–605.

Ellis, L.L., and Carney, G.E. (2010a). Mating alters gene expression patterns in Drosophila melanogaster male heads. *BMC Genomics* **11**, 558.

Ellis, L.L., and Carney, G.E. (2010b). Socially-responsive gene expression in male Drosophila melanogaster is influenced by the sex of the interacting partner. *Genetics* **187**, 157–169.

Ferveur, J.F. (1997). The pheromonal role of cuticular hydrocarbons in Drosophila melanogaster. *Bioessays* **19**, 353–358.

Ferveur, J.F. (2005). Cuticular hydrocarbons: their evolution and roles in Drosophila pheromonal communication. *Behav. Genet.* **35**, 279–295.

Ferveur, J.F. (2010). Drosophila female courtship and mating behaviors: sensory signals, genes, neural structures and evolution. *Curr. Opin. Neurobiol.* **20**, 764–769.

Ferveur, J.F., Stortkuhl, K.F., Stocker, R.F., and Greenspan, R.J. (1995). Genetic feminization of brain structures and changed sexual orientation in male Drosophila [see comments]. *Science* **267**, 902–905.

Fujii, S., and Amrein, H. (2002). Genes expressed in the Drosophila head reveal a role for fat cells in sex-specific physiology. *EMBO J.* **21**, 5353–5363.

Gailey, D.A., and Hall, J.C. (1989). Behavior and cytogenetics of fruitless in Drosophila melanogaster: different courtship defects caused by separate, closely linked lesions. *Genetics* **121**, 773–785.

Galikova, M., and Flatt, T. (2010). Dietary restriction and other lifespan extending pathways converge at the activation of the downstream effector takeout. *Aging (Albany, NY)* **2**, 387–389.

Goldman, T.D., and Arbeitman, M.N. (2007). Genomic and functional studies of Drosophila sex hierarchy regulated gene expression in adult head and nervous system tissues. *PLoS Genet.* **3**, e216.

Greenspan, R.J. (1995). Understanding the genetic construction of behavior. *Sci. Am.* **272**, 72–78.

Griffith, L.C., and Ejima, A. (2009a). Courtship learning in Drosophila melanogaster: diverse plasticity of a reproductive behavior. *Learn Mem.* **16**, 743–750.

Griffith, L.C., and Ejima, A. (2009b). Multimodal sensory integration of courtship stimulating cues in Drosophila melanogaster. *Ann. N. Y. Acad. Sci.* **1170**, 394–398.

Hall, J.C. (1979). Control of male reproductive behavior by the central nervous system of Drosophila: dissection of a courtship pathway by genetic mosaics. *Genetics* **92**, 437–457.

Hall, J.C. (1994). The mating of a fly. *Science* **264**, 1702–1714.

Hotta, Y., and Benzer, S. (1972). Mapping of behaviour in Drosophila mosaics. *Nature* **240**, 527–535.

Ito, H., Fujitani, K., Usui, K., Shimizu-Nishikawa, K., Tanaka, S., and Yamamoto, D. (1996). Sexual orientation in Drosophila is altered by the satori mutation in the sex-determination gene fruitless that encodes a zinc finger protein with a BTB domain. *Proc. Natl. Acad. Sci. U S A* **93**, 9687–9692.

Kido, A., and Ito, K. (2002). Mushroom bodies are not required for courtship behavior by normal and sexually mosaic Drosophila. *J. Neurobiol.* **52**, 302–311.

Kimura, K. (2011). Role of cell death in the formation of sexual dimorphism in the Drosophila central nervous system. *Dev. Growth Differ.* **53**, 236–244.

Kimura, K., Hachiya, T., Koganezawa, M., Tazawa, T., and Yamamoto, D. (2008). Fruitless and doublesex coordinate to generate male-specific neurons that can initiate courtship. *Neuron* **59**, 759–769.

Kimura, K., Ote, M., Tazawa, T., and Yamamoto, D. (2005). Fruitless specifies sexually dimorphic neural circuitry in the Drosophila brain. *Nature* **438**, 229–233.

Kitamoto, T. (2002). Conditional disruption of synaptic transmission induces male-male courtship behavior in Drosophila. *Proc. Natl. Acad. Sci. U S A* **99**, 13232–13237.

Koganezawa, M., Haba, D., Matsuo, T., and Yamamoto, D. (2010). The shaping of male courtship posture by lateralized gustatory inputs to male-specific interneurons. *Curr. Biol.* **20**, 1–8.

Kohatsu, S., Koganezawa, M., and Yamamoto, D. (2011). Female contact activates male-specific interneurons that trigger stereotypic courtship behavior in Drosophila. *Neuron* **69**, 498–508.

Kubli, E. (1992). The sex-peptide. *Bioessays* **14**, 779–784.

Kurtovic, A., Widmer, A., and Dickson, B.J. (2007). A single class of olfactory neurons mediates behavioural responses to a Drosophila sex pheromone. *Nature* **446**, 542–546.

Kyriacou, C.P., and Hall, J.C. (1986). Interspecific genetic control of courtship song production and reception in Drosophila. *Science* **232**, 494–497.

Kyriacou, C.P., van den Berg, M.J., and Hall, J.C. (1990). Drosophila courtship song cycles in normal and period mutant males revisited. *Behav. Genet.* **20**, 617–644.

Lazareva, A.A., Roman, G., Mattox, W., Hardin, P.E., and Dauwalder, B. (2007). A role for the adult fat body in Drosophila male courtship behavior. *PLoS Genet.* **3**, e16.

Lee, G., Foss, M., Goodwin, S.F., Carlo, T., Taylor, B.J., and Hall, J.C. (2000). Spatial, temporal, and sexually dimorphic expression patterns of the fruitless gene in the Drosophila central nervous system. *J. Neurobiol.* **43**, 404–426.

Lee, G., Hall, J.C., and Park, J.H. (2002). Doublesex gene expression in the central nervous system of Drosophila melanogaster. *J. Neurogenet.* **16**, 229–248.

Manoli, D.S., Foss, M., Villella, A., Taylor, B.J., Hall, J.C., and Baker, B.S. (2005). Male-specific fruitless specifies the neural substrates of Drosophila courtship behaviour. *Nature* **436**, 395–400.

McRobert, S.P., and Tompkins, L. (1983). Courtship of young males is ubiquitous in Drosophila melanogaster. *Behav. Genet.* **13**, 517–523.

McRobert, S.P., and Tompkins, L. (1985). The effect of transformer, doublesex and intersex mutations on the sexual behavior of Drosophila melanogaster. *Genetics* **111**, 89–96.

Mellert, D.J., Knapp, J.M., Manoli, D.S., Meissner, G.W., and Baker, B.S. (2009). Midline crossing by gustatory receptor neuron axons is regulated by fruitless, doublesex and the Roundabout receptors. *Development* **137**, 323–332.

Miyamoto, T., and Amrein, H. (2008). Suppression of male courtship by a Drosophila pheromone receptor. *Nat. Neurosci.* **11**, 874–876.

Moon, S.J., Lee, Y., Jiao, Y., and Montell, C. (2009). A Drosophila gustatory receptor essential for aversive taste and inhibiting male-to-male courtship. *Curr. Biol.* **19**, 1623–1627.

Possidente, D.R., and Murphey, R.K. (1989). Genetic control of sexually dimorphic axon morphology in Drosophila sensory neurons. *Dev. Biol.* **132**, 448–457.

Rideout, E.J., Billeter, J.C., and Goodwin, S.F. (2007). The sex-determination genes fruitless and doublesex specify a neural substrate required for courtship song. *Curr. Biol.* **17**, 1473–1478.

Rideout, E.J., Dornan, A.J., Neville, M.C., Eadie, S., and Goodwin, S.F. (2010). Control of sexual differentiation and behavior by the doublesex gene in Drosophila melanogaster. *Nat. Neurosci.* **13**, 458–466.

Ringo, J., Werczberger, R., and Segal, D. (1992). Male sexual signaling is defective in mutants of the apterous gene of Drosophila melanogaster. *Behav. Genet.* **22**, 469–487.

Ruta, V., Datta, S.R., Vasconcelos, M.L., Freeland, J., Looger, L.L., and Axel, R. (2010). A dimorphic pheromone circuit in Drosophila from sensory input to descending output. *Nature* **468**, 686–690.

Ryner, L.C., Goodwin, S.F., Castrillon, D.H., Anand, A., Villella, A., Baker, B.S., Hall, J.C., Taylor, B. J., and Wasserman, S.A. (1996). Control of male sexual behavior and sexual orientation in Drosophila by the fruitless gene. *Cell* **87**, 1079–1089.

Sanders, L.E., and Arbeitman, M.N. (2008). Doublesex establishes sexual dimorphism in the Drosophila central nervous system in an isoform-dependent manner by directing cell number. *Dev. Biol.* **320**, 378–390.

Sarov-Blat, L., So, W.V., Liu, L., and Rosbash, M. (2000). The Drosophila takeout gene is a novel molecular link between circadian rhythms and feeding behavior. *Cell* **101**, 647–656.

Spieth, H.T. (1974). Courtship behavior in Drosophila. *Annu. Rev. Entomol.* **19**, 385–405.

Stockinger, P., Kvitsiani, D., Rotkopf, S., Tirian, L., and Dickson, B.J. (2005). Neural circuitry that governs Drosophila male courtship behavior. *Cell* **121**, 795–807.

Taylor, B.J., and Truman, J.W. (1992). Commitment of abdominal neuroblasts in Drosophila to a male or female fate is dependent on genes of the sex-determining hierarchy. *Development* **114**, 625–642.

Taylor, B.J., Villella, A., Ryner, L.C., Baker, B.S., and Hall, J.C. (1994). Behavioral and neurobiological implications of sex-determining factors in Drosophila. *Dev. Genet.* **15**, 275–296.

Villella, A., and Hall, J.C. (1996). Courtship anomalies caused by doublesex mutations in Drosophila melanogaster. *Genetics* **143**, 331–344.

Villella, A., and Hall, J.C. (2008). Neurogenetics of courtship and mating in Drosophila. *Adv. Genet.* **62**, 67–184.

von Philipsborn, A.C., Liu, T., Yu, J.Y., Masser, C., Bidaye, S.S., and Dickson, B.J. (2011). Neuronal control of Drosophila courtship song. *Neuron* **69**, 509–522.

Von Schilcher, F. (1976). The behavior of cacophony, a courtship song mutant in Drosophila melanogaster. *Behav. Biol.* **17**, 187–196.

von Schilcher, F., and Hall, J.C. (1979). Neural topography of courtship song in sex mosaics of Drosophila melanogaster. *J. Comp. Physiol. [A]* **129**, 85–89.

Vosshall, L.B. (2008). Scent of a fly. *Neuron* **59**, 685–689.

Wilson, T.G., DeMoor, S., and Lei, J. (2003). Juvenile hormone involvement in Drosophila melanogaster male reproduction as suggested by the methoprene-tolerant (27) mutant phenotype. *Insect. Biochem. Mol. Biol.* **33**, 1167–1175.

Wolfner, M.F. (2009). Battle and ballet: molecular interactions between the sexes in Drosophila. *J. Hered.* **100**, 399–410.

Yu, J.Y., Kanai, M.I., Demir, E., Jefferis, G.S., and Dickson, B.J. (2010). Cellular organization of the neural circuit that drives Drosophila courtship behavior. *Curr. Biol.* **20**, 1602–1614.

CIRCADIAN PLASTICITY: FROM STRUCTURE TO BEHAVIOR

Lia Frenkel and María Fernanda Ceriani

Laboratorio de Genética del Comportamiento, Fundación Instituto Leloir, Instituto de Investigaciones Bioquímicas de Buenos Aires, CONICET, Av. Patricias Argentinas 435 - Ciudad Autónoma de Buenos Aires, (C1405BWE) Argentina

Over the years it has become clear that the biological clock acts at different levels, ranging from the control of gene expression, protein stability, or subcellular localization of key proteins, to the fine tuning of network properties and modulation of input signals, ultimately ensuring that the organism will be best synchronized to a changing environment at the physiological and behavioral levels. The purpose of this chapter is to discuss the circadian control of clock outputs, spanning the most immediate ones within pacemaker neurons (i.e., membrane excitability, release of neurotransmitters, structural changes) to the circadian modulation of different behaviors (locomotor activity, learning and memory, social interaction), with a focus on the examples that shed light on the surprising degree of plasticity that characterizes the underlying circuits.

INTERNATIONAL REVIEW OF
NEUROBIOLOGY, VOL. 99
DOI: 10.1016/B978-0-12-387003-2.00005-7

I. Introduction

The circadian clock serves as a temporal filter to synchronize gene expression, cell metabolism, physiology, and behavior to the most critical moments in the day, thus contributing to the adaptation of the organism to a changing environment. Although in the last 15 years substantial progress has been made in elucidating the molecular processes that impart this temporal control at the cellular level in *Drosophila* and other organisms (Hardin, 2005), how this translates into rhythmic behavior is less clear (Nitabach and Taghert, 2008). At the cellular level, the circadian clock is based on self-sustaining, cell-autonomous transcriptional negative feedback loops, which ultimately give rise to rhythms in the abundance, phosphorylation state, and nuclear localization of key intracellular proteins, such as PERIOD (PER) and TIMELESS (TIM) (Stanewsky, 2002).

To date several neuronal clusters have been shown to include a molecular oscillator in the *Drosophila* adult brain. The one best understood encompasses the small ventral lateral neurons (sLNvs), which are located anteriorly in the accessory medulla at the level of the esophageal foramen and send projections dorsally and posteriorly toward the superior protocerebrum, ending close to other relevant groups: the dorsal neurons (DNs) clusters 1 and 2 (Helfrich-Forster, 2003). The sLNv cluster comprises five cells, of which four rhythmically release the neuropeptide pigment dispersing factor (PDF) at their dorsal terminals (Park *et al.*, 2000). Relatively close to the sLNvs, there is a second cluster of larger somata known as the large LNvs (lLNvs) that release PDF through ipsi- and contralateral projections in the optic lobe. Other oscillators within the fly brain include the dorsal lateral neurons (LNds), the lateral posterior neurons (LPNs), together with the DN3, which is a group of about 40 neurons relevant for entrainment under light-dark (LD) cycles (Veleri *et al.*, 2003). A thorough description of the different anatomical clusters was provided by Taghert and colleagues (Shafer *et al.*, 2006). Recently, significant progress has been made in the understanding of how different environmental cues (namely, light and temperature) manage to impact on this cellular mechanism (Dubruille and Emery, 2008). This chapter will delve into the circadian control of clock outputs, ranging from the most immediate one within pacemaker neurons to the circadian modulation of different behaviors.

II. Transmitting Molecular Oscillations to Neuronal Networks

A. Circadian Modulation of Membrane Electrical Properties

About a decade ago, the advent of high-density oligonucleotide arrays spanning whole genomes provided an ample picture of the extent of clock-controlled genes at the transcriptional level, and hence, of the potential diversity of processes that take place at regular intervals throughout the day in a given organism (as diverse as plants (Harmer *et al.*, 2000), flies (Ceriani *et al.*, 2002; Claridge-Chang *et al.*, 2001; Lin *et al.*, 2002; McDonald and Rosbash, 2001; Nagoshi *et al.*, 2010; Ueda *et al.*, 2002), or mammals (Akhtar *et al.*, 2002; Oishi *et al.*, 2003; Panda *et al.*, 2002)). Although cycling transcript levels do not necessarily reflect circadian fluctuations in activity of a given protein, in time it became clear that many aspects of metabolism, physiology, and behavior are structured to anticipate changes accompanying the 24-h cycles imparted by the rotating planet. In fact, clock-controlled transcription is strikingly widespread, and about half of all mammalian genes fluctuate with a 24-h rhythm in at least one tissue (Yan *et al.*, 2008). Although this issue has not been extensively explored in *Drosophila*, very little overlap was reported between cycling genes in the head and body parts (Ceriani *et al.*, 2002), implying that genes might cycle only on those tissues in which they play a circadian relevant function. Moreover, given that the molecular clock is essentially a transcriptional/translational phenomenon, it is not surprising that basic membrane constituents such as ion channels and receptors are regulated at the mRNA level (Ceriani *et al.*, 2002; Claridge-Chang *et al.*, 2001; Kula-Eversole *et al.*, 2010). Claridge-Chang and colleagues reported that genes known to be involved in vesicle recycling and transport oscillate at the transcript level. This observation led them to propose that there is circadian modulation of the synaptic vesicle pool and, hence, that synaptic function and plasticity could change throughout the day (Claridge-Chang *et al.*, 2001). In separate studies, a number of genes involved in neurotransmitter synthesis, transport, and recycling were found to change in a circadian fashion (Ceriani *et al.*, 2002; Ueda *et al.*, 2002), lending further support to the potential modulation of synaptic function exerted by the circadian clock.

One possible way to convey time of day information to locomotor centers and other processing areas in the central brain is to control the firing properties of different circadian clusters, which could be achieved in a straightforward manner through direct transcriptional regulation of ion channels and transporters (Ceriani *et al.*, 2002; Claridge-Chang *et al.*, 2001). Although at the protein level such oscillation has rarely been addressed (except for the Ca^{2+}-dependent, voltage-gated K^+ channel *slowpoke* (Ceriani *et al.*, 2002)), direct measurements of distinct membrane properties support this possibility (Cao and Nitabach, 2008; Park and Griffith, 2006; Sheeba *et al.*, 2008). Park and Griffith (2006) monitored the resting membrane potential and membrane resistance of one of the circadian clusters—the lLNvs—at two different time points under light–dark cycles and constant darkness (LD and DD, respectively). They observed that both the resting potential and membrane resistance fluctuate with the clock and light conditions. Later on, a thorough analysis of basic properties of the same pacemaker cluster indicated that, despite no apparent differences at the morphological level, the lLNvs show two distinct "modes" of spontaneous firing throughout the day, tonic or bursting, and that a proportion of the sampled neurons could switch between both states during the recording (Sheeba *et al.*, 2008). Holmes and colleagues also showed that spontaneous firing of action potentials rapidly increases in response to light, and such modulation depends on *cryptochrome* (*cry*), the photoreceptor expressed in most circadian cells (Emery *et al.*, 1998, 2000; Stanewsky *et al.*, 1998). In an elegant experiment, Sheeba, Gu, and colleagues showed that, under free-running conditions (DD15; 15 days upon transfer to constant darkness), the spontaneous firing rate maxima occurred during the subjective day and the minima occurred during the subjective night (Sheeba *et al.*, 2008). Accordingly, a larger proportion of lLNvs display burst firing activity during the early subjective day, which progressively decreased by mid-subjective day and was absent later on. Consistent with this behavior the resting membrane potential peaks in the early to mid-subjective day and reaches a minimum at the beginning of the subjective night, underscoring a clear circadian regulation on this intrinsic property of this pacemaker cluster. In parallel, Cao and Nitabach (2008), in a slightly different preparation, also investigated the temporal regulation of membrane excitability. They recorded from the lLNvs, which are the most accessible cluster among the PDF+ cells, and also from the sLNvs, which are the most relevant neurons in terms of rhythmic behavior. Under light–dark cycles they found that the resting membrane potential of the lLNvs becomes most hyperpolarized from dawn to dusk, whereas the firing rate and membrane resistance decreases. In the dark phase the membrane resting potential becomes more depolarized as the night progresses, despite the fact that the firing rate and membrane resistance stay relatively low throughout. As the authors point out, it is unclear why these properties do not change accordingly, and could result from the differential regulation of specific ion channel subtypes (Cao and Nitabach, 2008); alternatively, it could reflect the variability of the recording preparation. As expected, circadian modulation of these membrane properties was lost in a clockless fly (a *per*^{01} null mutant) and also under free-running conditions (on DD1) in the wild type. The latter is potentially accounted for by the loss of coherent molecular oscillations in this cluster during the first 1–2 days upon transfer to constant darkness (Shafer *et al.*, 2002; Yang and Sehgal, 2001), a potential indication of the cell autonomous nature of this property (Cao and Nitabach, 2008). Interestingly, perhaps as a result of the optimized recording procedure, Cao and Nitabach also detected action potentials of roughly half the amplitude, derived from arborizations within the optic lobe of the contralateral side, opening the possibility that each lLNv processes information received at both sites independently.

In addition, Cao and Nitabach (2008) performed for the first-time recordings from the sLNvs, and reported that these cells are most depolarized near lights on, becoming hyperpolarized as the day proceeds. Toward the end of the night period the resting membrane potential begins to increase, though not as steadily as during the day, and the cells become more depolarized again. As the sLNvs cluster is responsible for the increase in locomotor activity anticipating lights on, the reported changes in membrane excitability nicely correlate with overt behavior. No clear trend was observed during the 12-h window between transitions (Cao and Nitabach, 2008); whether this reflects a property of the sLNvs or results from an inherently difficult biological preparation is yet to be determined. Whether sLNvs firing rate or membrane resistance is modulated by the circadian clock remains to be explored. No information is yet available for the remainder of the circadian network.

The connection between the circadian clock and membrane properties has been explored in different model organisms. Free-running circadian rhythms in membrane conductance and delayed-rectifier K^+ channel current were reported in pacemaker neurons of the molluscan retina (Michel *et al.*, 1993, 1999) and in the suprachiasmatic nuclei (SCNs) in the mammalian brain (De Jeu *et al.*, 1998; Itri *et al.*, 2005, 2010; Kuhlman and McMahon, 2004). *per1* expressing SCN cells exhibit daily changes in membrane potential (Belle *et al.*, 2009) and even cytoplasmic calcium levels have been shown to oscillate in a circadian manner (Colwell, 2000; Pennartz *et al.*, 2002). The conservation of the molecular mechanisms underlying the cell-autonomous biological clock as well as in this subset of physiological properties underscore that very similar mechanisms convey time of day information to prelocomotor centers in evolutionary distant organisms (Welsh and colleagues have recently reviewed this topic (Welsh *et al.*, 2010)).

B. Adjustment in the Release of Signaling Molecules Throughout the Day

Among the properties of circadian pacemaker neurons that are particularly relevant in the control of rhythmic behavior (recently reviewed by Nitabach and Taghert (2008)) is the release of molecules that could trigger specific (i.e., resetting) signaling events in downstream targets. In *Drosophila*, the candidate molecule best suited to play this role is the neuropeptide PDF. Initially, PDF reactive neurons were singled out as potential pacemaker cells as they coexpressed the core component PER and were located in a region that fulfilled the anatomical criteria proposed for circadian pacemakers in insects (Helfrich-Forster, 1995, 1997; Helfrich-Forster and Homberg, 1993). Surprisingly, no oscillations at the mRNA level could be detected by Northern blot analysis (Park and Hall, 1998), opening the possibility that the clock controls other aspects of PDF synthesis. Park and colleagues reevaluated *pdf* expression by *in situ* hybridization and concluded that no temporal regulation of PDF expression by the circadian clock takes place at the mRNA level. Nonetheless, the cyclic expression of PDF is lost within the sLNvs in certain clock mutants, that is, $Clock^{jrk}$ and cyc^{02} (Park *et al.*, 2000). This observation is consistent with the view that, in this cluster, CLOCK and CYCLE are indirect positive regulators of *pdf* expression. Interestingly, upon closer examination they reported changes in PDF immunoreactivity throughout the day in the axonal projections of the sLNvs at the dorsal protocerebrum, with a peak during the early day or subjective day, and a trough at the beginning of the night, which are lost in two null clock mutants, *per*^{01} and *tim*^{01} (Blau and Young, 1999).

The relevance of the PDF neuropeptide in rhythmic control of locomotor behavior became clear as it was reported that a *pdf* null mutant becomes arrhythmic a few days upon transfer to constant conditions (Renn *et al.*, 1999). PDF signaling is mediated by a G-protein coupled receptor, PDFR (Hyun *et al.*, 2005; Lear *et al.*, 2005b; Mertens *et al.*, 2005). In response to exogenous PDF application, increased cAMP levels are observed in most circadian clusters, implying that most of the circadian network is able to respond to PDF, with the notable exception of the lLNvs (Shafer *et al.*, 2008). More recently, the extent of PDFR expression was refined through promoter analysis (Im and Taghert, 2010). Among clock outputs, PDF immunoreactivity at the dorsal projections of the sLNvs has been extensively explored and reported; although to date it is still uncertain whether it reflects circadian changes in peptide synthesis, processing, axonal transport, and/or release. Aberrant PDF levels were found in mutants with impaired neuronal function such as *narrow abdomen* (*na*^{har}) (Lear *et al.*, 2005a), *slowpoke* (*slo*^{4}) (Fernández *et al.*, 2007) or *shaw* (Hodge and Stanewsky, 2008), suggesting that membrane excitability directly affects PDF accumulation at the dorsal terminals. Despite extensive reports on the differential PDF immunoreactivity in the axonal terminals of the sLNvs, its relevance in the control of rhythmic locomotor activity has been questioned (Kula *et al.*, 2006). However, in the set of experiments that support their conclusion the control group *pdf*-Gal4 does not show cyclical changes in PDF levels as was repetitively reported (Blanchard *et al.*, 2010; Fernández *et al.*, 2007; Harrisingh *et al.*, 2007; Helfrich-Forster *et al.*, 2000; Lear *et al.*, 2005a; Park *et al.*, 2000), calling for a reexamination of this issue.

Despite the relevance of the PDF neuropeptide, it is possible that additional molecules could be employed by the circuit to convey time of day information to downstream targets. Early on, Kaneko and colleagues reported that expression of the tetanus toxin light chain (TeTxLC), which impairs fast synaptic transmission through cleavage of synaptobrevin, affects circadian behavior if expressed in the PER/TIM+ cells, but not when expressed in the PDF+ cells (Kaneko *et al.*, 2000). In addition, this observation implies that the sLNvs do not release other molecules relevant for circadian control of locomotor behavior, at least through a mechanism sensitive to TeTxLC. In fact, as the authors note, intercellular communication within the network could still be taking place through electrical coupling (gap junctions), thus explaining the rhythmic behavior in flies expressing TeTxLC within the PDF circuit. Evidence of the latter has recently been found in the lLNvs in physiological recordings of acute brain preparations (Cao and Nitabach, 2008), in line with what was previously described in the accessory medulla in the cockroach (Schneider and Stengl, 2006). More recently, complementary analyses were undertaken employing a temperature-sensitive, dominant-negative version of the *shibire* gene. The *shibire* gene encodes *Drosophila* dynamin. With the shi^{ts1} mutation, one can use temperature shifts to disrupt receptor-mediated endocytosis and vesicle trafficking (reviewed in Kitamoto, 2001). Interestingly, expressing SHI^{ts} in PDF+ cells lengthens the free-running period by almost 3 h at the restrictive temperature (Kilman *et al.*, 2009; Wulbeck *et al.*, 2009). Strikingly, PDF levels still cycle in the axonal terminals at the dorsal protocerebrum, leading Helfrich Forster and colleagues to discount potential effects on PDF transport, and to propose that the long period phenotype—which is lost in the absence of PDFR—derives from PDF acting on the sLNvs (Wulbeck *et al.*, 2009). In a parallel study, Allada and coauthors showed that period lengthening was largely dependent on the LNv cluster because it disappears when broad circadian expression is combined with Gal80-mediated repression exclusively in the PDF+ cells (Kilman *et al.*, 2009). As expected, at the behavioral level, the long period phenotype required PDF; however, at the molecular level, PER entry to the nucleus still showed a delay in the absence of the neuropeptide. These and other observations led them to propose that impaired cell membrane recycling may modulate the clock through an effect on the expression, stability or activity of the clock component CLK, likely through a PKA-signaling pathway (Kilman *et al.*, 2009).

PDF is not the only neurotransmitter of the LNv cells. At the ultrastructural level it was reported that small clear vesicles could be detected near synaptic output sites in the axonal terminals of the sLNvs, in addition to the PDF-containing dense core vesicles (Miskiewicz *et al.*, 2004; Yasuyama and Meinertzhagen, 2010), suggesting that additional fast neurotransmitter(s) could take part in this process. Nassel and colleagues (Johard *et al.*, 2009) have recently shown that the sLNv express small neuropeptide F (sNPF), suggesting that at least part of the effects derived from SHI^{ts} expression could derive from defective release of other neuropeptides; in fact, as PDF not only affects the sLNvs but also additional dorsal clusters (Lin *et al.*, 2004; Peng *et al.*, 2003) an alternative interpretation that accounts for the absence of long period phenotype in $pdfr^{O1}$ mutants is that the desynchronization of specific dorsal clusters contributes to the deconsolidated locomotor pattern, thus obscuring the contribution of additional molecules, whose effect on intercellular communication would be most likely subtle (Berni *et al.*, 2008).

Aside from sNPF—whose relevance to circadian biology is yet to be addressed—there is no clear candidate neurotransmitter in the sLNv cells. They were found not to be immunoreactive against several biogenic amines such as serotonin, dopamine, and histamine during the early morning (Hamasaka and Nassel, 2005), which does not rule out a potential accumulation later in the day. In addition, no GABA is expressed in LNv neurons, although they do receive gabaergic inputs relevant in the control of behavior (Chung *et al.*, 2009; Dahdal *et al.*, 2010; Parisky *et al.*, 2008). Less is known about the remaining circadian clusters (Hong *et al.*, 2006), with the notable exception of the work reported by Shafer *et al.*

Table I

Summary of the Neurotransmitters and Neuropeptides Expressed in Different Subsets of Circadian Neurons in the Adult *Drosophila* Brain.

molecule/cluster	DN1	DN2	DN3	sLNv	5th sLNv	lLNv	LNd (6)
PDF[1-3]	–	–	–	+ c	–	+ c	–
sNPF[4]	–	–	–	+	–	–	2(CRY+)/6
NPF[4]	–	–	–	–	–	–	1♀ or 3♂/6
ITP[4]	–	–	–	–	+	–	1(NPF+)/6
chA[4]	–	–	–	–	+	–	2(sNPF+)/6
glutamate[5]	+	–	+	–	–	–	–
NPLP1[6]	+	–	–	–	–	–	–
5HT[7]	–	–	–	–	–	–	–
TH (DA)[5]	–	–	–	–	-	–	–
HA[8-9]	–	–	–	–	–	–	–

Only those pertaining to the LNv cluster were described in the text. +, indicates presence within a subset or all neurons within that specific cluster, which was demonstrated by either direct immunocytochemistry or reporter expression guided by enhancer traps or specific promoter sequences. A lowercase c denotes confirmed cycling. Colocalization with additional circadian relevant markers is indicated in parenthesis. DA, dopamine; HA, histamine; ITP, ion transport peptide; NPLP1, neuropeptide-like precursor 1; 5HT, serotonin; TH, tyrosine hydroxylase; . References: [1]Helfrich-Forster (1995); [2]Renn *et al.* (1999); [3]Park *et al.* (2000); [4]Johard *et al.* (2009); [5]Hamasaka *et al.* (2007); [6]Shafer *et al.* (2006); [7]Hamasaka and Nassel (2006); [8]Hong *et al.* (2006); [9]Hamasaka and Nassel (2008).

(2006) as well as the Nässel laboratory (Hamasaka and Nassel, 2005; Hamasaka *et al.*, 2005, 2007; Johard *et al.*, 2009; Shafer *et al.*, 2006). A summary of potential neurotransmitters expressed in the different clusters is included in Table I.

There is evidence that synaptic interactions involving neuropeptides play critical roles in the maintenance and synchronization of circadian rhythms in the mammalian brain. In fact, vasoactive intestinal peptide (VIP) appears to play a similar role to PDF in the mammalian brain, synchronizing SCN neurons, modulating molecular oscillations within individual oscillators, and acting downstream of light cues in entrainment (reviewed in (Vosko *et al.*, 2007).

C. Daily Changes in Neural Structure: Another Output of the Clock

Over the years it has become increasingly evident that the biological clock acts on different layers within a cell, not only controlling the levels and activity of given proteins but also modulating the structural properties of the cell itself. An example of this structural remodeling can be found in the visual system of different species, which undergoes circadian morphological changes. Pyza's laboratory found circadian changes in the size and shape of fly interneurons within the first optic neuropil both in *Musca* (Pyza and Meinertzhagen, 1996) and *Drosophila* (Pyza and Meinertzhagen, 1999). By quantifying the axon caliber of L1 and L2 monopolar cells, the authors found that axons of adult flies swell and shrink rhythmically both under LD and DD and that axon size correlates with the crepuscular pattern of locomotor activity. Axons swell at the beginning of both day and night, and shrink throughout the interval in between (Pyza and Meinertzhagen, 1999). As axon girth is associated with conduction velocity (Matsumoto and Tasaki, 1977), the fact that axon size correlates with the crepuscular pattern of locomotor activity might be functionally linked to a need for increased speed in the processing of visual cues. In fact, in L2 neurons not only the size of the axon is changing but also the nuclei and dendrites (revised in Pyza and Gorska-Andrzejak (2008)). Pyza and coworkers recently reported that the size of the L2 dendritic tree and spine morphology change daily, and is regulated by a circadian clock, increasing at the beginning of the day and decreasing afterward; this pattern persists under DD, in agreement with changes in axonal girth; no rhythmic changes were observed in dendritic tree size under LL (Weber *et al.*, 2009). Moreover, while exploring dendritic size in *per* mutants they were not able to find a temporal pattern of dendritic changes either under LD or in constant conditions (DD and LL), leading them to conclude that the structural plasticity in the L2 dendritic tree is exclusively circadian-driven (a pure clock output) as it cannot be masked by light (in contrast to what happens to rhythmic locomotor activity). In addition, they found that spines in *per*^{01} flies were shorter than in wild type under LD, in line with what was reported in other examples of circadian structural plasticity (see below within this section (p. 116)). However, given that these analyses were performed in the same null mutant background, developmental defects cannot be ruled out. To address the source of oscillations that govern this type of plasticity they restored *per* expression in different clusters. Neither *per* expression in lateral neurons nor in lateral neurons and retina photoreceptors appears to completely rescue circadian plasticity, suggesting that additional clock cells could be involved, for example, glial cells (Pyza and Gorska-Andrzejak, 2008). In fact, Pyza and colleagues had already suggested that *Musca* homologues of *Drosophila* PER+ glial cells could contribute to the regulation of structural plasticity (Pyza and Gorska-Andrzejak, 2004). The relevance of glial cells in the regulation of this circadian remodeling has been extensively reviewed (Jackson, 2010). Weber *et al.* (2009) also examined structural plasticity in the L2 dendritic tree in *cry baby* (*cry*^{b}) mutants, which have nonfunctional peripheral circadian oscillators (so far confirmed in specific organs (Ivanchenko *et al.*, 2001; Krishnan *et al.*, 2001; Myers *et al.*, 2003)) but operative central pacemakers (Krishnan *et al.*, 2001) and found circadian changes in dendritic tree size in LD and DD, but not in LL. This result provides evidence that circadian structural plasticity in L2 dendrites is dependent on the central clock, perhaps within the lLNv because they extend their projections to the optic neuropils (Helfrich-Forster, 1997, 2003). Interestingly, a recent report from the Strauss laboratory shows that structural plasticity in the lamina, measured in cartridge cross-sections as the circumference of the membrane surrounding the photoreceptor terminals, the extension of epithelial glial cells (capitate projections) and the lamina volume, is controlled by the circadian clock but also depends on phototransduction mechanisms (Barth *et al.*, 2010). Additionally, PER expression restricted to photoreceptor cells in the null *per*^{01} mutant rescues volumetric plasticity of the lamina in both DD and LL (Barth *et al.*, 2010). Moreover, such restricted expression is enough to recover the day/night differences in optomotor responses that can be found in the wild-type strain. Thus, Strauss and colleagues concluded that the circadian system would optimally adapt the visual system to the ambient light environment, as a functional circadian clock in photoreceptor cells R1–6 is sufficient to control photoreceptor terminal plasticity, thereby regulating the changes in sensitivity of optomotor behavior (Barth *et al.*, 2010).

In sum, the visual system of *Drosophila* undergoes circadian-driven structural changes in the photoreceptor–interneuron L2 synapses; interestingly, oscillators controlling pre- (Barth *et al.*, 2010) and postsynaptic (Weber *et al.*, 2009) changes appear not necessarily to be the same; although at the postsynaptic level dendritic plasticity depends mostly on the central clock, presynaptic photoreceptor plasticity is driven by its own clock and modulated by light input. Thus, rhythmic structural plasticity in the visual system is a striking example of the complexity underlying this phenomenon.

A while ago our laboratory showed evidence that remodeling in the axonal terminals of the PDF circuit may be a mechanism complementary to that of rhythmic PDF release for the synchronization between circadian clusters (Lin *et al.*, 2004; Peng *et al.*, 2003). The measurement of the complexity of the arborizations within the dorsal protocerebrum at different *zeitgeber* times revealed the existence of daily plastic changes in the axonal termini (Fernández *et al.*, 2008). The degree of axonal arborization was quantified and showed that these projections are more complex during early morning than during early night, which coincides with peak and trough levels of PDF immunoreactivity at these terminals. Such remodeling is not light-driven as it can be found in animals that were kept under constant darkness for 2 days. In addition, this circadian structural plasticity seems to persist over days because it is not restricted to newly eclosed flies (even 2-week-old flies showed differential complexity in the dorsal PDF projections), and there is no sexual dimorphism. To address the involvement of the circadian clock in this axonal remodeling, the number of PDF branches was quantified in the early morning and night in *per*^{01} and *tim*^{01} strains. *per*^{01} and *tim*^{01}, which display constitutively lower and higher PDF levels respectively (Park *et al.*, 2000), show no changes in axonal complexity throughout the day (Fernández *et al.*, 2008).

Interestingly, a detailed analysis evidenced a contrasting phenotype with regard to the length and degree of arborization of these termini: short and hyperbranched in *tim*^{01}, and long and poorly developed in *per*^{01}. In that work it was proposed that structural plasticity could be a mechanism for transmitting clock information on a time-dependent manner, leading to a circadian-regulated variation in the establishment of synapses (Fernández *et al.*, 2008).

Another example of circadian-dependent neuronal plasticity was found in an identified motoneuron (MN5) that innervates two longitudinal indirect flight muscles in *Drosophila* (Mehnert *et al.*, 2007). When comparing the size of the boutons of MN5 terminals from flies entrained to a 12:12-h LD cycle, they reported that mean bouton size is bigger in the middle of the day than in the middle of the night, with a peak at ZT3 and a trough at ZT15 (ZT stands for *zeitgeber* time; ZT0/−12 indicates lights on/off). In spite of a decrement both during the subjective day and night, the difference in size at these two time points is preserved in animals that were kept under constant darkness, reminiscent of a circadian regulation (Mehnert *et al.*, 2007). No changes throughout the day were observed in various *tim* and *per* mutants, again indicative of a clock-controlled phenomenon. Their analysis uncovered another phenotype in these arrhythmic mutants: while *tim* alleles cause hyperbranching of the axonal projections, *per*^{01} displays significantly fewer branches compared to wild-type flies, as previously discussed. Interestingly, the double *per*^{01};*tim*^{01} mutant has a normal branching pattern but no circadian-dependent neuronal plasticity in the size of the boutons. Moreover, enduring plasticity was reported in 1-month-old flies but not in older ones, in line with the characteristic age dependency of neuronal plasticity. Later on, Mehnert and Cantera showed that rhythmic changes in the size of the MN5 terminal synaptic boutons depends on peripheral functional clocks because it remains intact in decapitated flies (Mehnert and Cantera, 2008). The authors suggested that the clock governing MN5 bouton plasticity lies within the PER+ glial cells in the thoracic ganglia, as they did not find PER or TIM expression either in flight motorneurons or in muscles. Surprisingly, rhythmic changes in the MN5 terminals do not seem to depend on locomotor activity itself because neither morning suppression nor evening increment of synaptic activity for 4 h (by paralyzing the flies and forcing locomotor activity, respectively) affect the underlying rhythm (Mehnert *et al.*, 2007; Mehnert and Cantera, 2008). However, synaptic activity disruption for longer bouts could lead to a different conclusion as is the case in zebrafish hypocretin neurons (Appelbaum *et al.*, 2010), in line with a previous report from Shaw and colleagues stating that, in *Drosophila*, a 6-h window of sleep deprivation is needed to see changes in synaptic markers (Donlea *et al.*, 2009). Hypocretin neurons in zebrafish larvae also exhibit circadian regulation of structural plasticity, measured by time-lapse imaging in living animals across day (Appelbaum *et al.*, 2010). An increased number of active synapses is found during the day in projections to the hindbrain and pineal gland, although the phase of such synaptic plasticity is not the same in both areas. A brief (3 h) period of sleep deprivation does not impinge on this form of circadian plasticity. However, a longer period of forced locomotor activity (6 h) disrupts the rhythm by increasing the number of active synapses near the pineal gland during the subjective night (Appelbaum *et al.*, 2010). The latter result suggests that, in addition to the circadian-driven fluctuation, there is also homeostatic control of the number of active synapses of hypocretin neurons. A similar mechanism could be in place in the fly MN5, where longer periods of paralysis and sleep deprivation could impinge upon synaptic bouton size. Daily changes in the size and distribution of synaptic vesicles in the MN5 terminals could provide a functional correlate of the structural plasticity (Ruiz *et al.*, 2010). Appelbaum *et al.* (2010) offered a possible mechanism responsible of this plasticity, whereby the circadian expression of neuronal pentraxin (NPTX2b) in the hypothalamus would command its secretion in the axon terminals promoting AMPA receptor clustering in postsynaptic neurons and, consequently, lead to the configuration of new synapses. Interestingly, there are two pentraxin-like molecules (*furrowed* and b6) in *Drosophila*, which have not been studied in great detail; although no connection to the clock has been established, that is, they do not appear to be particularly enriched in the PDF+ cells (Kula-Eversole *et al.*, 2010).

Although the mechanisms that underlie circadian plasticity (or plasticities?) are still elusive, one possibility is that they share components with the ones involved in circuit formation and establishment. During development, gradients of cell surface and secreted molecules (synaptotrophins) target axons and dendrites to approximately the correct brain region, and then synaptic transmission trigger events that either increase or decrease the likelihood that synapses would be maintained (reviewed in Cline and Haas (2008)). In that regard, our laboratory reported that a hypomorph mutant in a transmembrane receptor crucial for proper axon guidance during embryogenesis (*roundabout*^{hypomorph}, *robo*^{hy}) exhibits an unexpected earlier entrance of PER to the nucleus and, concomitantly, a short period in locomotor activity (Berni *et al.*, 2008). It is noteworthy, that the loss of rhythmicity of *pdf*^{01} null flies (in a *robo* wild type background) is rescued in the heterozygous *robo*^{hy} (*robo*^{hy}/+), suggesting that both molecules can affect the coherent output from the circadian network, and thus impinge upon behavioral rhythmicity. As ROBO is expressed in most neuropils in the adult brain (Berni *et al.*, 2008), it is likely playing a role in the maintenance of synaptic connections throughout life. Thus, if ROBO mediates development/stabilization of synaptic contacts (i.e., through the modulation of N-cadherin homophilic interactions (Rhee *et al.*, 2002, 2007), it follows that a possible outcome of reduced ROBO levels would be a less stable/more plastic (hence, more robust) circadian network, with higher ability to cope with the loss of synchronizing signals. Such a scenario would offer an explanation for the increased rhythmicity of *robo*^{hy}/+;*pdf*^{01} flies. However, it has yet to be demonstrated that ROBO plays a role in the circadian structural remodeling of

Table II
SUMMARY OF NEURONAL STRUCTURES THAT UNDERGO CIRCADIAN STRUCTURAL PLASTICITY IN ADULT *DROSOPHILA*.

Structure	Genotype	LD	DD	LL
L1 and L2– axon[1]	wt	+	+	ns
	tim^{01}	–	–	ns
	per^{01}	–	–	ns
	cry^{b}	ns	ns	ns
L2– dendrites[2]	wt	+	+	–
	tim^{01}	ns	ns	ns
	per^{01}	–	–	–
	cry^{b}	+	+	–
lamina volume[3]	wt	+	+	+
	tim^{01}	+	–	–
	per^{01}	+	–	–
	cry^{b}	ns	ns	ns
MN5[4]	wt	+	+	ns
	tim^{01}	–	–	ns
	per^{01}	–	–	ns
	cry^{b}	ns	ns	ns
sLNv [5]	wt	+	+	ns
	tim^{01}	–	ns	ns
	per^{01}	–	ns	ns
	cry^{b}	ns	ns	ns

L1 and L2 refer to lamina interneurons, MN5 to a motoneuron that innervates two longitudinal indirect flight muscles, and sLNv to the small ventral lateral neurons. +, represents occurrence of circadian structural plasticity; – , indicates the absence the phenomena and ns, not studied. References: [1]Pyza and Meinertzhagen (1999); [2]Weber *et al.* (2009); [3]Barth *et al.* (2010); [4]Mehnert *et al.* (2007); [5]Fernández *et al.* (2008).

the PDF circuit. A summary of the structural plasticity described in adult flies is included in Table II.

So far, there is evidence supporting the hypothesis that the mechanism triggering morphological plasticity might act nonautonomously, as appears to be the case in the MN5 (Mehnert and Cantera, 2008), as well as cell autonomously. Such examples include the sLNvs (although additional experiments are required to more directly address this issue) and the hypocretin terminals. In this case, circadian plasticity would depend to some extent on the secretion of a clock-regulated protein that promotes postsynaptic receptor clustering. However, in the latter case, the postsynaptic terminal could play a major role as terminals from specific neurons projecting to different targets cycle with different phases (Appelbaum *et al.*, 2010).

Circadian plasticity refers to changes in communication efficacy throughout the day, probably through the regulation of synaptic assembly. Hence, it could involve both pre- and postsynaptic adjustments; nonetheless, it is likely initiated by the presynaptic neuron. Supporting this notion: (1) during development, axon guidance and branching is independent of a functional synapse. In contrast, it depends on the highly regulated expression of attraction and rejection molecules (i.e., ROBO/SLIT) but not on the synaptic communication with the target. (2) Remodeling takes place within the axon terminals of the neuronal pacemakers, as it is the case of the sLNvs, a clear example of presynaptic initiation. Regarding the circadian structural changes in L2 and MN5, it has been suggested that the glial clock plays a preponderant role in commanding synaptic remodeling. Although retrograde signaling from muscles to motoneurons was demonstrated during larval development (Keshishian and Kim, 2004; Marques and Zhang, 2006), circadian plasticity in MN5 refers to changes in the presynaptic site. On the contrary, L2 daily remodeling is the only example that discloses synaptic plasticity at the postsynaptic site.

Daily changes in neuronal architecture were reported in the mammalian major circadian pacemaker, the SCN. Vasointestinal peptide (VIP) but not arginine-vasopressin (AVP) immunoreactive neurons in the SCN experience ultrastructural reorganization in the neuronal-glia network (Becquet *et al.*, 2008) and at glutamatergic and nonglutamatergic synaptic terminals (Girardet *et al.*, 2010). As these changes are light-driven, the authors suggested that the changes have a role in photic synchronization (reviewed in Bosler *et al.* (2009)). But this kind of plasticity is not restricted to the SCN, as dendritic architecture and spine density of rat neurons from medial prefrontal cortex increase in the dark period, when the animals are active, and decrease during day time, when they rest. When the authors analyzed several parameters that account for the morphology of pyramidal neurons located in layer III of the infralimbic cortex (which receives direct input from the SCN), they found greater changes in basilar dendrites compared to apical ones. Stress during the active period abolishes these daily differences in morphology (Perez-Cruz *et al.*, 2009). In line with these results, a very recent report shows that a circadian disruption protocol (in which mice were kept in 20-h LD cycles 10-h light:10-h dark) not only disrupt temperature and metabolic hormonal level rhythms but also decrease apical, but not basal, dendrite size and morphology complexity of layer III medial prefrontal cortex neurons. Moreover, at the behavioral level, circadian disruption is not correlated with a difficulty in learning but with a reduced capability of re-learning, suggesting reduced behavioral plasticity (Karatsoreos *et al.*, 2011).

Thus, circadian remodeling of synaptic terminals appears to be widespread (and not only restricted to the projections of circadian pacemakers) and could in itself encode timing information relevant for the coherent control of a variety of behaviors.

III. From Oscillating Networks to Coherent Circadian Behaviors

A. Peripheral Clocks Control Behavioral Responses Independently From the Central Clock

An increasing body of evidence suggests that multiple modes of sensory processes are modulated by the endogenous clock (Allada and Chung, 2010; Glossop and Hardin, 2002). These studies show that peripheral circadian oscillators are crucial for sustaining these rhythms. First, regulation of olfactory responses in the antennae by a circadian clock was demonstrated (Krishnan *et al.*, 1999; Tanoue *et al.*, 2004), and reported to operate independently from the central clock neurons that govern rhythmic eclosion and locomotor activity patterns during adulthood (the PDF+ sLNvs). Soon after, Guo and colleagues showed a circadian rhythm in attractive and repulsive olfactory responses that depends neither on PDF nor on the presence of the LNvs (Zhou *et al.*, 2005). When looking for the cellular and molecular mechanisms that drive this olfactory rhythm, it was reported that spontaneous spike amplitude fluctuate throughout the day in certain sensillae, probably due to rhythmically controlled abundance of a particular G-coupled protein kinase that controls odorant receptor-dependent ion channel activity/composition and localization (Krishnan *et al.*, 2008; Tanoue *et al.*, 2008). In line with previous reports, Krishnan *et al.* showed an increased sensory processing during either midnight or subjective midnight (ZT17 and CT17), which coincides with the peak performance in the olfactory responses (Zhou *et al.*, 2005).

More recently, Chatterjee *et al.* (2010) elegantly demonstrated that the major chemosensory organ in *Drosophila*, the proboscis, harbors a circadian oscillator that controls gustatory physiology and regulates appetitive behavior independently of the central brain (Chatterjee *et al.*, 2010). These results are in line with those of Barth *et al.* (2010), whose work focused on photoreceptor cells and the control of optomotor turning response. Both reports suggest that behaviorally relevant adaptive processing of visual and gustatory information is performed, in part, at the input level (Barth *et al.*, 2010).

B. The Locomotor Activity Pattern of *Drosophila melanogaster*: An Extensively Studied Circadian Behavioral Output

Drosophila constitutes an excellent animal model for learning about the neuronal clock that drives rhythmic behaviors. The very first observation of a rhythmic behavior in flies was the time of emergence of the adult from the pupal case. Since the 1960s and 1970s eclosion rhythms have been used to reveal mutations in clock genes (as an example, see Konopka and Benzer, 1971). The eventual development of devices capable of automated recording of locomotor activity patterns of individual adult flies opened the research to a second, and by far, best understood circadian-driven behavior that, in contrast to eclosion that takes place only once in a lifetime, can be analyzed during the entire adult life.

Crepuscular animals such as *Drosophila* show a bimodal pattern of activity (two peaks per cycle under light–dark conditions): around dawn and dusk activity increases, and at noon and during the night activity is reduced. In the laboratory, the complex natural illumination spanning changes in light intensity and wavelength content throughout the day are replaced by a square pulse of light that is either on (usually for 12-h nonstop) or off. Under this simplistic condition, at a constant temperature (20–25°C) and 12-h of light alternated with 12-h of darkness (LD 12:12), individually housed male flies display a bimodal activity pattern. While looking for sexual dimorphism in the pattern of rhythmic locomotor activity of several wild-type strains, Helfrich-Forster (2000) found that males tend to be active before lights-on and reached their peak of activity around lights-on; their activity decreases continuously and stays low for several hours until approximately 2 h before lights-off. After this, activity is reduced until the next day. On the contrary, females (both virgin and mated) do not anticipate the lights-on transition with an increase in locomotor activity. They reach their peak about 1-h after lights-on and remain active until later than males. While analyzing the phase of the female's evening activity bout, Helfrich-Forster found strain differences in the degree of sex-dependency, leading her to propose that the phase of the morning peak has a stronger dependency on sex than the evening one. After transferring the flies to constant darkness (DD), strain differences in activity pattern emerge more prominently than sex-specific ones. Half of the flies changed the organization of activity from a bimodal to a unimodal display. In these flies, the morning peak merged with the evening peak to form a band of activity. In persistently bimodal flies, the morning peak becomes weaker and the phase relationship between peaks of activity becomes smaller. The examination of locomotor patterns after prolonged time under constant darkness (25–40 days) uncovered a proportion of flies showing spontaneous changes in period that correlated with changes in phase relationship between morning and evening peaks; only a few showed sudden internal desynchronization. The author proposed that this pattern of activity could represent another example of hierarchical organization between oscillators: one behaving as the pacemaker and the other one as the slave or driven (reminiscent of the "morning" and "evening" oscillator model put forward by Pittendrigh and Daan (Pittendrigh, 1976)). Thus, differential regulation of morning and evening peaks and spontaneous splitting and disorganization of locomotor patterns would suggest the existence of two oscillators with a pacemaker–slave relationship that control circadian locomotor behavior (Helfrich-Forster, 2000).

The major role played by the LNvs in locomotor rhythms was first inferred by analyzing the performance of the *disconnected* mutant, which lacks this cluster

(Helfrich-Forster, 1998). Soon after, it was shown that *pdf*^{01} flies, lacking the main output from the LNvs (the PDF neuropeptide), become largely arrhythmic once they are transferred to DD (Renn *et al.*, 1999), lending further support to the crucial role of the LNvs. In addition, PER overexpression in the LNvs disrupts locomotor rhythmicity despite protein oscillations in specific dorsal clusters (Blanchardon *et al.*, 2001). In order to further dissect the neural basis of rhythmic locomotor activity two groups made use of different genetic strategies to elucidate which neurons control the morning and evening activity peaks. Rouyer and colleagues reinstated PER function in a subset of clock neurons in an otherwise *per* null mutant, employing different Gal4 drivers (Grima *et al.*, 2004). They focused on *pdf*-Gal4 (expressed in the LNvs), Mz520-Gal4 (LNvs), *cry*-Gal4 (DN1-3; LNds; LNvs), C929-Gal4 (lLNvs among other nonclock neurons), and Mai179-Gal4 (neurosecretory cells including sLNvs and the non PDF+ 5th sLNv, occasionally one lLNv and some LNds). Later, they evaluated the locomotor activity under LD and DD in a *per*^{01} null background carrying different drivers and UAS-*period* transgenes, and found morning but not evening anticipation in *pdf>per* and Mz520>*per* flies, but no anticipation at all in c929>*per* animals. These results indicated that PER expression in the sLNvs is sufficient to rescue the control of morning anticipation. Moreover, as Mai179>*per* flies anticipate both morning and evening transitions but *pdf>per*, Mz520>*per* and Mz520>*per* flies do not, they concluded that PER expression in the LNds is enough to restore the evening peak. Moreover, PER expression in the sLNvs supported robust PER cycling and rhythmic behavior in *pdf>per* and Mz520>*per* but not in c929>*per* flies, indicating that the PDF+ sLNvs are an autonomous oscillator capable of driving rhythmic behavior in the absence of any other functional clock. In parallel, a similar strategy (specific drivers controlling a proapoptotic transgene expression or rescue of PER function) enabled Rosbash and colleagues to arrive to basically the same conclusions: the LNvs would embody the morning oscillator and the DN1, LNds and the 5th sLNv would form the evening one. Their approach allowed them to conclude that (i) such oscillators are self-sustained, (ii) the LNvs are necessary and sufficient for anticipating the lights-on transition and are not responsible for the light-off response, and (iii) the LNvs are necessary and sufficient for sustaining rhythmicity in DD (Stoleru *et al.*, 2004). In addition, the leading role of the LNvs in the control of rhythmic behavior was evidenced from the observation of the progressively defective PER oscillations (in terms of amplitude and synchronization within the cluster) in the sLNvs and LNds clusters in *pdf*^{01} mutants after increasing bouts in constant darkness. From this work Taghert and colleagues concluded that PDF is required to coordinate the phase of the molecular rhythms within the sLNvs, and set the phase of the LNds (Lin *et al.*, 2004).

To more specifically address the relationship between the different clusters, Stoleru *et al.* shortened the period of molecular oscillations through the over-expression of a TIM kinase (SHAGGY, SGG) in different clock clusters, and examined its impact on locomotor activity under DD. Restricting SGG overexpression to the PDF circuit not only leads to an advance in the evening activity but also to an increased pace of the molecular oscillations—measured as *tim* RNA level—in neuronal clusters, including the ones responsible for the evening activity bout (LNds and DN1s). Surprisingly, *tim* oscillations are not affected in the lLNVs despite SHAGGY overexpression within this subset of neurons. In contrast, the period observed in the lLNvs is locked to that of the DN2s, and both structures are unresponsive to SHAGGY overexpression. Guiding SHAGGY overexpression to the DN1s, LNds and the 5th sLNv clusters does not affect the free-running period (which is determined mainly by the sLNvs) but shortens the length of the subjective day. All in all, the authors suggest that the evening phase within each cycle is a reflection of the endogenous rhythm of the evening oscillator but the period of the cycle correlates with the morning clock (Stoleru *et al.*, 2005).

The circadian clock of animals in the wild is faced with a variety of input signals at once, which might change the hierarchy between the oscillators described under laboratory conditions. Thus, more complex and/or subtler interactions among the different clusters could be uncovered under different environmental conditions. As an example, rest–activity cycles and PER and TIM oscillations can be driven by temperature cycles in LL (Glaser and Stanewsky, 2005; Matsumoto *et al.*, 1998; Yoshii *et al.*, 2005) despite the fact that constant light disrupts molecular and behavioral rhythms at a given temperature. The study of circadian locomotor rhythmicity under constant light in *cry*^{b} mutants, together with other evidence, led Rouyer and colleagues to ascribe a novel effect of light—through differential CRY degradation—, which would allow the control of rhythmic locomotor behavior to either the morning or evening oscillators, depending on ambient conditions. Thus, aside from its ability to synchronize to the environment, light controls the behavioral output from the morning and evening oscillators in an opposite fashion (Picot *et al.*, 2007). In a parallel study also exploring the response to constant light, Rosbash and colleagues concluded that the circadian network switches control between the morning and evening oscillators depending on day length, and further suggested that this switch could define the adjustment to a changing photoperiod (Stoleru *et al.*, 2007). An even more flexible interpretation of morning and evening oscillators was put forward from experiments carried out under dim light; Helfrich-Forster and colleagues found that under these conditions only four cells support PER cycling; two LNds behave as the morning oscillator, and the 5th sLNv and one additional LNd constitute the evening one; these observations prompted them to conclude that the morning/evening oscillator function may not be restricted to certain anatomically defined groups of clock neurons, but instead depends on the environmental conditions (Rieger *et al.*, 2009).

How environmental conditions affect the mutual relationship between clock clusters is still not understood; for example, experiments carried out under temperature cycles indicated that the LNvs seem to be preferentially light-entrainable, whereas the DNs and LPNs seem to be primarily temperature-entrainable (Miyasako *et al.*, 2007). Not only responsiveness to environmental conditions appears to differ among clock clusters, but also the mechanism leading to this response. Although light synchronization through CRY is mainly a cell-autonomous process, temperature seems to require signaling from specific sensory structures, the chordotonal organs, to the brain (Sehadova *et al.*, 2009). However, a highly overlapped set of cycling genes could be detected upon entrainment to light–dark or temperature cycles underscoring that irrespective of the *zeitgeber* there is a rather defined set of cycling genes, that is, clock outputs (Boothroyd *et al.*, 2007).

Although the underlying mechanism is not fully understood yet, behavioral output demonstrates a synergic effect of light and temperature cycles in entrainment. When flies are exposed to both *zeitgebers* the rhythm of locomotor activity is more robust than after synchronization to each *zeitgeber* by itself (Yoshii *et al.*, 2009). This result suggests that an even more complex set of conditions would modify the fine tuning of the rhythm, ensuring a better adjustment to the particular environment. A summary of the best known circadian outputs from the sLNvs are depicted in Fig. 1.

In sum, these results underscore the high degree of plasticity displayed by the connectivity within the circadian network, which would enable it to accommodate to variable environmental input signals and result in coherent behavior. The mechanisms underlying how biological clocks ensure rhythmic rest–activity cycles have been the object of extensive review and are beyond the scope of the present work (see Allada and Chung (2010); Nitabach and Taghert (2008); Peschel and Helfrich-Forster (2011)).

C. Circadian Modulation of Short-Term Memory in *Drosophila*

The projections of the sLNvs branch near the calyx of the mushroom bodies (Helfrich-Forster, 1997), which are essential centers for memory processing in insects (Dubnau *et al.*, 2001; Heisenberg, 1998; Heisenberg *et al.*, 1985); such a neuroanatomical proximity led to propose a functional relationship between these systems, even though the presence of PDF receptors in the calyx had not yet been demonstrated. Whether other clock clusters contact the mushroom bodies has yet to be explored. In 2009, Lyons and Roman showed circadian modulation of short-term memory (Lyons and Roman, 2009). Employing the T-maze device developed by Tully and Quinn (Tully and Quinn, 1985) young adult flies receive an electric shock while exposed to an air current transporting a first odor but not when exposed to a second odor. During testing, animals are placed at a decision point where they should move toward one of the odors. Memory is expressed by an avoidance behavior to the shock-associated odor 3 min after training. Lyons and Roman assayed two kinds of training protocols, a strong one (12 shocks, 1-min odor presentation) and a weak one (1 shock, 10-s odor presentation) at various circadian

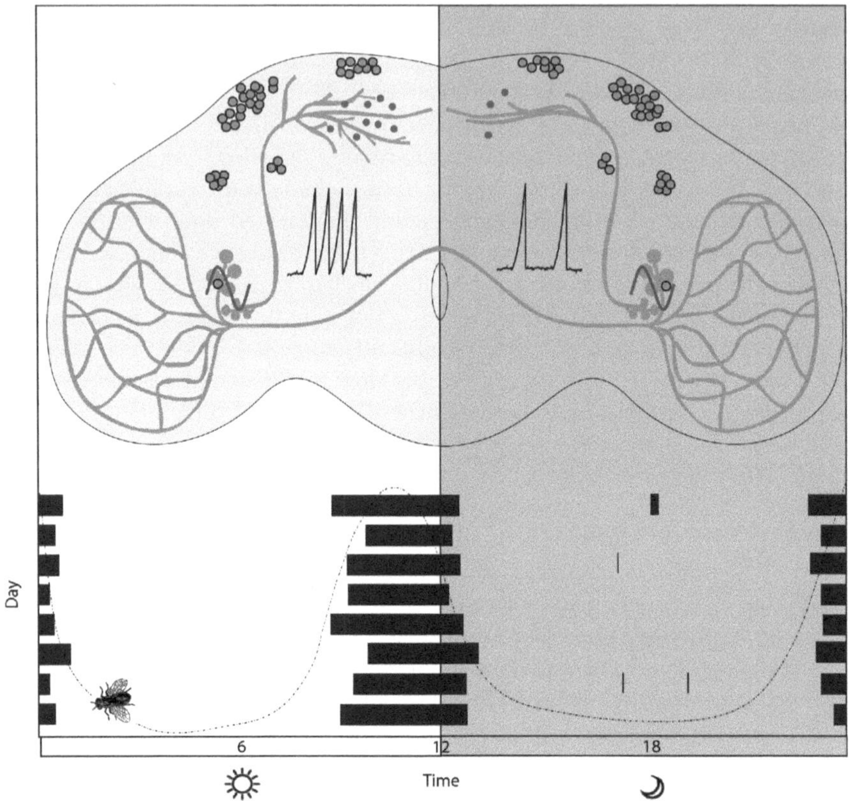

FIG. 1. From oscillating networks to coherent behavior. Schematic diagram displaying examples of circadian outputs in the adult *Drosophila* brain. The clock provides multiple layers of organization to provide a coherent response to the surrounding environment. Among them, electrical properties (indicated as action potentials) as well as the morphology of axonal terminals within PDF+ neurons (in green) are affected by the clock; in addition, variations in transmitter release (red circles) and complex behaviors (such as locomotor activity) are also examples of outputs modulated by the clock that change rhythmically accompanying the daily cycles. (For interpretation of the references to color in this figure legend, the reader is referred to the web version of this book.)

times, to find that the weaker stimulus uncovers a larger circadian effect. This would indicate that a "bad time for learning" could be overcome with a stronger training, suggesting that the time-of-day effect would contribute to improving rather than worsening memory formation, as (1) animals are capable of learning throughout the day and (2) both *per*^{01} and *tim*^{01} mutants, as well as animals exposed to constant-light (another means to disrupt the molecular clock) display memory levels similar to the worst time points in wild-type strains. In order to assess whether the rhythm on performance depends on modulation of sensory processing along the day, rhythms in shock and odor avoidance behaviors were characterized. Under the conditions used in their study they did not find any rhythm, suggesting a central clock involvement on circadian modulation of memory. But a circadian rhythm in the olfactory attractive and repulsive responses dependent on the antennal oscillator was described (Zhou *et al.*, 2005) with a peak response at CT17. Despite the coincidence between the time points for best olfaction responses (CT17) and memory performance (CT13 and CT17), the authors discounted a major antennal peripheral clock contribution on memory performance because *cry* mutant flies (*cry*^{b}), which have nonfunctional peripheral antennal circadian oscillators but operative central pacemakers, displayed circadian rhythms in memory.

Although not excluding the possibility that the circadian clock also affects memory recall, Lyons and Roman suggested that the circadian clock is regulating the strength of the association during memory formation probably due to a modulation of the availability of second messengers or other molecules taking part of the signaling cascade. In support of this possibility a diurnal rhythm of short-term synaptic plasticity was found in GABAergic synapses of the SCN (Gompf and Allen, 2004).

D. Circadian Modulation on Long-Term Memory

Although not yet studied in flies, evidence from other animal models suggests that long-term memory is also modulated by the circadian clock (Barbosa and Albuquerque, 2008; Decker *et al.*, 2007; Fernandez *et al.*, 2003; Gerstner and Yin, 2010; Hauber and Bareiss, 2001; Lyons *et al.*, 2005). For example, the long-term sensitization of the siphon withdrawal reflex in *Aplysia californica*—a non-associative form of learning—is influenced by the circadian clock. Moreover, authors demonstrated that the effect seems to depend on the time of training rather than the time of testing, suggesting that the circadian clock might play an important role in the acquisition/consolidation to form long-term memories *in vivo* (Fernandez *et al.*, 2003). A similar result was obtained with an associative form of learning in the cockroach *Leucophaea maderae* (Decker *et al.*, 2007), indicating that the circadian imposition is not linked to the type of learning but to the cellular mechanisms underlying them. In vertebrates—specifically birds and mammals—, the literature is ambiguous regarding a time-of-day effect, probably due to the diversity of paradigms and experimental procedures employed; nonetheless, a clear light-phase effect on memory was repeatedly reported (Chen and Wolpaw, 1995; Hauber and Bareiss, 2001; Moura *et al.*, 2009; Reijmers *et al.*, 2001; Valentinuzzi *et al.*, 1997, 2004). The mechanism underlying such phenomena is still not clear, although it was proposed that the dynamics of hippocampal clock gene expression imprints a temporal structure on memory processing and shapes, at the same time, the efficacy of behavioral learning (Jilg *et al.*, 2010; Wang *et al.*, 2009). Moreover, data coming from reports on hippocampus long-term potentiation (LTP), a well-established model for studying activity-dependent changes in synaptic strength, reveal a circadian action on neuronal responses that relies on the endogenous clock (Chaudhury *et al.*, 2005; Harris and Teyler, 1983). The current hypothesis states that there is an independent circadian pacemaker that controls time-of-day-dependent changes in hippocampal plasticity, and that the arousal state or sleep *per se* are not necessary for those circadian changes in LTP (Chaudhury *et al.*, 2005). A charming hypothesis is that consolidation—the ability to store memories, which requires synaptic plasticity—changes in response to circadian structural remodeling in the memory centers. In fact, structural plasticity was proposed as a mechanism that could trigger circadian changes in the number of active synapses (Fernández *et al.*, 2008). But it is not restricted solely to pacemaker cells because dendritic architecture and spine density of pyramidal neurons in the rat infralimbic cortex display daily rhythms (Perez-Cruz *et al.*, 2009). As the infralimbic cortex is involved in higher order cognitive functions (Bach *et al.*, 2008; DeSteno and Schmauss, 2008; Vertes, 2006; Wall *et al.*, 2004), it is legitimate to expect that memory consolidation would be modulated by the endogenous clock through, among others, changes in the degree of structural plasticity in memory centers. This potential mechanism would give alternative explanations to, for example, the susceptibility of certain types of memories to the disruption of the clock by constant illumination (Ma *et al.*, 2007), rapid changes in light/dark cycle (Loh *et al.*, 2010) or prolonged 20-h light/dark cycles (Karatsoreos *et al.*, 2011).

E. Circadian Rhythms and Social Interactions

Honey bees display complex social interactions that provide the opportunity to study the relationship between endogenous rhythms and social contact. When studying the molecular oscillations on clock genes in worker bees that switch between nursing (around the clock activity) and foraging (fine-tuned circadian activity), strong oscillations were found only in foragers (Bloch *et al.*, 2001). Furthermore, the lack of oscillation in the nurses is due to their contact with the brood demonstrating a social imposition to the central clock in individual bees (Shemesh *et al.*, 2010).

In flies, such a social interaction was studied at the behavioral level: while the addition of arrhythmic animals (per^{01}) to a group of wild-type flies disperses the phase of the host flies, introducing per^{S} mutants advances it. This effect is dependent on the proportion of visitors, the time-of-day when the animals are inserted, and on a functional olfactory system, as anosmic hosts do not change their behavior when arrhythmic flies are introduced (Levine *et al.*, 2002). In a 2D arena, Fujii *et al.* (2007) described a novel rhythmic locomotor output based on the observation of the social interaction between a male and female fly, termed "close proximity." Close proximity is defined as the percentage of time spent within

≤5 mm of each other. They established that these pairs display an activity pattern that stays rhythmic over days, linked to courtship and copulation. Opposite to the classical crepuscular activity and nocturnal inactivity exhibited by isolated male flies under LD and DD, the rhythmic "close proximity" behavior, reliant mainly on male activity, is elevated during most of the subjective night and morning and minimal at the approximate time of anticipated dusk. This rhythm is not entrainable, as it depends on the continuous presence of the female and is conditional to a functional male central circadian clock. The authors suggested that external cues perceived by the olfactory and other sensory systems feed into the male central pacemakers to cause a shift in the circadian activity. This would imply that a female presence acts as a peripheral stimulus that resets the central pacemaker only in males, as the female activity pattern is not influenced by the presence of a second animal (Fujii *et al.*, 2007). It is yet to be determined whether the relative hierarchy of the different clock clusters is modified by the presence of a social cue. More recent experiments suggest that functional LNvs are essential for male rhythms and lack of PDF disrupts it. In addition, DN1s are required to synchronize the trough at dusk in male proximity behavior (Fujii and Amrein, 2010). The important role of DN1s is supported by another report suggesting that neurons responsible for close proximity rhythms are among the neuropeptide F negative (NPF–) LNds and DN1s, as *npf*-Gal4 directed ablation does not impair this rhythm (Hamasaka *et al.*, 2010). However, Ishida and colleagues also found that disruption of the evening, but not the morning, oscillator caused arrhythmic male proximity behavior, ruling out that the LNvs would be the primary determinants. The former proposed that under the influence of PDF, the DN1 cluster would signal whether the male fly should generate a sex drive or, in contrast, an isolated rhythmic locomotor response (Fujii and Amrein, 2010). On the contrary, Hamasaka *et al.* suggested that, similarly to the notion that morning and evening cells act differently depending on the environmental conditions (Rieger *et al.*, 2009), the NPF– DN1s could also change their behavioral pattern from isolated locomotor rhythms to courtship rhythms triggered by the presence of the female (Hamasaka *et al.*, 2010).

IV. Conclusions

How do molecular clocks manage to transmit time of day information to physiology and behavior? Although the molecular mechanisms are yet to be defined, the notion that the biological clock directly controls output at different levels, from the most immediate one within the nucleus (i.e., gene expression) and cellular properties (i.e., axonal girth, synaptic strength) to a variety of behaviors has consolidated over the years. Perhaps one of the most striking discoveries is the degree of plasticity within the circadian network *per se*, as different neuronal clusters dictate the pace of overt behavior in response to a changing environment.

Do the different cues that synchronize the clock impinge upon different clusters changing their relative weight in the final output? No definite answer is possible today, although initial observations included in this review point in that direction. A striking example of network plasticity is the stomatogastric ganglion of crustaceans, where a small number of neurons exhibit a broad range of outputs depending on which input they receive (Marder and Bucher, 2007). In brief, this ganglion produces several rhythmic outputs that trigger different motor programs finally leading to the processing of different types of food; neuromodulatory molecules can reconfigure circuit dynamics by altering synaptic strength- and voltage-dependent conductances. Moreover, individual neurons can switch among different functional circuits. It would be tempting to speculate that the circadian network operates in a similar fashion, where distinct *zeitgebers* could reconfigure circuit dynamics by uniquely affecting each neuronal cluster (or even specific neurons). Certainly, in this scenario the role of PDF should not be underestimated.

How pervasive is the circadian control of brain functions? An interaction between the PDF circuit and arousal has already been established (Lebestky *et al.*, 2009; Shang *et al.*, 2008). Thus, it is appealing to propose that the clock would exert its modulatory effects not only in complex brain functions such as arousal but also in motivational states, behavioral flexibility, and time awareness among others.

Acknowledgments

We thank Nara Muraro and Ana Depetris-Chauvin for critical comments on the manuscript and Axel Gorostiza for help in assembling the figure. L.F. holds a postoctoral fellowship and M.F.C. is a member of the Argentine Research Council (CONICET). This work was supported by a grant from the ANPCyT, Argentina (PICT2006-1249) and by a FIRCA-NIH grant (1R03TW008342) to M.F.C.

References

Akhtar, R.A., Reddy, A.B., Maywood, E.S., Clayton, J.D., King, V.M., Smith, A.G., Gant, T.W., Hastings, M.H., and Kyriacou, C.P. (2002). Circadian cycling of the mouse liver transcriptome, as revealed by cDNA microarray, is driven by the suprachiasmatic nucleus. *Curr. Biol.* **12**, 540–550.

Allada, R., and Chung, B.Y. (2010). Circadian organization of behavior and physiology in Drosophila. *Annu. Rev. Physiol.* **72**, 605–624.

Appelbaum, L., Wang, G., Yokogawa, T., Skariah, G.M., Smith, S.J., Mourrain, P., and Mignot, E. (2010). Circadian and homeostatic regulation of structural synaptic plasticity in hypocretin neurons. *Neuron* **68**, 87–98.

Bach, M.E., Simpson, E.H., Kahn, L., Marshall, J.J., Kandel, E.R., and Kellendonk, C. (2008). Transient and selective overexpression of D2 receptors in the striatum causes persistent deficits in conditional associative learning. *Proc. Natl. Acad. Sci. U S A* **105**, 16027–16032.

Barbosa, F.F., and Albuquerque, F.S. (2008). Effect of the time-of-day of training on explicit memory. *Braz. J. Med. Biol. Res.* **41**, 477–481.

Barth, M., Schultze, M., Schuster, C.M., and Strauss, R. (2010). Circadian plasticity in photoreceptor cells controls visual coding efficiency in Drosophila melanogaster. *PLoS One* **5**, e9217.

Becquet, D., Girardet, C., Guillaumond, F., Francois-Bellan, A.M., and Bosler, O. (2008). Ultrastructural plasticity in the rat suprachiasmatic nucleus. Possible involvement in clock entrainment. *Glia* **56**, 294–305.

Belle, M.D., Diekman, C.O., Forger, D.B., and Piggins, H.D. (2009). Daily electrical silencing in the mammalian circadian clock. *Science* **326**, 281–284.

Berni, J., Beckwith, E.J., Fernandez, M.P., and Ceriani, M.F. (2008). The axon-guidance roundabout gene alters the pace of the Drosophila circadian clock. *Eur. J. Neurosci.* **27**, 396–407.

Blanchard, F.J., Collins, B., Cyran, S.A., Hancock, D.H., Taylor, M.V., and Blau, J. (2010). The transcription factor Mef2 is required for normal circadian behavior in Drosophila. *J. Neurosci.* **30**, 5855–5865.

Blanchardon, E., Grima, B., Klarsfeld, A., Chelot, E., Hardin, P.E., Preat, T., and Rouyer, F. (2001). Defining the role of Drosophila lateral neurons in the control of circadian rhythms in motor activity and eclosion by targeted genetic ablation and PERIOD protein overexpression. *Eur. J. Neurosci.* **13**, 871–888.

Blau, J., and Young, M.W. (1999). Cycling vrille expression is required for a functional Drosophila clock. *Cell* **99**, 661–671.

Bloch, G., Toma, D.P., and Robinson, G.E. (2001). Behavioral rhythmicity, age, division of labor and period expression in the honey bee brain. *J. Biol. Rhythms* **16**, 444–456.

Boothroyd, C.E., Wijnen, H., Naef, F., Saez, L., and Young, M.W. (2007). Integration of light and temperature in the regulation of circadian gene expression in Drosophila. *PLoS Genet.* **3**, e54.

Bosler, O., Girardet, C., Sage-Ciocca, D., Jacomy, H., Francois-Bellan, A.M., and Becquet, D. (2009). Mechanisms of structural plasticity associated with photic synchronization of the circadian clock within the suprachiasmatic nucleus. *J. Soc. Biol.* **203**, 49–63.

Cao, G., and Nitabach, M.N. (2008). Circadian control of membrane excitability in Drosophila melanogaster lateral ventral clock neurons. *J. Neurosci.* **28**, 6493–6501.

Ceriani, M.F., Hogenesch, J.B., Yanovsky, M., Panda, S., Straume, M., and Kay, S.A. (2002). Genome-wide expression analysis in Drosophila reveals genes controlling circadian behavior. *J. Neurosci.* **22**, 9305–9319.

Claridge-Chang, A., Wijnen, H., Naef, F., Boothroyd, C., Rajewsky, N., and Young, M.W. (2001). Circadian regulation of gene expression systems in the Drosophila head. *Neuron* **32**, 657–671.

Cline, H., and Haas, K. (2008). The regulation of dendritic arbor development and plasticity by glutamatergic synaptic input: a review of the synaptotrophic hypothesis. *J. Physiol.* **586**, 1509–1517.

Chatterjee, A., Tanoue, S., Houl, J.H., and Hardin, P.E. (2010). Regulation of gustatory physiology and appetitive behavior by the Drosophila circadian clock. *Curr. Biol.* **20**, 300–309.

Chaudhury, D., Wang, L.M., and Colwell, C.S. (2005). Circadian regulation of hippocampal long-term potentiation. *J. Biol. Rhythms* **20**, 225–236.

Chen, X.Y., and Wolpaw, J.R. (1995). Operantly conditioned plasticity and circadian rhythm in rat H-reflex are independent phenomena. *Neurosci. Lett.* **195**, 109–112.

Chung, B.Y., Kilman, V.L., Keath, J.R., Pitman, J.L., and Allada, R. (2009). The GABA(A) receptor RDL acts in peptidergic PDF neurons to promote sleep in Drosophila. *Curr. Biol.* **19**, 386–390.

Colwell, C.S. (2000). Circadian modulation of calcium levels in cells in the suprachiasmatic nucleus. *Eur. J. Neurosci.* **12**, 571–576.

Dahdal, D., Reeves, D.C., Ruben, M., Akabas, M.H., and Blau, J. (2010). Drosophila pacemaker neurons require g protein signaling and GABAergic inputs to generate twenty-four hour behavioral rhythms. *Neuron* **68**, 964–977.

De Jeu, M., Hermes, M., and Pennartz, C. (1998). Circadian modulation of membrane properties in slices of rat suprachiasmatic nucleus. *Neuroreport* **9**, 3725–3729.

Decker, S., McConnaughey, S., and Page, T.L. (2007). Circadian regulation of insect olfactory learning. *Proc. Natl. Acad. Sci. U S A* **104**, 15905–15910.

DeSteno, D.A., and Schmauss, C. (2008). Induction of early growth response gene 2 expression in the forebrain of mice performing an attention-set-shifting task. *Neuroscience* **152**, 417–428.

Donlea, J.M., Ramanan, N., and Shaw, P.J. (2009). Use-dependent plasticity in clock neurons regulates sleep need in Drosophila. *Science* **324**, 105–108.

Dubnau, J., Grady, L., Kitamoto, T., and Tully, T. (2001). Disruption of neurotransmission in Drosophila mushroom body blocks retrieval but not acquisition of memory. *Nature* **411**, 476–480.

Dubruille, R., and Emery, P. (2008). A plastic clock: how circadian rhythms respond to environmental cues in Drosophila. *Mol. Neurobiol.* **38**, 129–145.

Emery, P., So, W.V., Kaneko, M., Hall, J.C., and Rosbash, M. (1998). CRY, a Drosophila clock and light-regulated cryptochrome, is a major contributor to circadian rhythm resetting and photosensitivity. *Cell* **95**, 669–679.

Emery, P., Stanewsky, R., Hall, J.C., and Rosbash, M. (2000). A unique circadian-rhythm photoreceptor. *Nature* **404**, 456–457.

Fernández, M.P., Berni, J., and Ceriani, M.F. (2008). Circadian remodeling of neuronal circuits involved in rhythmic behavior. *PLoS Biol.* **6**, e69.

Fernández, M.P., Chu, J., Villella, A., Atkinson, N., Kay, S.A., and Ceriani, M.F. (2007). Impaired clock output by altered connectivity in the circadian network. *Proc. Natl. Acad. Sci. U S A* **104**, 5650–5655.

Fernandez, R.I., Lyons, L.C., Levenson, J., Khabour, O., and Eskin, A. (2003). Circadian modulation of long-term sensitization in Aplysia. *Proc. Natl. Acad. Sci. U S A* **100**, 14415–14420.

Fujii, S., and Amrein, H. (2010). Ventral lateral and DN1 clock neurons mediate distinct properties of male sex drive rhythm in Drosophila. *Proc. Natl. Acad. Sci. U S A* **107**, 10590–10595.

Fujii, S., Krishnan, P., Hardin, P., and Amrein, H. (2007). Nocturnal male sex drive in Drosophila. *Curr. Biol.* **17**, 244–251.

Gerstner, J.R., and Yin, J.C. (2010). Circadian rhythms and memory formation. *Nat. Rev. Neurosci.* **11**, 577–588.

Girardet, C., Blanchard, M.P., Ferracci, G., Leveque, C., Moreno, M., Francois-Bellan, A.M., Becquet, D., and Bosler, O. (2010). Daily changes in synaptic innervation of VIP neurons in the rat suprachiasmatic nucleus: contribution of glutamatergic afferents. *Eur. J. Neurosci.* **31**, 359–370.

Glaser, F.T., and Stanewsky, R. (2005). Temperature synchronization of the Drosophila circadian clock. *Curr. Biol.* **15**, 1352–13563.

Glossop, N.R., and Hardin, P.E. (2002). Central and peripheral circadian oscillator mechanisms in flies and mammals. *J. Cell Sci.* **115**, 3369–3377.

Gompf, H.S., and Allen, C.N. (2004). GABAergic synapses of the suprachiasmatic nucleus exhibit a diurnal rhythm of short-term synaptic plasticity. *Eur. J. Neurosci.* **19**, 2791–2798.

Grima, B., Chelot, E., Xia, R., and Rouyer, F. (2004). Morning and evening peaks of activity rely on different clock neurons of the Drosophila brain. *Nature* **431**, 869–873.

Hamasaka, Y., and Nassel, D.R. (2005). Mapping of serotonin, dopamine, and histamine in relation to different clock neurons in the brain of Drosophila. *J. Comp. Neurol.* **494**, 314–330.

Hamasaka, Y., Rieger, D., Parmentier, M.L., Grau, Y., Helfrich-Forster, C., and Nassel, D.R. (2007). Glutamate and its metabotropic receptor in Drosophila clock neuron circuits. *J. Comp. Neurol.* **505**, 32–45.

Hamasaka, Y., Suzuki, T., Hanai, S., and Ishida, N. (2010). Evening circadian oscillator as the primary determinant of rhythmic motivation for Drosophila courtship behavior. *Genes Cells* **15**, 1240–1248.

Hamasaka, Y., Wegener, C., and Nassel, D.R. (2005). GABA modulates Drosophila circadian clock neurons via GABAB receptors and decreases in calcium. *J. Neurobiol.* **65**, 225–240.

Hardin, P.E. (2005). The circadian timekeeping system of Drosophila. *Curr. Biol.* **15**, R714–R722.

Harmer, S.L., Hogenesch, J.B., Straume, M., Chang, H.S., Han, B., Zhu, T., Wang, X., Kreps, J.A., and Kay, S.A. (2000). Orchestrated transcription of key pathways in Arabidopsis by the circadian clock. *Science* **290**, 2110–2113.

Harris, K.M., and Teyler, T.J. (1983). Age differences in a circadian influence on hippocampal LTP. *Brain Res.* **261**, 69–73.

Harrisingh, M.C., Wu, Y., Lnenicka, G.A., and Nitabach, M.N. (2007). Intracellular Ca^{2+} regulates free-running circadian clock oscillation in vivo. *J. Neurosci.* **27**, 12489–12499.

Hauber, W., and Bareiss, A. (2001). Facilitative effects of an adenosine A1/A2 receptor blockade on spatial memory performance of rats: selective enhancement of reference memory retention during the light period. *Behav. Brain Res.* **118**, 43–52.

Heisenberg, M. (1998). What do the mushroom bodies do for the insect brain? an introduction. *Learn Mem.* **5**, 1–10.

Heisenberg, M., Borst, A., Wagner, S., and Byers, D. (1985). Drosophila mushroom body mutants are deficient in olfactory learning. *J. Neurogenet.* **2**, 1–30.

Helfrich-Forster, C. (1995). The period clock gene is expressed in central nervous system neurons which also produce a neuropeptide that reveals the projections of circadian pacemaker cells within the brain of Drosophila melanogaster. *Proc. Natl. Acad. Sci. U S A* **92**, 612–616.

Helfrich-Forster, C. (1997). Development of pigment-dispersing hormone-immunoreactive neurons in the nervous system of Drosophila melanogaster. *J. Comp. Neurol.* **380**, 335–354.

Helfrich-Forster, C. (1998). Robust circadian rhythmicity of Drosophila melanogaster requires the presence of lateral neurons: a brain-behavioral study of disconnected mutants. *J. Comp. Physiol. [A]* **182**, 435–453.

Helfrich-Forster, C. (2000). Differential control of morning and evening components in the activity rhythm of Drosophila melanogaster—sex-specific differences suggest a different quality of activity. *J. Biol. Rhythms* **15**, 135–154.

Helfrich-Forster, C. (2003). The neuroarchitecture of the circadian clock in the brain of Drosophila melanogaster. *Microsc. Res. Tech.* **62**, 94–102.

Helfrich-Forster, C., and Homberg, U. (1993). Pigment-dispersing hormone-immunoreactive neurons in the nervous system of wild-type Drosophila melanogaster and of several mutants with altered circadian rhythmicity. *J. Comp. Neurol.* **337**, 177–190.

Helfrich-Forster, C., Tauber, M., Park, J.H., Muhlig-Versen, M., Schneuwly, S., and Hofbauer, A. (2000). Ectopic expression of the neuropeptide pigment-dispersing factor alters behavioral rhythms in Drosophila melanogaster. *J. Neurosci.* **20**, 3339–3353.

Hodge, J.J., and Stanewsky, R. (2008). Function of the Shaw potassium channel within the Drosophila circadian clock. *PLoS One* **3**, e2274.

Hong, S.T., Bang, S., Paik, D., Kang, J., Hwang, S., Jeon, K., Chun, B., Hyun, S., Lee, Y., and Kim, J. (2006). Histamine and its receptors modulate temperature-preference behaviors in Drosophila. *J. Neurosci.* **26**, 7245–7256.

Hyun, S., Lee, Y., Hong, S.T., Bang, S., Paik, D., Kang, J., Shin, J., Lee, J., Jeon, K., and Hwang, S et al., (2005). Drosophila GPCR Han is a receptor for the circadian clock neuropeptide PDF. *Neuron* **48**, 267–278.

Im, S.H., and Taghert, P.H. (2010). PDF receptor expression reveals direct interactions between circadian oscillators in Drosophila. *J. Comp. Neurol.* **518**, 1925–1945.

Itri, J.N., Michel, S., Vansteensel, M.J., Meijer, J.H., and Colwell, C.S. (2005). Fast delayed rectifier potassium current is required for circadian neural activity. *Nat. Neurosci.* **8**, 650–656.

Itri, J.N., Vosko, A.M., Schroeder, A., Dragich, J.M., Michel, S., and Colwell, C.S. (2010). Circadian regulation of a-type potassium currents in the suprachiasmatic nucleus. *J. Neurophysiol.* **103**, 632–640.

Ivanchenko, M., Stanewsky, R., and Giebultowicz, J.M. (2001). Circadian photoreception in Drosophila: functions of cryptochrome in peripheral and central clocks. *J. Biol. Rhythms* **16**, 205–215.

Jackson, F.R. (2011). Glial cell modulation of circadian rhythms. *Glia.* **59**, 1341–1350.

Jilg, A., Lesny, S., Peruzki, N., Schwegler, H., Selbach, O., Dehghani, F., and Stehle, J.H. (2010). Temporal dynamics of mouse hippocampal clock gene expression support memory processing. *Hippocampus* **20**, 377–388.

Johard, H.A., Yoishii, T., Dircksen, H., Cusumano, P., Rouyer, F., Helfrich-Forster, C., and Nassel, D. R. (2009). Peptidergic clock neurons in Drosophila: ion transport peptide and short neuropeptide F in subsets of dorsal and ventral lateral neurons. *J. Comp. Neurol.* **516**, 59–73.

Kaneko, M., Park, J.H., Cheng, Y., Hardin, P.E., and Hall, J.C. (2000). Disruption of synaptic transmission or clock-gene-product oscillations in circadian pacemaker cells of Drosophila cause abnormal behavioral rhythms. *J. Neurobiol.* **43**, 207–233.

Karatsoreos, I.N., Bhagat, S., Bloss, E.B., Morrison, J.H., and McEwen, B.S. (2011). Disruption of circadian clocks has ramifications for metabolism, brain, and behavior. *Proc. Natl. Acad. Sci. U S A* **108**, 1657–1662.

Keshishian, H., and Kim, Y.S. (2004). Orchestrating development and function: retrograde BMP signaling in the Drosophila nervous system. *Trends Neurosci.* **27**, 143–147.

Kilman, V.L., Zhang, L., Meissner, R.A., Burg, E., and Allada, R. (2009). Perturbing dynamin reveals potent effects on the Drosophila circadian clock. *PLoS One* **4**, e5235.

Kitamoto, T. (2001). Conditional modification of behavior in Drosophila by targeted expression of a temperature-sensitive shibire allele in defined neurons. *J Neurobiol.* **47**, 81–92.

Konopka, R.J., and Benzer, S. (1971). Clock mutants of Drosophila melanogaster. *Proc Natl Acad Sci. U S A* **68**, 2112–2116.

Krishnan, B., Dryer, S.E., and Hardin, P.E. (1999). Circadian rhythms in olfactory responses of Drosophila melanogaster. *Nature* **400**, 375–378.

Krishnan, B., Levine, J.D., Lynch, M.K., Dowse, H.B., Funes, P., Hall, J.C., Hardin, P.E., and Dryer, S.E. (2001). A new role for cryptochrome in a Drosophila circadian oscillator. *Nature* **411**, 313–317.

Krishnan, P., Chatterjee, A., Tanoue, S., and Hardin, P.E. (2008). Spike amplitude of single-unit responses in antennal sensillae is controlled by the Drosophila circadian clock. *Curr. Biol.* **18**, 803–807.

Kuhlman, S.J., and McMahon, D.G. (2004). Rhythmic regulation of membrane potential and potassium current persists in SCN neurons in the absence of environmental input. *Eur. J. Neurosci.* **20**, 1113–1117.

Kula-Eversole, E., Nagoshi, E., Shang, Y., Rodriguez, J., Allada, R., and Rosbash, M. (2010). Surprising gene expression patterns within and between PDF-containing circadian neurons in Drosophila. *Proc. Natl. Acad. Sci. U S A* **107**, 13497–134502.

Kula, E., Levitan, E.S., Pyza, E., and Rosbash, M. (2006). PDF cycling in the dorsal protocerebrum of the Drosophila brain is not necessary for circadian clock function. *J. Biol. Rhythms* **21**, 104–117.

Lear, B.C., Lin, J.M., Keath, J.R., McGill, J.J., Raman, I.M., and Allada, R. (2005 a) The ion channel narrow abdomen is critical for neural output of the Drosophila circadian pacemaker. *Neuron* **48**, 965–976.

Lear, B.C., Merrill, C.E., Lin, J.M., Schroeder, A., Zhang, L., and Allada, R. (2005 b) A G protein-coupled receptor, groom-of-PDF, is required for PDF neuron action in circadian behavior. *Neuron* **48**, 221–227.

Lebestky, T., Chang, J.S., Dankert, H., Zelnik, L., Kim, Y.C., Han, K.A., Wolf, F.W., Perona, P., and Anderson, D.J. (2009). Two different forms of arousal in Drosophila are oppositely regulated by the dopamine D1 receptor ortholog DopR via distinct neural circuits. *Neuron* **64**, 522–536.

Levine, J.D., Funes, P., Dowse, H.B., and Hall, J.C. (2002). Resetting the circadian clock by social experience in Drosophila melanogaster. *Science* **298**, 2010–2012.

Lin, Y., Han, M., Shimada, B., Wang, L., Gibler, T.M., Amarakone, A., Awad, T.A., Stormo, G.D., Van Gelder, R.N., and Taghert, P.H. (2002). Influence of the period-dependent circadian clock on diurnal, circadian, and aperiodic gene expression in Drosophila melanogaster. *Proc. Natl. Acad. Sci. U S A* **99**, 9562–9567.

Lin, Y., Stormo, G.D., and Taghert, P.H. (2004). The neuropeptide pigment-dispersing factor coordinates pacemaker interactions in the Drosophila circadian system. *J. Neurosci.* **24**, 7951–7957.

Loh, D.H., Navarro, J., Hagopian, A., Wang, L.M., Deboer, T., and Colwell, C.S. (2010). Rapid changes in the light/dark cycle disrupt memory of conditioned fear in mice. *PLoS One* **5**(9); pii: e12546.

Lyons, L.C., Rawashdeh, O., Katzoff, A., Susswein, A.J., and Eskin, A. (2005). Circadian modulation of complex learning in diurnal and nocturnal Aplysia. *Proc. Natl. Acad. Sci. U S A* **102**, 12589–12594.

Lyons, L.C., and Roman, G. (2009). Circadian modulation of short-term memory in Drosophila. *Learn Mem.* **16**, 19–27.

Ma, W.P., Cao, J., Tian, M., Cui, M.H., Han, H.L., Yang, Y.X., and Xu, L. (2007). Exposure to chronic constant light impairs spatial memory and influences long-term depression in rats. *Neurosci. Res.* **59**, 224–230.

Marder, E., and Bucher, D. (2007). Understanding circuit dynamics using the stomatogastric nervous system of lobsters and crabs. *Annu. Rev. Physiol.* **69**, 291–316.

Marques, G., and Zhang, B. (2006). Retrograde signaling that regulates synaptic development and function at the Drosophila neuromuscular junction. *Int. Rev. Neurobiol.* **75**, 267–285.

Matsumoto, A., Matsumoto, N., Harui, Y., Sakamoto, M., and Tomioka, K. (1998). Light and temperature cooperate to regulate the circadian locomotor rhythm of wild type and period mutants of Drosophila melanogaster. *J. Insect. Physiol.* **44**, 587–596.

Matsumoto, G., and Tasaki, I. (1977). A study of conduction velocity in nonmyelinated nerve fibers. *Biophys. J.* **20**, 1–13.

McDonald, M.J., and Rosbash, M. (2001). Microarray analysis and organization of circadian gene expression in Drosophila. *Cell* **107**, 567–578.

Mehnert, K.I., Beramendi, A., Elghazali, F., Negro, P., Kyriacou, C.P., and Cantera, R. (2007). Circadian changes in Drosophila motor terminals. *Dev. Neurobiol.* **67**, 415–421.

Mehnert, K.I., and Cantera, R. (2008). A peripheral pacemaker drives the circadian rhythm of synaptic boutons in Drosophila independently of synaptic activity. *Cell. Tissue Res.* **334**, 103–109.

Mertens, I., Vandingenen, A., Johnson, E.C., Shafer, O.T., Li, W., Trigg, J.S., De Loof, A., Schoofs, L., and Taghert, P.H. (2005). PDF receptor signaling in Drosophila contributes to both circadian and geotactic behaviors. *Neuron* **48**, 213–219.

Michel, S., Geusz, M.E., Zaritsky, J.J., and Block, G.D. (1993). Circadian rhythm in membrane conductance expressed in isolated neurons. *Science* **259**, 239–241.

Michel, S., Manivannan, K., Zaritsky, J.J., and Block, G.D. (1999). A delayed rectifier current is modulated by the circadian pacemaker in Bulla. *J. Biol. Rhythms* **14**, 141–150.

Miskiewicz, K., Pyza, E., and Schurmann, F.W. (2004). Ultrastructural characteristics of circadian pacemaker neurones, immunoreactive to an antibody against a pigment-dispersing hormone in the fly's brain. *Neurosci. Lett.* **363**, 73–77.

Miyasako, Y., Umezaki, Y., and Tomioka, K. (2007). Separate sets of cerebral clock neurons are responsible for light and temperature entrainment of Drosophila circadian locomotor rhythms. *J. Biol. Rhythms* **22**, 115–126.

Moura, P.J., Gimenes-Junior, J.A., Valentinuzzi, V.S., and Xavier, G.F. (2009). Circadian phase and intertrial interval interfere with social recognition memory. *Physiol. Behav.* **96**, 51–56.

Myers, E.M., Yu, J., and Sehgal, A. (2003). Circadian control of eclosion: interaction between a central and peripheral clock in Drosophila melanogaster. *Curr. Biol.* **13**, 526–533.

Nagoshi, E., Sugino, K., Kula, E., Okazaki, E., Tachibana, T., Nelson, S., and Rosbash, M. (2010). Dissecting differential gene expression within the circadian neuronal circuit of Drosophila. *Nat. Neurosci.* **13**, 60–68.

Nitabach, M.N., and Taghert, P.H. (2008). Organization of the Drosophila circadian control circuit. *Curr. Biol.* **18**, R84–R93.

Oishi, K., Miyazaki, K., Kadota, K., Kikuno, R., Nagase, T., Atsumi, G., Ohkura, N., Azama, T., Mesaki, M., and Yukimasa, S et al., (2003). Genome-wide expression analysis of mouse liver reveals CLOCK-regulated circadian output genes. *J. Biol. Chem.* **278**, 41519–41527.

Panda, S., Antoch, M.P., Miller, B.H., Su, A.I., Schook, A.B., Straume, M., Schultz, P.G., Kay, S.A., Takahashi, J.S., and Hogenesch, J.B. (2002). Coordinated transcription of key pathways in the mouse by the circadian clock. *Cell* **109**, 307–320.

Parisky, K.M., Agosto, J., Pulver, S.R., Shang, Y., Kuklin, E., Hodge, J.J., Kang, K., Liu, X., Garrity, P. A., and Rosbash, M et al., (2008). PDF cells are a GABA-responsive wake-promoting component of the Drosophila sleep circuit. *Neuron* **60**, 672–682.

Park, D., and Griffith, L.C. (2006). Electrophysiological and anatomical characterization of PDF-positive clock neurons in the intact adult Drosophila brain. *J. Neurophysiol.* **95**, 3955–3960.

Park, J.H., and Hall, J.C. (1998). Isolation and chronobiological analysis of a neuropeptide pigment-dispersing factor gene in Drosophila melanogaster. *J. Biol. Rhythms* **13**, 219–228.

Park, J.H., Helfrich-Forster, C., Lee, G., Liu, L., Rosbash, M., and Hall, J.C. (2000). Differential regulation of circadian pacemaker output by separate clock genes in Drosophila. *Proc. Natl. Acad. Sci. U S A* **97**, 3608–3613.

Peng, Y., Stoleru, D., Levine, J.D., Hall, J.C., and Rosbash, M. (2003). Drosophila free-running rhythms require intercellular communication. *PLoS. Biol.* **1**, E13.

Pennartz, C.M., de Jeu, M.T., Bos, N.P., Schaap, J., and Geurtsen, A.M. (2002). Diurnal modulation of pacemaker potentials and calcium current in the mammalian circadian clock. *Nature* **416**, 286–290.

Perez-Cruz, C., Simon, M., Flugge, G., Fuchs, E., and Czeh, B. (2009). Diurnal rhythm and stress regulate dendritic architecture and spine density of pyramidal neurons in the rat infralimbic cortex. *Behav. Brain Res.* **205**, 406–413.

Peschel, N., and Helfrich-Forster, C. (2011). Setting the clock—by nature: circadian rhythm in the fruitfly Drosophila melanogaster. *FEBS Lett* **585**, 1435–1442.

Picot, M., Cusumano, P., Klarsfeld, A., Ueda, R., and Rouyer, F. (2007). Light activates output from evening neurons and inhibits output from morning neurons in the Drosophila circadian clock. *PLoS Biol.* **5**, e315.

Pittendrigh, C.S.a.D.S. (1976). A functional analysis of circadian pacemakers in nocturnal rodents: V. Pacemaker structure: A clock for all seasons. *J. Comp. Physiol. [A]* **106**, 333–355.

Pyza, E., and Gorska-Andrzejak, J. (2004). Involvement of glial cells in rhythmic size changes in neurons of the housefly's visual system. *J. Neurobiol.* **59**, 205–215.

Pyza, E., and Gorska-Andrzejak, J. (2008). External and internal inputs affecting plasticity of dendrites and axons of the fly's neurons. *Acta Neurobiol. Exp. (Wars)* **68**, 322–333.

Pyza, E., and Meinertzhagen, I.A. (1996). Neurotransmitters regulate rhythmic size changes amongst cells in the fly's optic lobe. *J. Comp. Physiol. A* **178**, 33–45.

Pyza, E., and Meinertzhagen, I.A. (1999). Daily rhythmic changes of cell size and shape in the first optic neuropil in Drosophila melanogaster. *J. Neurobiol.* **40**, 77–88.

Reijmers, L.G., Leus, I.E., Burbach, J.P., Spruijt, B.M., and van Ree, J.M. (2001). Social memory in the rat: circadian variation and effect of circadian rhythm disruption. *Physiol. Behav.* **72**, 305–309.

Renn, S.C., Park, J.H., Rosbash, M., Hall, J.C., and Taghert, P.H. (1999). A pdf neuropeptide gene mutation and ablation of PDF neurons each cause severe abnormalities of behavioral circadian rhythms in Drosophila. *Cell* **99**, 791–802.

Rhee, J., Buchan, T., Zukerberg, L., Lilien, J., and Balsamo, J. (2007). Cables links Robo-bound Abl kinase to N-cadherin-bound beta-catenin to mediate Slit-induced modulation of adhesion and transcription. *Nat. Cell Biol.* **9**, 883–892.

Rhee, J., Mahfooz, N.S., Arregui, C., Lilien, J., Balsamo, J., and VanBerkum, M.F. (2002). Activation of the repulsive receptor roundabout inhibits N-cadherin-mediated cell adhesion. *Nat. Cell Biol.* **4**, 798–805.

Rieger, D., Wulbeck, C., Rouyer, F., and Helfrich-Forster, C. (2009). Period gene expression in four neurons is sufficient for rhythmic activity of Drosophila melanogaster under dim light conditions. *J. Biol. Rhythms* **24**, 271–282.

Ruiz, S., Ferreiro, M.J., Casanova, G., Olivera, A., and Cantera, R. (2010). Synaptic vesicles in motor synapses change size and distribution during the day. *Synapse* **64**, 14–19.

Schneider, N.L., and Stengl, M. (2006). Gap junctions between accessory medulla neurons appear to synchronize circadian clock cells of the cockroach Leucophaea maderae. *J. Neurophysiol.* **95**, 1996–2002.

Sehadova, H., Glaser, F.T., Gentile, C., Simoni, A., Giesecke, A., Albert, J.T., and Stanewsky, R. (2009). Temperature entrainment of Drosophila's circadian clock involves the gene nocte and signaling from peripheral sensory tissues to the brain. *Neuron* **64**, 251–266.

Shafer, O.T., Helfrich-Forster, C., Renn, S.C., and Taghert, P.H. (2006). Reevaluation of Drosophila melanogaster's neuronal circadian pacemakers reveals new neuronal classes. *J. Comp. Neurol.* **498**, 180–193.

Shafer, O.T., Kim, D.J., Dunbar-Yaffe, R., Nikolaev, V.O., Lohse, M.J., and Taghert, P.H. (2008). Widespread receptivity to neuropeptide PDF throughout the neuronal circadian clock network of Drosophila revealed by real-time cyclic AMP imaging. *Neuron* **58**, 223–237.

Shafer, O.T., Rosbash, M., and Truman, J.W. (2002). Sequential nuclear accumulation of the clock proteins period and timeless in the pacemaker neurons of Drosophila melanogaster. *J. Neurosci.* **22**, 5946–5954.

Shang, Y., Griffith, L.C., and Rosbash, M. (2008). Light-arousal and circadian photoreception circuits intersect at the large PDF cells of the Drosophila brain. *Proc. Natl. Acad. Sci. U S A* **105**, 19587–19594.

Sheeba, V., Gu, H., Sharma, V.K., O'Dowd, D.K., and Holmes, T.C. (2008). Circadian- and light-dependent regulation of resting membrane potential and spontaneous action potential firing of Drosophila circadian pacemaker neurons. *J. Neurophysiol.* **99**, 976–988.

Shemesh, Y., Eban-Rothschild, A., Cohen, M., and Bloch, G. (2010). Molecular dynamics and social regulation of context-dependent plasticity in the circadian clockwork of the honey bee. *J. Neurosci.* **30**, 12517–12525.

Stanewsky, R. (2002). Clock mechanisms in Drosophila. *Cell Tissue Res.* **309**, 11–26.

Stanewsky, R., Kaneko, M., Emery, P., Beretta, B., Wager-Smith, K., Kay, S.A., Rosbash, M., and Hall, J.C. (1998). The cryb mutation identifies cryptochrome as a circadian photoreceptor in Drosophila. *Cell* **95**, 681–692.

Stoleru, D., Nawathean, P., Fernandez, M.L., Menet, J.S., Ceriani, M.F., and Rosbash, M. (2007). The Drosophila circadian network is a seasonal timer. *Cell* **129**, 207–219.

Stoleru, D., Peng, Y., Agosto, J., and Rosbash, M. (2004). Coupled oscillators control morning and evening locomotor behaviour of Drosophila. *Nature* **431**, 862–868.

Stoleru, D., Peng, Y., Nawathean, P., and Rosbash, M. (2005). A resetting signal between Drosophila pacemakers synchronizes morning and evening activity. *Nature* **438**, 238–242.

Tanoue, S., Krishnan, P., Chatterjee, A., and Hardin, P.E. (2008). G protein-coupled receptor kinase 2 is required for rhythmic olfactory responses in Drosophila. *Curr. Biol.* **18**, 787–794.

Tanoue, S., Krishnan, P., Krishnan, B., Dryer, S.E., and Hardin, P.E. (2004). Circadian clocks in antennal neurons are necessary and sufficient for olfaction rhythms in Drosophila. *Curr. Biol.* **14**, 638–649.

Tully, T., and Quinn, W.G. (1985). Classical conditioning and retention in normal and mutant Drosophila melanogaster. *J. Comp. Physiol. A* **157**, 263–277.

Ueda, H.R., Matsumoto, A., Kawamura, M., Iino, M., Tanimura, T., and Hashimoto, S. (2002). Genome-wide transcriptional orchestration of circadian rhythms in Drosophila. *J. Biol. Chem.* **277**, 14048–14052.

Valentinuzzi, V.S., Menna-Barreto, L., and Xavier, G.F. (2004). Effect of circadian phase on performance of rats in the Morris water maze task. *J. Biol. Rhythms* **19**, 312–324.

Valentinuzzi, V.S., Scarbrough, K., Takahashi, J.S., and Turek, F.W. (1997). Effects of aging on the circadian rhythm of wheel-running activity in C57BL/6 mice. *Am. J. Physiol.* **273**, R1957–R1964.

Veleri, S., Brandes, C., Helfrich-Forster, C., Hall, J.C., and Stanewsky, R. (2003). A self-sustaining, light-entrainable circadian oscillator in the Drosophila brain. *Curr. Biol.* **13**, 1758–1767.

Vertes, R.P. (2006). Interactions among the medial prefrontal cortex, hippocampus and midline thalamus in emotional and cognitive processing in the rat. *Neuroscience* **142**, 1–20.

Vosko, A.M., Schroeder, A., Loh, D.H., and Colwell, C.S. (2007). Vasoactive intestinal peptide and the mammalian circadian system. *Gen. Comp. Endocrinol.* **152**, 165–175.

Wall, P.M., Blanchard, R.J., Markham, C., Yang, M., and Blanchard, D.C. (2004). Infralimbic D1 receptor agonist effects on spontaneous novelty exploration and anxiety-like defensive responding in CD-1 mice. *Behav. Brain Res.* **152**, 67–79.

Wang, L.M., Dragich, J.M., Kudo, T., Odom, I.H., Welsh, D.K., O'Dell, T.J., and Colwell, C.S. (2009). Expression of the circadian clock gene Period2 in the hippocampus: possible implications for synaptic plasticity and learned behaviour. *ASN Neuro* **1**(3); pii: e00012.

Weber, P., Kula-Eversole, E., and Pyza, E. (2009). Circadian control of dendrite morphology in the visual system of Drosophila melanogaster. *PLoS One* **4**, e4290.

Welsh, D.K., Takahashi, J.S., and Kay, S.A. (2010). Suprachiasmatic nucleus: cell autonomy and network properties. *Annu. Rev. Physiol.* **72**, 551–577.

Wulbeck, C., Grieshaber, E., and Helfrich-Forster, C. (2009). Blocking endocytosis in Drosophila's circadian pacemaker neurons interferes with the endogenous clock in a PDF-dependent way. *Chronobiol. Int.* **26**, 1307–1322.

Yan, J., Wang, H., Liu, Y., and Shao, C. (2008). Analysis of gene regulatory networks in the mammalian circadian rhythm. *PLoS Comput. Biol.* **4**, e1000193.

Yang, Z., and Sehgal, A. (2001). Role of molecular oscillations in generating behavioral rhythms in Drosophila. *Neuron* **29**, 453–467.

Yasuyama, K., and Meinertzhagen, I.A. (2010). Synaptic connections of PDF-immunoreactive lateral neurons projecting to the dorsal protocerebrum of Drosophila melanogaster. *J. Comp. Neurol.* **518**, 292–304.

Yoshii, T., Heshiki, Y., Ibuki-Ishibashi, T., Matsumoto, A., Tanimura, T., and Tomioka, K. (2005). Temperature cycles drive Drosophila circadian oscillation in constant light that otherwise induces behavioural arrhythmicity. *Eur. J. Neurosci.* **22**, 1176–1184.

Yoshii, T., Vanin, S., Costa, R., and Helfrich-Forster, C. (2009). Synergic entrainment of Drosophila's circadian clock by light and temperature. *J. Biol. Rhythms* **24**, 452–464.

Zhou, X., Yuan, C., and Guo, A. (2005). Drosophila olfactory response rhythms require clock genes but not pigment dispersing factor or lateral neurons. *J. Biol. Rhythms* **20**, 237–244.

LEARNING AND MEMORY IN *DROSOPHILA*: BEHAVIOR, GENETICS, AND NEURAL SYSTEMS

Lily Kahsai and Troy Zars

University of Missouri, Division of Biological Sciences, 114 Lefevre Hall, Columbia, MO 65211, USA

The rich behavioral repertoire that *Drosophila* use to navigate in their natural environment suggests that flies can use memories to inform decisions. Development of paradigms to examine memories that restrict behavioral choice was essential in furthering our understanding of the genetics and neural systems of memory formation in the fly. Olfactory, visual, and place memory paradigms have proven influential in determining principles for the mechanisms of memory formation. Several parts of the nervous system have been shown to be important for different types of memories, including the mushroom bodies and the central complex. Thus far, about 40 genes have been linked to normal olfactory short-term memory. A subset of these genes have also been tested for a role in visual and place memory. Some genes have a common function in memory formation, specificity of action comes from where in the nervous system these genes act. Alternatively, some genes have a more restricted role in different types of memories.

DOI: 10.1016/B978-0-12-387003-2.00006-9

0074-7742/10 $35.00

Watching wild fruit flies quickly suggests that they move through a sensory-rich world filled with both rewards and dangers. The attraction of these little flies to overripe bananas or a glass of beer is a common experience in the summer. While landing and feeding on a banana is likely a treat for the fly, a last bitter swim would surely be considered a danger. If one spends long enough watching flies, there is also the chance to see what looks like flies chasing each other. Sometimes it is a male chasing a female fly, and courtship can end in copulation. Other times there is a fight, over food or a potential mate. The implication of these natural behaviors is that flies can have goals (e.g., eating, finding a mate), use multiple sensory cues to orient in space (e.g., olfactory, visual), and achieve outcomes that could inform future strategies in reaching those goals. Feedback from a good or bad outcome suggests that flies can learn from experience, and use memory to alter behavior over time.

To understand the behavioral, neural, and genetic bases of learning and memory, paradigms have been developed that restrict the number of stimuli or behaviors that might be used by a fly to predict good or bad outcomes. Indeed, the need to restrict behavioral options was recognized early on in the quest to understand the neural basis of behavior (M. Heisenberg, personal communication). For example, first attempts to examine visually guided behavior on a bench top were frustrated by a complex visual environment and the problem that a fly seemed to do whatever it wanted to do, including behaviors apparently not related to the experimenters view of the task (e.g., wing cleaning, standing in place). Thus, testing of "simpler" behaviors such as object fixation and release benefited tremendously from simplifying the testing arena (Benzer, 1967; Buelthoff *et al.*, 1982; Heisenberg, 1983). Design of paradigms that reduced behavioral choice was a requisite step in examining memory formation in the fly. So then, how does a fly form a new memory?

This chapter will concentrate on learning and memory in *Drosophila melanogaster*. We use the definition of learning as a lasting behavioral change due to experience and memory as the retention over time of learned information (Dudai, 2002). Discussion of the mechanisms of learning and memory will be restricted to those memories formed in the minutes to hours range after training. Memories tested in three paradigms will be examined: olfactory conditioning, visual pattern recognition memory, and place learning. The reader is referred to several excellent reviews of long-term memory mechanisms for studies that examine memory in the 24-h range and beyond, and other learning paradigms (Bolduc and Tully, 2009; Keene and Waddell, 2007; McGuire *et al.*, 2005; Pitman *et al.*, 2009). Furthermore, the mechanisms that will be discussed here will concentrate on the genetic and neural bases of olfactory, visual, and place memory.

I. Behavioral Measures of Learning and Memory in *Drosophila*

Fruit flies can be trained to memorize odors, visual patterns, and tactile/spatial cues through classical and/or operant learning. In classical conditioning an organism is trained to associate a conditioned stimulus (CS) with a rewarding or punishing unconditioned stimulus (US) (Pavlov, 1927). In operant conditioning, a behavior of an organism is either rewarded or punished (Skinner, 1950). Classical and operant memory with olfactory, visual, and spatial cues in adult and larval (when appropriate) *Drosophila* will be discussed.

A. Olfactory Memory

Adult *Drosophila* can form an olfactory memory through classical conditioning using aversive electric shock punishment or appetitive sugar reward (Heisenberg *et al.*, 1985; Quinn *et al.*, 1974; Tempel *et al.*, 1983; Tully and Quinn, 1985). In the most popular version for testing aversive olfactory conditioning, a group of about 100 flies are trained to associate one odor (CS) with electric shock punishment (US) and a second odor without punishment (Tully and Quinn, 1985). After a period of rest (from minutes to hours to days), the flies are forced to choose between the two odors in a T-maze choice point (Fig. 1A). Importantly, the experiment is done in a

Fig. 1. Three memory paradigms: olfactory, visual pattern, and place. (A) The T-maze olfactory conditioning assay. During training flies are presented with two odors, one of which is associated with an aversive or appetitive stimulus. Memory is tested at a choice point, where two air streams with the odors used in training converge through long acrylic tubes. Trained flies typically avoid an odor arm associated with, for example, shock or approach an odor associated with sugar. (B) The flight simulator for visual pattern recognition. A single fly attached to a hook between its head and thorax is suspended in the middle of this LED-based arena. Different patterns can be displayed on the arena wall, in this case an upright "T" is shown. One of two visual patterns is associated with aversive high temperature (an upside-down T is typically used to complement the upright T). Visual pattern preference is determined after training (and randomization of pattern location). Flies typically avoid the pattern associated with high temperature. (C) Heat box place memory. Individual flies are loaded into a single chamber (entrance hidden by the black knob). Typically, one half of a chamber is associated with a high temperature, the other half with a preferred cool temperature. Place memory is measured as the change in spatial preference after training when the chamber is kept at the cool temperature. This single chamber was removed from an array of multiple chambers that condition up to 16 flies in parallel. (For color version of this figure, the reader is referred to the web version of this book.)

reciprocal fashion. That is, a second group of flies has the "other" odor associated with the US before the memory is tested. The number of flies choosing both odors are counted and used to generate one half of a performance index (PI) score (i.e., the number of flies choosing the US-associated odor is subtracted from the number choosing the other odor, and this number is divided by the total number of flies). The half-PIs from each US-associated odor are then averaged to make a single PI measure. This conditioning is done in a reciprocal fashion to rule out nonassociative changes leading to altered odor preferences (Gerber *et al.*, 2004b). Appetitive olfactory conditioning is similar in principle to the aversive conditioning except that flies are food-deprived before training, and odorants are associated with sugar reward (Schwaerzel *et al.*, 2003; Tempel *et al.*, 1983).

It was recently shown that adding an operant component to aversive olfactory learning increases memory levels (Claridge-Chang *et al.*, 2009). In this new assay individual flies walk in a narrow chamber, one side of the chamber is filled with an odor that is associated with electric shock; the other side of the chamber is filled with a different odor not associated with shock. Memory from this action contingent conditioning is tested in the chambers with odors converging on the midline, and a fly will typically move into the odor that was not associated with shock. The chamber used for this mixed operant and classical olfactory conditioning is similar to chambers that do not use olfactory cues and focus on testing place memory (Fig. 1C) (Wustmann *et al.*, 1996; Zars, 2009). Nevertheless, addition of the operant conditioning component improves olfactory memory levels compared to classical conditioning in the same chambers.

Both aversive and appetitive olfactory conditioning in larval *Drosophila* with reciprocal training is also possible (Aceves-Pina and Quinn, 1979; Heisenberg *et al.*, 1985; Scherer *et al.*, 2003; Schipanski *et al.*, 2008). In this case, single or multiple larvae are placed on Petri plates and are exposed to one of two odors (held in tiny odor cups that are not accessible to the larvae). For aversive conditioning, an odor is paired with agar that contains either salt or quinine. In appetitive olfactory conditioning, larvae are exposed to an odor when placed on agar containing a sugar (e.g., fructose). In both cases, the larvae are presented with a second odor in the absence of the salt or sugar. The olfactory memory test is measured some time after training, when larvae are placed midway between the two odor sources and preference within minutes is determined as the number of larvae close to either of the two odors. That is, separate groups of animals have one of the two odors paired with sugar or salt, and the combined difference in odor preference is a measure of associative memory. Interestingly, it appears that for aversive olfactory conditioning, presentation of the odor in the presence of the previously used aversive cue is required for memory recall (Gerber and Hendel, 2006).

B. Visual Memory

Visual learning and memory have been examined in several ways, centering on tests of flies to remember different visual patterns, colors, and illumination intensity. Visual pattern recognition has been extensively examined in the flight simulator (Dill *et al.*, 1993; Heisenberg *et al.*, 2001; Liu *et al.*, 2006; Tang *et al.*, 2004). In this assay, an individual fly is held in a single stationary position suspended in the center of an arena that provides a visual surround (Fig. 1B). The fly is connected by a hook glued on the thorax to a torque meter that measures the yaw force it generates as it tries to rotate in the arena. The paradigm is set up such that when a fly exerts counterclockwise yaw torque the visual patterns on the arena wall moves in a clockwise fashion. Individual patterns are presented in one of four quadrants of the arena wall, very often with two pairs of patterns. If one pattern type is in front of the fly, an identical pattern is directly behind. At 90° to the left or right of the fly is the second pair of patterns. Conditioning depends on the pairing of one of the pattern types with an aversive cue. The negative reinforcer in this paradigm is most often high-temperature (infrared light source) or a rapid back-and-forth rotation of the visual panorama (Dill *et al.*, 1993; Wolf *et al.*, 1998). Finally, a memory test comes after the visual patterns are moved, in the dark, to random positions and pattern preference is tested by the amount of time a fly keeps the pattern not associated with punishment in the front quadrant.

Tests for visual memory based on forced choice paradigms between lights of different wavelengths have been established (Schnaitmann *et al.*, 2010; Spatz *et al.*, 1974; van Swinderen *et al.*, 2009). Older versions of this sort of paradigm associated physical shaking or electric shock with exposure to one of two colored lights (e.g., blue or green). Behavioral tests in a T-maze, where flies could choose between two arms bathed in light of either type, examined visual memory. Newer versions of this type of memory test either improve on the speed of the experiment or increase the behavioral freedom for the flies. In one version, flies are presented with one of two lights in a V-shaped choice apparatus. Shaking of the tubes in the presence of one of two colored lights provides for aversive conditioning. The clever design of the apparatus immediately starts flies in the choice point after the shock exposure, with the choice point tubes pointing slightly upward to provide a strong negative geotaxis drive for flies to move away from the start point (van Swinderen *et al.*, 2009). Experience-dependent changes in colored light preference last for 10s of seconds. In a second new apparatus, flies are exposed to one of two colored lights in the presence or absence of sugar as a reward (or two aversive chemicals) (Schnaitmann *et al.*, 2010). Flies are allowed to choose between the two colors in a test phase when they are free to distribute over a pattern of lights in four quadrants. Memory after rewarded conditioning lasts at least 3 h. Aversive conditioning with the chemicals used to date either do not support aversive visual memory or induce a weak one.

Similar to olfactory learning in larvae, a larval visual learning assay has also been developed (Gerber *et al.*, 2004a). In this case, brightness and darkness are associated with sugar or quinine/salt as positive and negative stimuli. Behavioral tests of visual memory take place on Petri plates lined with agar (as in the conditioning phase) with bright and dark sectors.

C. Place Memory

Place memory tests flies' persistent preference for a position in an arena. The arena ranges in complexity from quite simple, the heat box, to those that are rich with visual landmarks (Foucaud *et al.*, 2010; Neuser *et al.*, 2008; Putz and Heisenberg, 2002; Wustmann *et al.*, 1996).

Probably the simplest means of examining place memory is in the heat box (Fig. 1C). The heat box is a long narrow dark chamber in which the temperature can be linked to spatial position of a fly (Wustmann and Heisenberg, 1997; Zars *et al.*, 2000b). Conditioning in the heat box is considered a mixed operant and classical paradigm because it is the activity of a fly as it explores the chamber that leads to the rise of a temperature (operant), but also that a particular place in the chamber is associated with unpleasantness (classical) (Zars, 2010a). Specifically, during training if a fly enters the half of the chamber associated with high temperature, the whole chamber heats up; when it returns to the other half the chamber cools to preferred temperatures (Sitaraman *et al.*, 2007; Sitaraman and Zars, 2010; Zars, 2001; Zars and Zars, 2006, 2009; Zars *et al.*, 2000b). In a test phase, the chamber is maintained at the preferred temperature of 24°C. Persistent place preference for the half of the chamber associated with cool temperatures measures place memory in this paradigm. That flies can orient in the dark and show persistent place preferences probably indicates that they use ideothetic, or self-derived, information about the dimensions and their place within the chamber (Wustmann and Heisenberg, 1997; Zars, 2009). Memory of the conditioning parameters in the heat box lasts about 2 h with extended intermittent training (Putz and Heisenberg, 2002).

In more complex arenas that test place memory, the visual environment can be used by the fly to aid in orientation. In the detour paradigm visible targets can be moved and turned off to test an orientation memory (Neuser *et al.*, 2008). An individual fly is allowed to move back and forth between two vertical stripes that cannot be reached. This clever design flashes a new vertical stripe 90° to the left or right of the visual target for a few seconds, and then all landmarks are turned off. The fly reorients toward the original target with high accuracy, suggesting that flies have an orientation memory. Flies have also been shown to be able to orient with respect to conspicuous distal landmarks (Foucaud *et al.*, 2010; Ofstad *et al.*, 2011; Zars, 2011). This paradigm uses landmarks that can be rotated 90° and a heated floor with a single cool place (Peltier tile). In this paradigm, based on the Morris water maze concept, flies search for the cool tile in the heated floor. Repeated experience with the complex visual landscape and position of the cool tile greatly decreases the escape latency to that tile. That flies follow the movement of the visual environment and the cool tile by 90° suggests that flies make an association between the distal landmarks and the place where the cool tile can be found. Thus, even in more complex environments flies can form place memories.

II. Neural Structures Implicated in Memory

The central fly brain contains approximately 200,000 neurons (Heisenberg and Wolf, 1984). These neurons are part of different structures, some of which have been implicated in memory formation. Major distinct neuropil structures in the fly brain include the mushroom bodies (MBs), the central complex, the antennal lobes, and the optic lobes. The form and function of the MBs and central complex will be discussed below. Furthermore, biogenic amine containing neuromodulatory systems will be explored in the reinforcement pathway section.

A. Structure of the Mushroom Bodies

The MBs are paired structures in the insect brain (Fig. 2) (Strausfeld *et al.*, 1998; Technau and Heisenberg, 1982). There are two types of neurons, intrinsic and extrinsic, in the MBs. About 2000 *intrinsic* neurons, largely made up of the Kenyon cells, have cell bodies clustered in the posterior lateral part in the adult

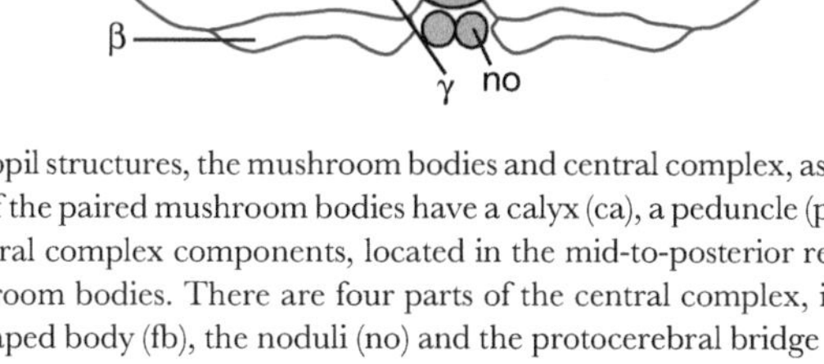

FIG. 2. Two neuropil structures, the mushroom bodies and central complex, associated with learning and memory. Each of the paired mushroom bodies have a calyx (ca), a peduncle (ped) and an α, α′, β, β′, and γ lobe. The central complex components, located in the mid-to-posterior region of the brain, are framed by the mushroom bodies. There are four parts of the central complex, including the ellipsoid body (eb), the fan-shaped body (fb), the noduli (no) and the protocerebral bridge (pb) (Fahrbach, 2006; Hanesch *et al.*, 1989). (For color version of this figure, the reader is referred to the web version of this book.)

Drosophila brain. The Kenyon cells have dendrites that form the MB calyx and axons that course anteriorly through the peduncle to form the MB lobes (Fig. 2) (Aso *et al.*, 2009; Technau and Heisenberg, 1982). By definition, there are no arborization fields of MB intrinsic neurons outside of the MBs. There are three known MB intrinsic neuron types that are not Kenyon cells, the best known are called the "MB-DPM" neurons, which will be described in more detail below (Ito *et al.*, 1998; Tanaka *et al.*, 2008; Waddell *et al.*, 2000). The extrinsic neurons, probably in the range of 20–30 cell types, have cell bodies outside of the Kenyon cell cluster. These extrinsic neurons project fibers into the MBs and have neurite projections to other regions of the brain (including other parts of the MB) (Ito *et al.*, 1998; Tanaka *et al.*, 2008).

The MB lobes can be divided into different functional regions. The lobes of the MBs project toward the midline and dorsally (Crittenden *et al.*, 1998; Lee *et al.*, 1999). Some of the Kenyon cells send a single neurite toward the midline as it reaches the front of the brain, making the γ lobe. Another set of Kenyon cells have a split neurite in the front of the brain, sending one branch toward the midline and the other branch dorsally. These split Kenyon cells populate the α/β and α'/β' lobes. Detailed anatomical examination of the MBs suggests that the medial and vertical lobes can be further subdivided, in addition to the characterization of the spur region (at the junction of the lobes and peduncle) and an accessory calyx (Tanaka *et al.*, 2008). This chapter will restrict discussion to the five primary lobe systems because most information on MB function is restricted to these structures.

B. Some Memory Functions for the Mushroom Bodies

Flies without MBs are, perhaps, surprisingly normal. Although this structure takes up a central position in the fly brain and is seemingly connected with most regions of the brain, altering this structure by chemical ablation or genetic mutation spares many behaviors, including flight, locomotion, chemorepulsion, and reproduction (Heisenberg *et al.*, 1985; Wolf *et al.*, 1998). Furthermore, although the MBs are sometimes called the "learning center" in the fly brain, several forms of learning are independent of the MBs (e.g., straightforward visual pattern recognition memory and place memory in the heat box) (Liu *et al.*, 1999; Wolf *et al.*, 1998; Zars *et al.*, 2000b). Further degrading the learning center label, the MBs are also important for several nonlearning behaviors (Besson and Martin, 2005; Martin *et al.*, 1998; Serway *et al.*, 2009; Zars, 2000). For example, thigmotaxis and fine-tuning of walking activity are altered by manipulation of the MBs.

The MBs have been implicated in olfactory memory formation. The first indication that the MBs are important for olfactory memory came from genetic mutations that altered the gross structure of the MBs. Mutations of the *mushroom body miniature* (*mbm*) and *mushroom body deranged* (*mbd*) genes have abnormal MBs and are deficient in olfactory memory formation (Heisenberg *et al.*, 1985). Important experiments that removed the MBs of adult flies by feeding very young larvae the drug hydroxyurea, killing the MB neuroblasts and one other neuroblast dividing at this time in development, provided a key link between the MBs and olfactory memory (de Belle and Heisenberg, 1994). Moreover, transgenic approaches altering G-protein signaling in the Kenyon cells showed that this type of signaling is necessary in the MBs for olfactory memory (Connolly *et al.*, 1996). The localization of the *rutabaga* type-1 adenlyl cyclase (*rut*-AC)-dependent plasticity to the MBs for olfactory memory formation provided first evidence for an olfactory memory trace in these neurons (Mao *et al.*, 2004; McGuire *et al.*, 2003; Schwaerzel *et al.*, 2002; Zars *et al.*, 2000a). Several additional genes have been implicated in olfactory memory, and they also function within the MBs for this task (discussed in more detail below) (Table I).

The different lobes of the MBs have been implicated in different components of olfactory memory. The α/β, α'/β', and γ lobes can be addressed with different sets of GAL4 drivers. The α/β and γ lobes have been implicated in *rut*-AC-dependent olfactory memory, and regulated cAMP signaling is involved in different phases within these lobes (Blum *et al.*, 2009; Mao *et al.*, 2004; McGuire *et al.*, 2003; Zars *et al.*, 2000a). Furthermore, an α'/β' lobe function has been identified for olfactory memory consolidation (Krashes *et al.*, 2007). Remarkably, visualization of cAMP and PKA activity in the MBs with paired application of acetylcholine and dopamine or octopamine (the biogenic amines will be discussed further in the next section) leads to synergistic increases in these signals in different lobe patterns (Gervasi *et al.*, 2010; Tomchik and Davis, 2009; Zars, 2010b). With acetylcholine/dopamine presentation, the α lobes show an increase in PKA activity; acetylcholine/octopamine presentation leads to increases of PKA activity in the α, β, and γ lobes.

C. Structure of the Central Complex

The central complex in *Drosophila* is composed of four major structures: the ellipsoid body (eb), the fan-shaped body (fb), the noduli (no), and the protocerebral bridge (pb) (Fig. 2) (Hanesch *et al.*, 1989; Strausfeld, 1976). There are two accessory regions that are also associated with the central complex: the lateral triangle (ltr) and the ventral bodies (vbo). The ellipsoid body is the most anterior substructure in the complex and can be divided into four concentric rings. The fan-shaped body is composed of several horizontal layers (perhaps six) and vertical segments (eight) (Hanesch *et al.*, 1989; Renn *et al.*, 1999). The protocerebral bridge is located posterior to the fan-shaped body and can be separated into 16 segments. The noduli are spherical structures ventral to the ellipsoid body. Neurons of the central complex are classified into three major groups: tangential, columnar, and pontine neurons (Hanesch *et al.*, 1989; Young and Armstrong, 2010a, 2010b). Tangential

Table I
GENES IMPLICATED IN OLFACTORY MEMORY FORMATION.

Gene	Protein Product	Genetic Evidence	Acute Versus Development	Neural System	References
AKAP	A kinase anchoring protein	$TI_{peptide}$	Acute	MB	(Schwaerzel *et al.*, 2007)
amnesiac (*amn*)	Neuropeptide	MA; R	ND	DPM neurons	(DeZazzo *et al.*, 1999; Feany and Quinn, 1995; Keene *et al.*, 2006; Quinn *et al.*, 1979; Waddell *et al.*, 2000)
crammer (*cer*)	Cysteine protease inhibitor	OA; R	ND	Glial cells	(Colomb *et al.*, 2009; Comas *et al.*, 2004; Krashes and Waddell, 2008)
DA1	Dopamine receptor	MA	Acute	MB	(Kim *et al.*, 2007)
DCO	PKA catalytic subunit	MA; $TI_{peptide}$; R	Acute	MB	(Drain *et al.*, 1991; Horiuchi *et al.*, 2008; Li *et al.*, 1996; Skoulakis *et al.*, 1993)
downstream of receptor kinase (*drk*)	Receptor tyrosine kinase linker protein	MA, TI_{RNAi}; R	Acute	MB	(Moressis *et al.*, 2009)
dunce (*dnc*)	cAMP phosphodiesterase	MA; R	Acute	ND	(Byers *et al.*, 1981; Dauwalder and Davis, 1995; Dudai *et al.*, 1976; Nighorn *et al.*, 1991; Qiu and Davis, 1993)
ethanol sens. low mem (*elm*)	CG2185	OA	ND	ND	(Laferriere *et al.*, 2008)
fasciclin II (*fas2*)	NCAM	MA; R	Acute	ND	(Cheng *et al.*, 2001)
foraging	cGMP-dependent protein kinase	MA; R	Acute	MB	(Mery *et al.*, 2007; Wang *et al.*, 2008)
fragileX mental retardation (*fmr*)	Fragile-X protein	MA, TI_{RNAi}; R	Acute	MB	(Bolduc *et al.*, 2008)
$G_{o\alpha}$[47a]	G-protein α subunit, O-type	$TI_{peptide}$	Acute	MB	(Ferris *et al.*, 2006)
$G_{s\alpha}$[60a]	G-protein α subunit, S-type	$TI_{peptide}$	ND	MB	(Connolly *et al.*, 1996)
gilgamesh (*gish*)	Casein kinase	MA; R	Acute	MB	(Tan *et al.*, 2010)

ignorant (ign)	S6 kinase Type 2	MA; R	ND	ND	(Putz *et al.*, 2004)
latheo (lat)	Origin recognition complex	MA; R	Development	ND	(Boynton and Tully, 1992; Pinto *et al.*, 1999)
leonardo (leo)	14-3-3 zeta	MA; R	Acute	ND	(Philip *et al.*, 2001; Skoulakis and Davis, 1996)
linotte/ derailed (drl)	Receptor tyrosine kinase	MA; Re; R	Development	ND	(Bolwig *et al.*, 1995; Bolwig *et al.*, 2002; Dura *et al.*, 1993; Dura *et al.*, 1995; Moreau-Fauvarque *et al.*, 2002)
mushroom body deranged (mbd)	ND	OA	Development	ND	(Heisenberg *et al.*, 1985)
mushroom body miniature (mbm)	Transcription factor	OA	Development	ND	(Heisenberg *et al.*, 1985; Raabe *et al.*, 2004)
nalyot (Adf1)	ADF1 transcription factor	OA; R	Acute and Development	ND	(DeZazzo *et al.*, 2000)
nebula/ sarah (sra)	Calcineurin inhibitor	MA; $TI_{peptide}$; R	Acute	MB	(Chang *et al.*, 2003)
neurofibromin (nf1)	Ras-GAP	MA; R	Acute	MB	(Buchanan and Davis, 2010; Guo *et al.*, 2000; Ho *et al.*, 2007)
NMDAR1	Glutamate receptor	MA; TI_{RNAi}; R	Acute	MB and CC	(Wu *et al.*, 2007; Xia *et al.*, 2005)
NMDAR2	Glutamate receptor	TI_{RNAi}	Acute	MB and CC	(Wu *et al.*, 2007; Xia *et al.*, 2005)
npf-receptor (npfr)	Neuropeptide Y receptor ortholog	OA; TI_{RNAi}	ND	Dopam. neurons	(Krashes *et al.*, 2009)
PKA-RI	PKA regulatory subunit	MA	ND	ND	(Goodwin *et al.*, 1997)
PQBP1	PolyQ binding protein type 1	OA; TI_{RNAi}; R	Acute	PN of AL	(Tamura *et al.*, 2010)
rac	Rho family - GTPase	$TI_{peptide}$	Acute	MB	(Shuai *et al.*, 2010)
radish (rsh)	CG42628	OA; R	Acute	ND	(Folkers *et al.*, 1993; Folkers *et al.*, 2006; Krashes and Waddell, 2008; Tully *et al.*, 1994)
Resistance to dieldrin (Rdl)	GABA-A receptor	$TI_{peptide}$; TI_{RNAi}	Acute	MB	(Liu *et al.*, 2009; Liu *et al.*, 2007)

(*Continued*)

Table I (*Continued*)

Gene	Protein Product	Genetic Evidence	Acute Versus Development	Neural System	References
rutabaga (*rut*)	Type 1 adenylyl cyclase	MA, R	Acute	MB	(Levin *et al.*, 1992; Mao *et al.*, 2004; McGuire *et al.*, 2003; Tully and Quinn, 1985; Zars *et al.*, 2000a, 2000b)
synapsin (*syn*)	Synapsin	OA	ND	ND	(Godenschwege *et al.*, 2004; Knapek *et al.*, 2010)
TβH	Tyramine beta hyroxylase	OA; R	Acute	Octop./ Tyram. neurons	(Schwaerzel *et al.*, 2003; Sitaraman *et al.*, 2010; Yarali and Gerber, 2010)
tequila (*teq*)	Neurotrypsin	OA; TI_{RNAi}	Acute	MB	(Colomb *et al.*, 2009; Didelot *et al.*, 2006; Krashes and Waddell, 2008)
tribbles	kinase	OA	ND	ND	(Laferriere *et al.*, 2008)
volado (*scab*)	α-Integrin	MA; R	Acute	ND	(Grotewiel *et al.*, 1998)
white (*w*)	ABC transporter	OA	ND	ND	(Diegelmann *et al.*, 2006a, 2006b; Yarali *et al.*, 2009)

Gene names, protein product and genetic evidence for the role of a gene are presented (MA, multiple alleles; OA, one allele; $TI_{peptide}$, transgenic interference peptide; TI_{RNAi}, transgenic interference RNAi; R, transgenic rescue; Re, reversion alleles). Whether a gene has been shown to be important for development or acutely is shown. Finally, the neural systems in which a gene acts is also shown (MB, mushroom bodies; MB-DPM, MB dorsal paired medial intrinsic neurons, CC, central complex; PN of AL, projection neurons of the antennal lobes). ND, not determined.

neurons provide connection to other adjacent brain regions and arborize in each of the horizontal layers of the fan-shaped body (known as F neurons [F1–F6]) or each of the rings of the ellipsoid body (known as R neurons [R1–R4]). The columnar neurons either innervate the vertical segments of the fan-shaped body or interconnect two or three central complex substructures. Pontine neurons are intrinsic to a central complex structure and integrate different layers.

D. Some Memory Functions for the Central Complex

The fan-shaped body and the ellipsoid body have been recently implicated in visual pattern memory (Liu *et al.*, 2006; Pan *et al.*, 2009; Wang *et al.*, 2008). Previous studies have shown that *Drosophila* can recognize and memorize five types of visual pattern parameters: color, elevation, size, contour orientation, and vertical compactness of the panorama (Ernst and Heisenberg, 1999; Tang *et al.*, 2004). Using *rut*-AC local rescue experiments in combination with targeted expression of central complex enhancer trap lines, memory traces for three visual pattern parameters (elevation, size, and contour orientation) were localized to tangential neurons innervating distinct layers in the fan-shaped body (Liu *et al.*, 2006). Neurons innervating the fifth (F5) and first (F1) layer of the fan-shaped body were found to be specifically important for visual pattern memories: elevation of the center of gravity in an object and contour orientation. This conclusion is somewhat blurred by the ability to locally rescue the *rut*-AC visual pattern memory with expression in the ring neurons of the ellipsoid body for additional pattern parameters (Pan *et al.*, 2009). Interestingly, using immunocytochemistry some fan-shaped body neurons were also shown to express some neuropeptides, such as tachykinin and short neuropeptide-f (Kahsai *et al.*, 2010; Kahsai and Winther, 2011). However, whether these neuromodulators are involved in visual pattern recognition memory remains to be investigated.

The central complex has also been shown to be important for olfactory and orientation memory (Heisenberg *et al.*, 1985; Neuser *et al.*, 2008; Wu *et al.*, 2007). Mutations that altered the structure of the central complex had a large effect on early studies of olfactory memory (Heisenberg *et al.*, 1985). Furthermore, *N*-methyl-D-aspartate (NMDA) receptor activity in some of the ring neurons of the ellipsoid body were recently shown to have an important role in consolidation of olfactory memory (Wu *et al.*, 2007). There is evidence to suggest that altering the function of the ring neurons implicated in olfactory memory are confoundingly important for locomotion and olfactory acuity based on an avoidance paradigm (Krashes and Waddell, 2008). Thus, the current evidence for a role of the ellipsoid body in olfactory memory is inconclusive. Finally, manipulation of the central complex shows that some of the ring neurons of the ellipsoid body (R3/R4d) are important for a seconds-long orientation memory (Neuser *et al.*, 2008).

E. Reinforcement Pathways

The dopaminergic, octopaminergic, and serotonergic neural systems have been implicated in reinforcement of different types of memory. The role of these systems in olfactory, visual, and place memory will be discussed.

The dopaminergic system mediates negative reinforcement in aversive olfactory memory formation (Claridge-Chang *et al.*, 2009; Schroll *et al.*, 2006; Schwaerzel *et al.*, 2003). Of about 200 dopamine-containing neurons in the adult fly brain, eight types of neurons grouped in three clusters are known to project to the MBs (Mao and Davis, 2009; Nassel and Elekes, 1992). These neurons include the PPL1 and PAM clusters that project to different regions of the vertical and horizontal lobes as well as the PPL2a cluster that projects to the calyx. Using calcium-imaging techniques, different types of dopaminergic cells in the PPL1 cluster were demonstrated to respond differently to unpaired odor and odor paired with electric shock stimulation, indicating participation of these neurons in both CS and US pathways (Mao and Davis, 2009; Riemensperger *et al.*, 2005). Indeed, paired activation of a specific cell type in the PPL1 cluster (MB-MP1) with an odor is sufficient to induce an aversive olfactory memory (Aso *et al.*, 2010; Claridge-Chang *et al.*, 2009). Furthermore, blocking output from another group of MB innervating dopamine neurons (MB-M3) impairs the formation of aversive olfactory memory (Aso *et al.*, 2010). However, when activation of these neurons is paired with an odor an aversive olfactory memory was induced that lasted only minutes. Thus, it seems that different dopaminergic neurons are capable of inducing olfactory memories with different stabilities. Importantly, one cannot conclude that the dopaminergic neurons are only important for negative reinforcement of aversive olfactory memory. MB-MP1 neurons are also critical in gating the motivational state of hungry flies in appetitive olfactory memory formation (Krashes *et al.*, 2009). The dopaminergic system is not necessary for negatively reinforced visual pattern memory or place memory in the heat box (Sitaraman *et al.*, 2008; Zhang *et al.*, 2007).

Octopamine is necessary and sufficient for appetitive olfactory memory (Schroll *et al.*, 2006; Schwaerzel *et al.*, 2003). That is, blocking of octopaminergic neuron function reduces appetitive olfactory memory in the adult fly, and pairing activation of octopaminergic neuron activity with an odor induces appetitive memory in larvae. Although much effort has been invested in anatomically characterizing the octopaminergic system, the circuits involved in the formation of appetitive memory in *Drosophila* are not yet known (Busch *et al.*, 2009; Sinakevitch and Strausfeld, 2006). The octopaminergic system is not necessary for heat box based place memory (Sitaraman *et al.*, 2010). Although evidence from crickets suggests that there is a role for the octopaminergic system in appetitive visual learning, this has not yet been examined in the fly (Unoki *et al.*, 2006).

Tools to unambiguously examine the function of the serotonergic system in *Drosophila* memory formation have only recently been developed (Alekseyenko *et al.*, 2010; Park *et al.*, 2006; Sitaraman *et al.*, 2008). Thus far, using transgenic and pharmacological manipulation, the serotonergic system has been shown to be necessary for normal place memory (Sitaraman *et al.*, 2008). Investigation into the role of serotonin in olfactory and visual memory in the fly is ongoing.

III. Genetics of Olfactory Memory Formation

The classical olfactory conditioning paradigm has been the primary means of identifying new genes with a role in *Drosophila* memory formation. There are about 40 genes, with various levels of knowledge about where and when in the nervous system they act, that are critical for memory formation in the range of minutes to hours of training (Table I). Although most knowledge about memory formation mechanisms centers on the cAMP/PKA-signaling pathway, there are now several genes that function outside of this cascade or provide starting points to investigate novel memory-critical signaling mechanisms. Finally, a relatively small number of genes have been identified that provide critical developmental functions for normal memory.

A. Olfactory Memory Genes Altering the cAMP/PKA Cascade

Several of the earliest mutations, identified in unbiased screens for flies with abnormal learning and memory, ultimately linked gene products critical in regulating the cAMP-signaling cascade with memory formation. The *rutabaga* and *dunce* genes encode a type 1 adenylyl cyclase (*rut*-AC) and cAMP-phosphodiesterase (*dnc*-PDE), respectively (Han *et al.*, 1992; Levin *et al.*, 1992; Nighorn *et al.*, 1991) (Table I). Because the *rut*-AC has the properties predicted for a coincidence detector (Dudai *et al.*, 1984; Lechner and Byrne, 1998), that is requiring both G-protein and Ca^{2+}/calmodulin input to increase cAMP levels, this gene has received much attention in memory investigations. Through restricted expression of the wild-type *rut*-AC gene in the MBs of *rutabaga* mutant flies, olfactory memory formation can be restored to normal levels. Thus, the neuronal plasticity deficits of *rutabaga* mutant flies was restored in the Kenyon cells of the MBs, and that *rut*-AC-dependent plasticity supports the formation of an olfactory memory (i.e., the identification of a memory trace or engram) (Mao *et al.*, 2004; McGuire *et al.*, 2003; Zars *et al.*, 2000a). Imaging of cAMP and PKA activity in the MBs of wild type and *rut*-AC mutant flies strengthens the conclusion that the *rut*-AC-dependent plasticity in the MBs supports olfactory memory formation (Gervasi *et al.*, 2010; Tomchik and Davis, 2009; Zars, 2010b). The *dnc*-PDE has the enzymatic activity that is expected to complement the *rut*-AC function. Although the *dnc*-PDE is necessary for normal olfactory memory formation, there is as yet little indication of where this product acts in promoting memory formation. Based on PKA-imaging studies, this protein could be acting within the MBs to restrict the cAMP increases to specific parts of this structure under learning conditions (Gervasi *et al.*, 2010). In contrast, using a set of mutations that alter specific versions of the *dnc*-PDE, expression domains outside of the MBs have been implicated for memory formation for this gene (Qiu and Davis, 1993). Localized rescue experiments with the *dnc*-PDE gene in a fashion similar to the *rut*-AC rescue approach should resolve this issue.

Several gene products influence the cAMP/PKA cascade, and the cAMP/PKA cascade influences several proteins, to regulate memory formation within the MB Kenyon cells. A G-protein S-type α (Gsa) subunit has been shown to be necessary within the MBs for normal olfactory memory formation, as have the catalytic subunit of PKA and the *AKAP* protein (critical for subcellular targeting of PKA) (Connolly *et al.*, 1996; Horiuchi *et al.*, 2008; Schwaerzel *et al.*, 2007; Skoulakis *et al.*, 1993). A regulatory subunit of PKA has also been implicated in olfactory memory, but where it acts in the central nervous system for this function is not known (Goodwin *et al.*, 1997). Mutation of the *neurofibromin* (*nf1*) gene, implicated in the human disease neurofibromatosis type 1, has strong effects on olfactory memory (Guo *et al.*, 2000). The Nf1 protein has been proposed to act as a GTPase-activating protein on the Gsa, and mutation of *nf1* alters *rut*-AC-dependent synthesis of cAMP. Double mutant *rut*-AC and *nf1* flies have similar memory levels as flies with mutations in either one of the two genes. NF1 expression is sufficient in the MBs for *nf1*-dependent olfactory memory formation, although the set of neurons that have been implicated only partially overlap the neurons needed for *rut*-AC-dependent memory (Buchanan and Davis, 2010). Mutation of a calcineurin inhibitor in the *nebula/sarah* (*sra*) gene has profound effects on olfactory memory formation (Chang *et al.*, 2003). Flies mutant in *sra* have a 50% reduction in PKA activity compared to wild-type flies. This suggests that *sra* inhibition of calcineurin phosphatase activity somehow influences PKA function, which could be linked to the olfactory memory deficit. Finally, a synapse-associated protein termed *synapsin* (*syn*), with potential PKA phosphorylation sites, has been implicated as a downstream target of cAMP-dependent memory formation (Diegelmann *et al.*, 2006a; Godenschwege *et al.*, 2004; Knapek *et al.*, 2010). That the loss of *syn* does not further reduce olfactory memory in *rut*-AC double mutant flies has been used to link the cAMP pathway to *syn*-dependent changes critical for memory.

There are several neurotransmitters/modulators that are thought to influence cAMP/PKA-dependent memory formation through receptor action on the MB

Kenyon cells. As previously discussed, the dopaminergic and octopaminergic neurons provide critical instructive roles in olfactory memory formation. One of the known dopamine receptors, DA1 is required for normal aversive and appetitive olfactory memory formation (Kim *et al.*, 2007). The DA1 receptor has been shown to function within the MB Kenyon cells through local rescue experiments. Little is known about how the octopaminergic signal is received at the MBs. The GABAergic system has been implicated in the olfactory pathway that is important for memory formation. Reducing expression of the GABA-A receptor encoded by the *Resistance to dieldrin* (*Rdl*) gene enhances memory levels, while increasing Rdl expression lowers memory (Liu *et al.*, 2007). The link of *Rdl* action with *rut*-AC-dependent memory comes from the failure of the reduction of *Rdl* expression in the MBs to enhance olfactory memory levels in *rut*-AC mutant flies, suggesting that *rut*-AC acts downstream of GABA-A receptor activity (Liu *et al.*, 2009). Neuropeptides have also been implicated in olfactory memory formation. The *amnesiac* gene encodes putative neuropeptides that may influence the cAMP/PKA cascade via MB extrinsic neurons. The *amnesiac* gene is predicted to encode three neuropeptides, one of which has some homology to the PACAP peptide (Feany and Quinn, 1995; Moore *et al.*, 1998). Although not a direct measure of the *amnesiac* peptides, biochemical studies of a mammalian PACAP have been shown to influence *rut*-AC activity (Guo *et al.*, 1997; Zhong, 1995). The *amnesiac* products are expressed in a rather small set of neurons in the adult fly brain. Remarkably, expressing a normal version of the *amnesiac* gene in a set of neurons called the MB-DPM cells can return *amnesiac* memory deficits to normal levels (Keene *et al.*, 2006; Waddell *et al.*, 2000). The DPM neurons ramify throughout the lobes of the MBs, and the current model is that the *amnesiac* product(s) are released onto the MBs during training, bind to as-yet-unidentified receptors on the MB Kenyon cells, and act to prolong *rut*-AC and PKA activity in those cells. Whether it is the PACAP-like peptide that is responsible for the *amnesiac* activity to be normal is not known. Use of potentially rescuing transgenes encoding each of the peptides could resolve this issue.

B. Olfactory Memory Genes That Implicate Novel Mechanisms

1. Genes With Demonstrated Function Outside of the cAMP/PKA Cascade

It should be expected that there will be some genes that are critical for olfactory memory formation that do not fall in the cAMP/PKA-signaling cascade because strong or null mutations in several of these genes do not reduce memory levels to zero. Thus far the strategy for identifying genes that complement the PKA cascade is to test new mutant genes in a *rut*-AC mutant background. If the *rut*-AC phenotype is enhanced in the double mutant, the new gene function is interpreted as acting outside of the PKA cascade. There has been some progress in identifying the cAMP/PKA-independent genes. Signaling through the G-protein O-type α subunit (Goα) has been implicated in memory formation outside of the PKA cascade. Expression of a pertussis toxin transgene in the MBs, altering Goα activity, strongly reduces olfactory memory levels. The low level of *rut*-AC mutant olfactory memory is further reduced in Goα altered flies. The *gilgamesh* (*gish*) gene has also been implicated in non-PKA-dependent olfactory learning (Tan *et al.*, 2010). The *gish* gene encodes a casein kinase 1 γ orthologue. Mutation of the *gish* gene reduces olfactory 3 min memory to about 50% levels in homozygous mutant flies and a still obvious but weaker phenotype in heterozygous flies. Manipulation of *gish* expression in the MBs shows that the olfactory memory function of this gene product can be restricted to the adult Kenyon cells. The reduction of *gish* with a single-mutant allele significantly enhanced *rut*-AC mutant flies memory performance.

A forgetting function in olfactory memory has been linked to the small rho-family GTPase *rac*, a gene product critical for regulating cytoskeleton dynamics among other cellular functions (Shuai *et al.*, 2010). Reduction of *rac* activity with expression of a dominant negative version of the gene in the MBs gives rise to higher memory levels after training. Furthermore, increases in *rac* activity with expression of a dominant active version speeds the decline in memory performance. With a nonacquisition function, one might expect that low *rut*-AC memory levels (*rut*-AC being important for acquisition) would not influence the *rac* manipulated memory enhancement, which is indeed the case.

2. *Genes With Unknown Linkage to cAMP/PKA Signaling*

Several genes have been identified as having critical roles in olfactory memory formation, implicating novel mechanisms, but a link or independence from the cAMP/PKA pathway is not clear. Processes that have been implicated include receptor tyrosine kinase (RTK) and NMDA-receptor signaling, regulation of protein expression, and neuronal connectivity.

Three genes have been identified that implicate RTK signaling in olfactory memory formation. The *linotte/derailed* gene encodes an RTK, and flies with a mutation in this gene have reduced olfactory memory. Furthermore, the *leonardo* 14-3-3zeta (*leo*) gene is important for memory formation. Gene interaction studies in developmental assays suggest *leo* signals through RAS and RAF proteins (Li *et al.*, 1997; Philip *et al.*, 2001; Skoulakis and Davis, 1996). Finally, the *downstream of receptor kinase* (*drk*) gene, providing a "linker" function between an RTK and signaling proteins recruited to the receptor, has also been implicated in MB-dependent olfactory memory formation (Moressis *et al.*, 2009). Indeed, expression of active mutant versions of RAS and RAF proteins in the MBs can compensate for a reduction in *drk* levels, suggesting that *drk* acts upstream of the RAS- and RAF-signaling components in memory formation. It should be pointed out, however, that in each of these three cases it is not clear what the relationship is between the different genes (e.g., is there an interaction of *drk* with *leo*?). Thus, although RTK signaling is important for olfactory memory formation, the complete picture of these events is far from realized.

The NMDA receptors have been implicated in olfactory memory formation (Wu *et al.*, 2007; Xia *et al.*, 2005). There are two NMDA receptors in the *Drosophila* genome, *DNR1* and *DNR2*. Mutation and RNAi-mediated reduction in *DNR1* in the MBs lead to a reduction in olfactory memory within hours of training (long-term memory requires *DNR1* in parts of the central complex). Similarly, reduction in *DNR2* in the MBs has a temporally restricted role in olfactory memory. Mutation of the gene *polyglutamine tract-binding protein-1* (*PQBP1*) also reduces olfactory memory, and is thought to do so by altering *NMDAR*-dependent function (Tamura *et al.*, 2010). The *PQBP1* gene alters memory by action in the projection neurons of the antennal lobe, a set of neurons that innervate the calyces of the MBs. Interestingly, overexpressing *DNR1* in *PQBP1* mutant flies (which otherwise have low DNR1 expression in the PN cells) reverted the olfactory memory deficit to normal levels. The discordant conclusions for where NMDA receptors function in memory formation in the minutes to hours range after training in the PNs and MBs have not been addressed (Tamura *et al.*, 2010). A finer temporal resolution of requirement of these receptors in the PNs and MB after training could resolve this issue.

Several genes have been identified that represent first insights into mechanisms of memory formation. The *volado* (an α integrin) and *fasciclin2* (an NCAM) genes implicate acute regulation of cell–cell interactions in memory (Cheng *et al.*, 2001; Grotewiel *et al.*, 1998). Several genes implicate acute but so far general regulation of protein stability or protein levels, including the *crammer* cysteine protease inhibitor, *fragile X mental retardation* (*fmr*), and the *tequila* neurotrypsin (Bolduc *et al.*, 2008; Comas *et al.*, 2004; Didelot *et al.*, 2006). The *ignorant* S6 type 2 and the *tribbles* kinases implicate new kinase cascades in memory formation (Laferriere *et al.*, 2008; Putz *et al.*, 2004). Despite the long history for the role of the *radish* (*rsh*) gene in olfactory memory, this gene falls into the class for which it is not clear how it acts in a cell or neural circuit (Folkers *et al.*, 2006; Krashes and Waddell, 2008). Finally, several genes implicate the normal function of neurmodulatory systems for memory, including the dopaminergic, octopaminergic, and serotonergic systems (Diegelmann *et al.*, 2006b; Krashes *et al.*, 2009; Schwaerzel *et al.*, 2003; Sitaraman *et al.*, 2008, 2010). A deeper knowledge of how this class of proteins interact with other signaling proteins and where they function in the nervous system (in some cases) will provide a better understanding of memory formation mechanisms in the fly.

C. Olfactory Memory Genes With a Role in Development

Although relatively underrepresented, several genes (~15% thus far) have been identified as critical in establishing the systems important for olfactory memory formation. Probably the most straightforward class of genes are those that when mutated lead to grossly altered brain structures. The genes *mushroom body miniature* (*mbm*) and *mushroom body deranged* (*mbd*) both alter the structure of the adult MBs and give rise to flies with severely disrupted olfactory memory (Heisenberg *et al.*, 1985; Raabe *et al.*, 2004). Several additional genes fall into the class for which mutation leads to either minor or nonobvious changes in the nervous system. These genes are "developmental" because expression of rescuing transgene copies must occur in early life stages to restore olfactory memory in the adult fly. Within this class, products that regulate gene expression (*latheo* and *nalyot*) are critical (Boynton and Tully, 1992; DeZazzo *et al.*, 2000; Pinto *et al.*, 1999). New tools to examine single neuron or synaptic morphology could be used to better understand the role of this class of gene in establishing memory forming systems.

IV. Common and Specific Genetic Mechanisms in Olfactory, Visual, and Spatial Memory Formation

Although not nearly as comprehensive, genetic mechanisms of olfactory, visual, and spatial memory have identified some common genetic and specific elements that are critical for memory formation. When common genetic mechanisms have been identified, specificity for gene/behavior action comes from the parts of the nervous system that are critical for different memories.

In addition to the role of the cAMP/PKA pathway in olfactory memory formation, this signaling pathway is important for visual pattern recognition memory and place memory. Mutation of the *rut*-AC and *dnc*-PDE, as well as the *amnesiac* peptides, shows that this pathway is important for place memory (Diegelmann *et al.*, 2006b; Putz, 2002; Wustmann *et al.*, 1996; Zars, 2010a; Zars *et al.*, 2000b). Less is known about the genes important for visual pattern recognition memory, but the *rut*-AC gene is necessary for at least a component of the visual pattern memory (Brembs and Plendl, 2008; Liu *et al.*, 2006).

Specificity for the function of the *rut*-AC, and therefore perhaps the cAMP/PKA pathway, in memory formation comes from the set of neurons in which it acts for memory formation. Sufficient *rut*-AC expression in subsets of the MB localizes regulated cAMP-dependent plasticity for olfactory memory formation (Blum *et al.*, 2009; Mao *et al.*, 2004; McGuire *et al.*, 2003; Zars *et al.*, 2000a). Expression of the same transgenes in the same mutant background does not restore *rut*-AC mutant visual pattern memory or place memory (Liu *et al.*, 2006; Zars *et al.*, 2000b). Instead, expression of a rescuing transgene in F-type neurons that innervate specific layers of the fan-shaped body restores visual pattern recognition memory for different pattern parameters. Furthermore, expression of a rescuing transgene in neurons of the median bundle, a structure that innervates the dorsal medial protocerebrum restores place memory to normal levels in otherwise *rut*-AC mutant flies.

In addition to PKA, several additional kinases have been implicated in olfactory, visual, and place memory. Specificity of action for different memories in this set of kinases is a function of the kinases themselves. A protein kinase C (PKC) is important for visual pattern memory and a modified version that is "active" (PKM) enhances olfactory memory (Brembs and Plendl, 2008; Drier *et al.*, 2002). Furthermore, natural variant alleles for a cGMP-dependent protein kinase (PKG) show that increased PKG activity is correlated with higher olfactory and visual pattern recognition memory (Mery *et al.*, 2007; Wang *et al.*, 2008). Specificity for PKG activity in olfactory and visual memory again depends on the neural circuit in which this protein acts. The same alleles that show differential roles in olfactory and visual memory do not alter place memory (Gioia and Zars, 2009). Finally, an S6 type-2 kinase (S6K2) was identified in a screen for mutant flies in the heat box place memory paradigm. Semidominant alleles have an effect on place memory but not olfactory memory; loss of function alleles alter olfactory memory but not place memory (Putz *et al.*, 2004; Zars, 2009). Interestingly, the loss of function alleles of S6K2 have an impact on the seconds-long orientation memory, and selective expression of the S6K2 product in ring neurons of S6K2 mutant flies in the ellipsoid body restores the orientation memory defect to normal (Neuser *et al.*, 2008). It is not yet clear whether or not the S6K2 has a role in visual pattern recognition memory. Finally, mutation of a *tribbles* encoded kinase has been shown to decrease place memory, but increase olfactory memory (Laferriere *et al.*, 2008). Where in the nervous system this kinase acts to support normal memory in both paradigms is under investigation, as is the potential role of this gene in visual pattern recognition memory.

V. Conclusion

The rich environment and behavioral repertoire of *Drosophila* suggest that flies should be able to form memories to inform future decisions. Development of learning paradigms with a restricted set of behavioral choice in receiving reward or avoiding punishment was essential for furthering our understanding of the genetic and neural system mechanisms of memory formation in the fly. Three paradigms have proven influential in determining principles of gene and neural action in memory: olfactory, visual, and place. Anatomical and genetic investigation of the nervous system suggested several parts of the nervous system would be important for memory. About 40 genes have been identified that alter olfactory short-term memory, a subset of these genes have been tested in multiple paradigms. Thus far, specificity for the role of a gene or pathway in memory formation comes either from the selective recruitment of a kinase-type to a subset of memory types or a function within different parts of the nervous system.

Acknowledgments

We thank Reinhard Wolf for providing the picture of the fly flying in the visual learning paradigm. This work was supported by grants from the National Science Foundation and the University of Missouri Research Board.

Note in proof: The *arouser* EPS8L3 has been recently shown to be necessary for normal olfactory and place memory (LaFerriere *et al.*, 2011).

References

Aceves-Pina, E.O., and Quinn, W.G. (1979). Learning in normal and mutant *Drosophila* larvae. *Science* **206**, 93–96.

Alekseyenko, O.V., Lee, C., and Kravitz, E.A. (2010). Targeted manipulation of serotonergic neurotransmission affects the escalation of aggression in adult male *Drosophila melanogaster*. *PLoS ONE* **5**, e10806.

Aso, Y., Grubel, K., Busch, S., Friedrich, A.B., Siwanowicz, I., and Tanimoto, H. (2009). The mushroom body of adult *Drosophila* characterized by GAL4 drivers. *J. Neurogenet.* **23**, 156–172.

Aso, Y., Siwanowicz, I., Bracker, L., Ito, K., Kitamoto, T., and Tanimoto, H. (2010). Specific dopaminergic neurons for the formation of labile aversive memory. *Curr. Biol.* **20**, 1445–1451.

Benzer, S. (1967). Behavioral mutants of *Drosophila* isolated by countercurrent distribution. *Proc. Natl. Acad. Sci. U S A* **58**, 1112–1119.

Besson, M., and Martin, J.R. (2005). Centrophobism/thigmotaxis, a new role for the mushroom bodies in *Drosophila*. *J. Neurobiol.* **62**, 386–396.

Blum, A.L., Li, W., Cressy, M., and Dubnau, J. (2009). Short- and long-term memory in *Drosophila* require cAMP signaling in distinct neuron types. *Curr. Biol.* **19**, 1341–1350.

Bolduc, F.V., Bell, K., Cox, H., Broadie, K.S., and Tully, T. (2008). Excess protein synthesis in *Drosophila fragile X* mutants impairs long-term memory. *Nat. Neurosci.* **11**, 1143–1145.

Bolduc, F.V., and Tully, T. (2009). Fruit flies and intellectual disability. *Fly* **3**, 91–104.

Bolwig, G.M., Del Vecchio, M., Hannon, G., and Tully, T. (1995). Molecular cloning of *linotte* in *Drosophila*: a novel gene that functions in adults during associative learning. *Neuron* **15**, 829–842.

Bolwig, G.M., Del Vecchio, M., Hannon, G., and Tully, T. (2002). Erratum: Molecular cloning of *linotte* in *Drosophila*: a novel gene that functions in adults during associative learning. *Neuron* **34**, 667.

Boynton, S., and Tully, T. (1992). *latheo*, a new gene involved in associative learning and memory in *Drosophila melanogaster*, identified from P element mutagenesis. *Genetics* **131**, 655–672.

Brembs, B., and Plendl, W. (2008). Double dissociation of PKC and AC manipulations on operant and classical learning in *Drosophila*. *Curr. Biol.* **18**, 1168–1171.

Buchanan, M.E., and Davis, R.L. (2010). A distinct set of *Drosophila* brain neurons required for *neurofibromatosis type 1*-dependent learning and memory. *J. Neurosci.* **30**, 10135–10143.

Buelthoff, H., Goetz, K.G., and Herre, M. (1982). Recurrent inversion of visual orientation in the walking fly *Drosophila melanogaster*. *J. Comp. Physiol. A* **148**, 471–482.

Busch, S., Selcho, M., Ito, K., and Tanimoto, H. (2009). A map of octopaminergic neurons in the *Drosophila* brain. *J. Comp. Neurol.* **513**, 643–667.

Byers, D., Davis, R.L., and Kiger Jr., J.A. (1981). Defect in cyclic AMP phosphodiesterase due to the *dunce* mutation of learning in *Drosophila melanogaster*. *Nature* **289**, 79–81.

Chang, K.T., Shi, Y.J., and Min, K.T. (2003). The *Drosophila* homolog of Down's syndrome critical region 1 gene regulates learning: implications for mental retardation. *Proc. Natl. Acad. Sci. U S A* **100**, 15794–15799.

Cheng, Y., Endo, K., Wu, K., Rodan, A.R., Heberlein, U., and Davis, R.L. (2001). *Drosophila fasciclinII* is required for the formation of odor memories and for normal sensitivity to alcohol. *Cell* **105**, 757–768.

Claridge-Chang, A., Roorda, R.D., Vrontou, E., Sjulson, L., Li, H., Hirsh, J., and Miesenbock, G. (2009). Writing memories with light-addressable reinforcement circuitry. *Cell* **139**, 405–415.

Colomb, J., Kaiser, L., Chabaud, M.A., and Preat, T. (2009). Parametric and genetic analysis of *Drosophila* appetitive long-term memory and sugar motivation. *Genes Brain Behav.* **8**, 407–415.

Comas, D., Petit, F., and Preat, T. (2004). *Drosophila* long-term memory formation involves regulation of *cathepsin* activity. *Nature* **430**, 460–463.

Connolly, J.B., Roberts, I.J., Armstrong, J.D., Kaiser, K., Forte, M., Tully, T., and O'Kane, C.J. (1996). Associative learning disrupted by impaired Gs signaling in *Drosophila* mushroom bodies. *Science* **274**, 2104–2107.

Crittenden, J.R., Skoulakis, E.M., Han, K.A., Kalderon, D., and Davis, R.L. (1998). Tripartite mushroom body architecture revealed by antigenic markers. *Learn. Mem.* **5**, 38–51.

Dauwalder, B., and Davis, R.L. (1995). Conditional rescue of the *dunce* learning/memory and female fertility defects with *Drosophila* or rat transgenes. *J. Neurosci.* **15**, 3490–3499.

de Belle, J.S., and Heisenberg, M. (1994). Associative odor learning in *Drosophila* abolished by chemical ablation of mushroom bodies. *Science* **263**, 692–695.

DeZazzo, J., Xia, S., Christensen, J., Velinzon, K., and Tully, T. (1999). Developmental expression of an *amn*(+) transgene rescues the mutant memory defect of *amnesiac* adults. *J. Neurosci.* **19**, 8740–8746.

DeZazzo, J., Sandstrom, D., de Belle, S., Velinzon, K., Smith, P., Grady, L., DelVecchio, M., Ramaswami, M., and Tully, T. (2000). *nalyot*, a mutation of the *Drosophila myb*-related *Adf1* transcription factor, disrupts synapse formation and olfactory memory. *Neuron* **27**, 145–158.

Didelot, G., Molinari, F., Tchenio, P., Comas, D., Milhiet, E., Munnich, A., Colleaux, L., and Preat, T. (2006). *Tequila*, a neurotrypsin ortholog, regulates long-term memory formation in *Drosophila*. *Science* **313**, 851–853.

Diegelmann, S., Nieratschker, V., Werner, U., Hoppe, J., Zars, T., and Buchner, E. (2006 a) The conserved protein kinase-A target motif in synapsin of *Drosophila* is effectively modified by pre-mRNA editing. *BMC Neurosci.* **7**, 76.

Diegelmann, S., Zars, M., and Zars, T. (2006 b) Genetic dissociation of acquisition and memory strength in the heat-box spatial learning paradigm in *Drosophila*. *Learn. Mem.* **13**, 72–83.

Dill, M., Wolf, R., and Heisenberg, M. (1993). Visual pattern recognition in *Drosophila* involves retinotopic matching. *Nature* **365**, 751–753.

Drain, P., Folkers, E., and Quinn, W.G. (1991). cAMP-dependent protein kinase and the disruption of learning in transgenic flies. *Neuron* **6**, 71–82.

Drier, E.A., Tello, M.K., Cowan, M., Wu, P., Blace, N., Sacktor, T.C., and Yin, J.C. (2002). Memory enhancement and formation by atypical PKM activity in *Drosophila melanogaster*. *Nat. Neurosci.* **5**, 316–324.

Dudai, Y. (2002). *Memory from A to Z.* Oxford University Press, Oxford, New York.

Dudai, Y., Jan, Y.N., Byers, D., Quinn, W.G., and Benzer, S. (1976). *dunce*, a mutant of *Drosophila* deficient in learning. *Proc. Natl. Acad. Sci. U S A* **73**, 1684–1688.

Dudai, Y., Zvi, S., and Segel, S. (1984). A defective conditioned behavior and a defective adenylate cyclase in the *Drosophila* mutant *rutabaga*. *J. Comp. Physiol. A* **155**, 569–576.

Dura, J.M., Preat, T., and Tully, T. (1993). Identification of *linotte*, a new gene affecting learning and memory in *Drosophila melanogaster*. *J. Neurogenet.* **9**, 1–14.

Dura, J.M., Taillebourg, E., and Preat, T. (1995). The *Drosophila* learning and memory gene *linotte* encodes a putative receptor tyrosine kinase homologous to the human RYK gene product. *FEBS Lett.* **370**, 250–254.

Ernst, R., and Heisenberg, M. (1999). The memory template in *Drosophila* pattern vision at the flight simulator. *Vision Res.* **39**, 3920–3933.

Fahrbach, S.E. (2006). Structure of the mushroom bodies of the insect brain. *Annu. Rev. Entomol.* **51**, 209–232.

Feany, M.B., and Quinn, W.G. (1995). A neuropeptide gene defined by the *Drosophila* memory mutant *amnesiac*. *Science* **268**, 869–873.

Ferris, J., Ge, H., Liu, L., and Roman, G. (2006). G(o) signaling is required for *Drosophila* associative learning. *Nat. Neurosci.* **9**, 1036–1040.

Folkers, E., Drain, P., and Quinn, W.G. (1993). *radish*, a *Drosophila* mutant deficient in consolidated memory. *Proc. Natl. Acad. Sci. U S A* **90**, 8123–8127.

Folkers, E., Waddell, S., and Quinn, W.G. (2006). The *Drosophila radish* gene encodes a protein required for anesthesia-resistant memory. *Proc. Natl. Acad. Sci. U S A* **103**, 17496–17500.

Foucaud, J., Burns, J.G., and Mery, F. (2010). Use of spatial information and search strategies in a water maze analog in *Drosophila melanogaster*. *PLoS ONE* **5**, e15231.

Gerber, B., and Hendel, T. (2006). Outcome expectations drive learned behaviour in larval *Drosophila*. *Proc. Biol. Sci.* **273**, 2965–2968.

Gerber, B., Scherer, S., Neuser, K., Michels, B., Hendel, T., Stocker, R.F., and Heisenberg, M. (2004 a) Visual learning in individually assayed *Drosophila* larvae. *J. Exp. Biol.* **207**, 179–188.

Gerber, B., Tanimoto, H., and Heisenberg, M. (2004 b) An engram found? Evaluating the evidence from fruit flies. *Curr. Opin. Neurobiol.* **14**, 737–744.

Gervasi, N., Tchenio, P., and Preat, T. (2010). PKA dynamics in a *Drosophila* learning center: coincidence detection by *rutabaga* adenylyl cyclase and spatial regulation by *dunce* phosphodiesterase. *Neuron* **65**, 516–529.

Gioia, A., and Zars, T. (2009). Thermotolerance and place memory in adult *Drosophila* are independent of natural variation at the *foraging* locus. *J. Comp. Physiol. A* **195**, 777–782.

Godenschwege, T.A., Reisch, D., Diegelmann, S., Eberle, K., Funk, N., Heisenberg, M., Hoppe, V., Hoppe, J., Klagges, B.R., Martin, J.R., Nikitina, E.A., Putz, G., Reifegerste, R., Reisch, N., Rister, J., Schaupp, M., Scholz, H., Schwarzel, M., Werner, U., Zars, T.D., Buchner, S., and Buchner, E. (2004). Flies lacking all *synapsins* are unexpectedly healthy but are impaired in complex behaviour. *Eur. J. Neurosci.* **20**, 611–622.

Goodwin, S.F., Del Vecchio, M., Velinzon, K., Hogel, C., Russell, S.R., Tully, T., and Kaiser, K. (1997). Defective learning in mutants of the *Drosophila* gene for a regulatory subunit of cAMP-dependent protein kinase. *J. Neurosci.* **17**, 8817–8827.

Grotewiel, M.S., Beck, C.D., Wu, K.H., Zhu, X.R., and Davis, R.L. (1998). Integrin-mediated short-term memory in *Drosophila*. *Nature* **391**, 455–460.

Guo, H.F., The, I., Hannan, F., Bernards, A., and Zhong, Y. (1997). Requirement of *Drosophila NF1* for activation of adenylyl cyclase by PACAP38-like neuropeptides. *Science* **276**, 795–798.

Guo, H.F., Tong, J., Hannan, F., Luo, L., and Zhong, Y. (2000). A *neurofibromatosis-1*-regulated pathway is required for learning in *Drosophila*. *Nature* **403**, 895–898.

Han, P.L., Levin, L.R., Reed, R.R., and Davis, R.L. (1992). Preferential expression of the *Drosophila rutabaga* gene in mushroom bodies, neural centers for learning in insects. *Neuron* **9**, 619–627.

Hanesch, U., Fischbach, K.-F., and Heisenberg, M. (1989). Neuronal architecture of the central complex in *Drosophila melanogaster*. *Cell Tissue Res.* **257**, 343–366.

Heisenberg, M. (1983). Initial activity and voluntary behavior in animals. *Naturwissenschaften* **70**, 70–78.

Heisenberg, M., Borst, A., Wagner, S., and Byers, D. (1985). *Drosophila* mushroom body mutants are deficient in olfactory learning. *J. Neurogenet.* **2**, 1–30.

Heisenberg, M., and Wolf, R. (1984). *Vision in Drosophila; Genetics of microbehavior*. Springer-Verlag, Berlin, Heidelberg, New York, Tokyo.

Heisenberg, M., Wolf, R., and Brembs, B. (2001). Flexibility in a single behavioral variable of *Drosophila*. *Learn. Mem.* **8**, 1–10.

Ho, I.S., Hannan, F., Guo, H.F., Hakker, I., and Zhong, Y. (2007). Distinct functional domains of *neurofibromatosis type 1* regulate immediate versus long-term memory formation. *J. Neurosci.* **27**, 6852–6857.

Horiuchi, J., Yamazaki, D., Naganos, S., Aigaki, T., and Saitoe, M. (2008). Protein kinase A inhibits a consolidated form of memory in *Drosophila*. *Proc. Natl. Acad. Sci. U S A* **105**, 20976–20981.

Ito, K., Suzuki, K., Estes, P., Ramaswami, M., Yamamoto, D., and Strausfeld, N.J. (1998). The organization of extrinsic neurons and their implications in the functional roles of the mushroom bodies in *Drosophila melanogaster* Meigen. *Learn. Mem.* **5**, 52–77.

Kahsai, L., Martin, J.R., and Winther, A.M. (2010). Neuropeptides in the *Drosophila* central complex in modulation of locomotor behavior. *J. Exp. Biol.* **213**, 2256–2265.

Kahsai, L., and Winther, A.M. (2011). Chemical neuroanatomy of the *Drosophila* central complex: distribution of multiple neuropeptides in relation to neurotransmitters. *J. Comp. Neurol.* **519**, 290–315.

Keene, A.C., Krashes, M.J., Leung, B., Bernard, J.A., and Waddell, S. (2006). *Drosophila* dorsal paired medial neurons provide a general mechanism for memory consolidation. *Curr. Biol.* **16**, 1524–1530.

Keene, A.C., and Waddell, S. (2007). *Drosophila* olfactory memory: single genes to complex neural circuits. *Nat. Rev. Neurosci.* **8**, 341–354.

Kim, Y.C., Lee, H.G., and Han, K.A. (2007). D1 dopamine receptor dDA1 is required in the mushroom body neurons for aversive and appetitive learning in *Drosophila*. *J. Neurosci.* **27**, 7640–7647.

Knapek, S., Gerber, B., and Tanimoto, H. (2010). Synapsin is selectively required for anesthesia-sensitive memory. *Learn. Mem.* **17**, 76–79.

Krashes, M.J., Dasgupta, S., Vreede, A., White, B., Armstrong, J.D., and Waddell, S. (2009). A neural circuit mechanism integrating motivational state with memory expression in *Drosophila*. *Cell* **139**, 416–427.

Krashes, M.J., Keene, A.C., Leung, B., Armstrong, J.D., and Waddell, S. (2007). Sequential use of mushroom body neuron subsets during *Drosophila* odor memory processing. *Neuron* **53**, 103–115.

Krashes, M.J., and Waddell, S. (2008). Rapid consolidation to a *radish* and protein synthesis-dependent long-term memory after single-session appetitive olfactory conditioning in *Drosophila*. *J. Neurosci.* **28**, 3103–3113.

Laferriere, H., Guarnieri, D.J., Sitaraman, D., Diegelmann, S., Heberlein, U., and Zars, T. (2008). Genetic dissociation of ethanol sensitivity and memory formation in *Drosophila melanogaster*. *Genetics* **178**, 1895–1902.

Laferriere, H., Ostrowski, D., Guarnieri, D.J., and Zars, T. (2011). The arouser EPS8L3 gene is critical for normal memory in Drosophila. *PLoS ONE* **6**(7); e22867.

Lechner, H.A., and Byrne, J.H. (1998). New perspectives on classical conditioning: a synthesis of Hebbian and non-Hebbian mechanisms. *Neuron* **20**, 355–358.

Lee, T., Lee, A., and Luo, L. (1999). Development of the *Drosophila* mushroom bodies: sequential generation of three distinct types of neurons from a neuroblast. *Development* **126**, 4065–4076.

Levin, L.R., Han, P.L., Hwang, P.M., Feinstein, P.G., Davis, R.L., and Reed, R.R. (1992). The *Drosophila* learning and memory gene *rutabaga* encodes a Ca2+/Calmodulin-responsive adenylyl cyclase. *Cell* **68**, 479–489.

Li, W., Skoulakis, E.M., Davis, R.L., and Perrimon, N. (1997). The *Drosophila* 14-3-3 protein *Leonardo* enhances *Torso* signaling through D-Raf in a Ras 1-dependent manner. *Development* **124**, 4163–4171.

Li, W., Tully, T., and Kalderon, D. (1996). Effects of a conditional *Drosophila* PKA mutant on olfactory learning and memory. *Learn. Mem.* **2**, 320–333.

Liu, G., Seiler, H., Wen, A., Zars, T., Ito, K., Wolf, R., Heisenberg, M., and Liu, L. (2006). Distinct memory traces for two visual features in the *Drosophila* brain. *Nature* **439**, 551–556.

Liu, L., Wolf, R., Ernst, R., and Heisenberg, M. (1999). Context generalization in *Drosophila* visual learning requires the mushroom bodies. *Nature* **400**, 753–756.

Liu, X., Buchanan, M.E., Han, K.A., and Davis, R.L. (2009). The GABAA receptor *RDL* suppresses the conditioned stimulus pathway for olfactory learning. *J. Neurosci.* **29**, 1573–1579.

Liu, X., Krause, W.C., and Davis, R.L. (2007). GABAA receptor *RDL* inhibits *Drosophila* olfactory associative learning. *Neuron* **56**, 1090–1102.

Mao, Z., and Davis, R.L. (2009). Eight different types of dopaminergic neurons innervate the *Drosophila* mushroom body neuropil: anatomical and physiological heterogeneity. *Front. Neural Circuits* **3**, 5.

Mao, Z., Roman, G., Zong, L., and Davis, R.L. (2004). Pharmacogenetic rescue in time and space of the *rutabaga* memory impairment by using Gene-Switch. *Proc. Natl. Acad. Sci. U S A* **101**, 198–203.

Martin, J.R., Ernst, R., and Heisenberg, M. (1998). Mushroom bodies suppress locomotor activity in *Drosophila melanogaster*. *Learn. Mem.* **5**, 179–191.

McGuire, S.E., Le, P.T., Osborn, A.J., Matsumoto, K., and Davis, R.L. (2003). Spatiotemporal rescue of memory dysfunction in *Drosophila*. *Science* **302**, 1765–1768.

McGuire, S.E., Deshazer, M., and Davis, R.L. (2005). Thirty years of olfactory learning and memory research in *Drosophila melanogaster*. *Prog. Neurobiol.* **76**, 328–347.

Mery, F., Belay, A.T., So, A.K., Sokolowski, M.B., and Kawecki, T.J. (2007). Natural polymorphism affecting learning and memory in *Drosophila*. *Proc. Natl. Acad. Sci. U S A* **104**, 13051–13055.

Moore, M.S., DeZazzo, J., Luk, A.Y., Tully, T., Singh, C.M., and Heberlein, U. (1998). Ethanol intoxication in *Drosophila*: Genetic and pharmacological evidence for regulation by the cAMP signaling pathway. *Cell* **93**, 997–1007.

Moreau-Fauvarque, C., Taillebourg, E., Preat, T., and Dura, J.M. (2002). Mutation of *linotte* causes behavioral defects independently of pigeon in *Drosophila*. *Neuroreport* **13**, 2309–2312.

Moressis, A., Friedrich, A.R., Pavlopoulos, E., Davis, R.L., and Skoulakis, E.M. (2009). A dual role for the adaptor protein *DRK* in *Drosophila* olfactory learning and memory. *J. Neurosci.* **29**, 2611–2625.

Nassel, D.R., and Elekes, K. (1992). Aminergic neurons in the brain of blowflies and *Drosophila*: dopamine- and tyrosine hydroxylase-immunoreactive neurons and their relationship with putative histaminergic neurons. *Cell Tissue Res.* **267**, 147–167.

Neuser, K., Triphan, T., Mronz, M., Poeck, B., and Strauss, R. (2008). Analysis of a spatial orientation memory in *Drosophila*. *Nature* **453**, 1244–1247.

Nighorn, A., Healy, M.J., and Davis, R.L. (1991). The cyclic AMP phosphodiesterase encoded by the *Drosophila dunce* gene is concentrated in the mushroom body neuropil. *Neuron* **6**, 455–467.

Ofstad, T., Zuker, C.S., and Reiser, M.B. (2011). Visual place learning in *Drosophila melanogaster*. *Nature* **474**, 204–207.

Pan, Y., Zhou, Y., Guo, C., Gong, H., Gong, Z., and Liu, L. (2009). Differential roles of the fan-shaped body and the ellipsoid body in *Drosophila* visual pattern memory. *Learn. Mem.* **16**, 289–295.

Park, J., Lee, S.B., Lee, S., Kim, Y., Song, S., Kim, S., Bae, E., Kim, J., Shong, M., Kim, J.M., and Chung, J. (2006). Mitochondrial dysfunction in *Drosophila* PINK1 mutants is complemented by *parkin*. *Nature* **441**, 1157–1161.

Pavlov, I.P. (1927). *Conditioned reflexes*. Oxford University Press, London.

Philip, N., Acevedo, S.F., and Skoulakis, E.M. (2001). Conditional rescue of olfactory learning and memory defects in mutants of the 14-3-3zeta gene *leonardo*. *J. Neurosci.* **21**, 8417–8425.

Pinto, S., Quintana, D.G., Smith, P., Mihalek, R.M., Hou, Z.H., Boynton, S., Jones, C.J., Hendricks, M., Velinzon, K., Wohlschlegel, J.A., Austin, R.J., Lane, W.S., Tully, T., and Dutta, A. (1999). *latheo* encodes a subunit of the origin recognition complex and disrupts neuronal proliferation and adult olfactory memory when mutant. *Neuron* **23**, 45–54.

Pitman, J.L., DasGupta, S., Krashes, M.J., Leung, B., Perrat, P.N., and Waddell, S. (2009). There are many ways to train a fly. *Fly* **3**, 3–9.

Putz, G. (2002). *Characterization of memories and ignorant (S6KII) mutants in operant conditioning in the heat-box., 115 in Department of Genetics and Neurobiology*. Julius Maximillians Universitaet Wuerzburg, Wuerzburg.

Putz, G., Bertolucci, F., Raabe, T., Zars, T., and Heisenberg, M. (2004). The S6KII (*rsk*) gene of *Drosophila melanogaster* differentially affects an operant and a classical learning task. *J. Neurosci.* **24**, 9745–9751.

Putz, G., and Heisenberg, M. (2002). Memories in *Drosophila* heat-box learning. *Learn. Mem.* **9**, 349–359.

Qiu, Y., and Davis, R.L. (1993). Genetic dissection of the learning/memory gene *dunce* of *Drosophila melanogaster*. *Genes Dev.* **7**, 1447–1458.

Quinn, W.G., Harris, W.A., and Benzer, S. (1974). Conditioned behavior in *Drosophila melanogaster*. *Proc. Natl. Acad. Sci. U S A* **71**, 708–712.

Quinn, W.G., Sziber, P.P., and Booker, R. (1979). The *Drosophila* memory mutant *amnesiac*. *Nature* **277**, 212–214.

Raabe, T., Clemens-Richter, S., Twardzik, T., Ebert, A., Gramlich, G., and Heisenberg, M. (2004). Identification of *mushroom body miniature*, a zinc-finger protein implicated in brain development of *Drosophila*. *Proc. Natl. Acad. Sci. U S A* **101**, 14276–14281.

Renn, S.C., Armstrong, J.D., Yang, M., Wang, Z., An, X., Kaiser, K., and Taghert, P.H. (1999). Genetic analysis of the *Drosophila* ellipsoid body neuropil: organization and development of the central complex. *J. Neurobiol.* **41**, 189–207.

Riemensperger, T., Voller, T., Stock, P., Buchner, E., and Fiala, A. (2005). Punishment prediction by dopaminergic neurons in *Drosophila*. *Curr. Biol.* **15**, 1953–1960.

Scherer, S., Stocker, R.F., and Gerber, B. (2003). Olfactory learning in individually assayed *Drosophila* larvae. *Learn. Mem.* **10**, 217–225.

Schipanski, A., Yarali, A., Niewalda, T., and Gerber, B. (2008). Behavioral analyses of sugar processing in choice, feeding, and learning in larval *Drosophila*. *Chem. Senses* **33**, 563–573.

Schnaitmann, C., Vogt, K., Triphan, T., and Tanimoto, H. (2010). Appetitive and aversive visual learning in freely moving *Drosophila*. *Front. Behav. Neurosci.* **4**, 10.

Schroll, C., Riemensperger, T., Bucher, D., Ehmer, J., Voller, T., Erbguth, K., Gerber, B., Hendel, T., Nagel, G., Buchner, E., and Fiala, A. (2006). Light-induced activation of distinct modulatory neurons triggers appetitive or aversive learning in *Drosophila* larvae. *Curr. Biol.* **16**, 1741–1747.

Schwaerzel, M., Heisenberg, M., and Zars, T. (2002). Extinction antagonizes olfactory memory at the sub-cellular level. *Neuron* **35**, 951–960.

Schwaerzel, M., Jaeckel, A., and Mueller, U. (2007). Signaling at A-kinase anchoring proteins organizes anesthesia-sensitive memory in *Drosophila*. *J. Neurosci.* **27**, 1229–1233.

Schwaerzel, M., Monastirioti, M., Scholz, H., Friggi-Grelin, F., Birman, S., and Heisenberg, M. (2003). Dopamine and octopamine differentiate between aversive and appetitive olfactory memories in *Drosophila*. *J. Neurosci.* **23**, 10495–10502.

Serway, C.N., Kaufman, R.R., Strauss, R., and de Belle, J.S. (2009). Mushroom bodies enhance initial motor activity in *Drosophila*. *J. Neurogenet.* **23**, 173–184.

Shuai, Y., Lu, B., Hu, Y., Wang, L., Sun, K., and Zhong, Y. (2010). Forgetting is regulated through *Rac* activity in *Drosophila*. *Cell* **140**, 579–589.

Sinakevitch, I., and Strausfeld, N.J. (2006). Comparison of octopamine-like immunoreactivity in the brains of the fruit fly and blow fly. *J. Comp. Neurol.* **494**, 460–475.

Sitaraman, D., and Zars, T. (2010). Lack of prediction for high-temperature exposures enhances *Drosophila* place learning. *J. Exp. Biol.* **213**, 4018–4022.

Sitaraman, D., Zars, M., Laferriere, H., Chen, Y.C., Sable-Smith, A., Kitamoto, T., Rottinghaus, G.E., and Zars, T. (2008). Serotonin is necessary for place memory in *Drosophila*. *Proc. Natl. Acad. Sci. U S A* **105**, 5579–5584.

Sitaraman, D., Zars, M., and Zars, T. (2007). Reinforcement pre-exposure enhances spatial memory formation in *Drosophila*. *J. Comp. Physiol. [A]* **193**, 903–908.

Sitaraman, D., Zars, M., and Zars, T. (2010). Place memory formation in *Drosophila* is independent of proper octopamine signaling. *J. Comp. Physiol. [A]* **196**, 299–305.

Skinner, B.F. (1950). Are theories of learning necessary? *Psychol. Rev.* **57**, 193–216.

Skoulakis, E.M., and Davis, R.L. (1996). Olfactory learning deficits in mutants for *leonardo*, a *Drosophila* gene encoding a 14-3-3 protein. *Neuron* **17**, 931–944.

Skoulakis, E.M., Kalderon, D., and Davis, R.L. (1993). Preferential expression in mushroom bodies of the catalytic subunit of protein kinase A and its role in learning and memory. *Neuron* **11**, 197–208.

Spatz, H.C., Emanns, A., and Reichert, H. (1974). Associative learning of *Drosophila melanogaster*. *Nature* **248**, 359–361.

Strausfeld, N.J. (1976). *Atlas of an Insect Brain*. Springer-Verlag, Berlin, Heidelberg, New York.

Strausfeld, N.J., Hansen, L., Li, Y., Gomez, R.S., and Ito, K. (1998). Evolution, discovery, and interpretations of arthropod mushroom bodies. *Learn. Mem.* **5**, 11–37.

Tamura, T., Horiuchi, D., Chen, Y.C., Sone, M., Miyashita, T., Saitoe, M., Yoshimura, N., Chiang, A. S., and Okazawa, H. (2010). *Drosophila PQBP1* regulates learning acquisition at projection neurons in aversive olfactory conditioning. *J. Neurosci.* **30**, 14091–14101.

Tan, Y., Yu, D., Pletting, J., and Davis, R.L. (2010). *Gilgamesh* is required for *rutabaga*-independent olfactory learning in *Drosophila*. *Neuron* **67**, 810–820.

Tanaka, N.K., Tanimoto, H., and Ito, K. (2008). Neuronal assemblies of the *Drosophila* mushroom body. *J. Comp. Neurol.* **508**, 711–755.

Tang, S., Wolf, R., Xu, S., and Heisenberg, M. (2004). Visual pattern recognition in *Drosophila* is invariant for retinal position. *Science* **305**, 1020–1022.

Technau, G., and Heisenberg, M. (1982). Neural reorganization during metamorphosis of the corpora pedunculata in *Drosophila melanogaster*. *Nature* **295**, 405–407.

Tempel, B.L., Bonini, N., Dawson, D.R., and Quinn, W.G. (1983). Reward learning in normal and mutant *Drosophila*. *Proc. Natl. Acad. Sci. U S A* **80**, 1482–1486.

Tomchik, S.M., and Davis, R.L. (2009). Dynamics of learning-related cAMP signaling and stimulus integration in the *Drosophila* olfactory pathway. *Neuron* **64**, 510–521.

Tully, T., and Quinn, W.G. (1985). Classical conditioning and retention in normal and mutant *Drosophila melanogaster*. *J. Comp. Physiol. [A]* **157**, 263–277.

Tully, T., Preat, T., Boynton, S.C., and Del Vecchio, M. (1994). Genetic dissection of consolidated memory in *Drosophila*. *Cell* **79**, 35–47.

Unoki, S., Matsumoto, Y., and Mizunami, M. (2006). Roles of octopaminergic and dopaminergic neurons in mediating reward and punishment signals in insect visual learning. *Eur. J. Neurosci.* **24**, 2031–2038.

van Swinderen, B., McCartney, A., Kauffman, S., Flores, K., Agrawal, K., Wagner, J., and Paulk, A. (2009). Shared visual attention and memory systems in the *Drosophila* brain. *PLoS ONE* **4**, e5989.

Waddell, S., Armstrong, J.D., Kitamoto, T., Kaiser, K., and Quinn, W.G. (2000). The *amnesiac* gene product is expressed in two neurons in the *Drosophila* brain that are critical for memory. *Cell* **103**, 805–813.

Wang, Z., Pan, Y., Li, W., Jiang, H., Chatzimanolis, L., Chang, J., Gong, Z., and Liu, L. (2008). Visual pattern memory requires *foraging* function in the central complex of *Drosophila*. *Learn. Mem.* **15**, 133–142.

Wolf, R., Wittig, T., Liu, L., Wustmann, G., Eyding, D., and Heisenberg, M. (1998). *Drosophila* mushroom bodies are dispensible for visual, tactile, and motor learning. *Learn. Mem.* **5**, 166–178.

Wu, C.L., Xia, S., Fu, T.F., Wang, H., Chen, Y.H., Leong, D., Chiang, A.S., and Tully, T. (2007). Specific requirement of NMDA receptors for long-term memory consolidation in *Drosophila* ellipsoid body. *Nat. Neurosci.* **10**, 1578–1586.

Wustmann, G., and Heisenberg, M. (1997). Behavioral manipulation of retrieval in a spatial memory task for *Drosophila melanogaster*. *Learn. Mem.* **4**, 328–336.

Wustmann, G., Rein, K., Wolf, R., and Heisenberg, M. (1996). A new paradigm for operant conditioning of *Drosophila melanogaster*. *J. Comp. Physiol. [A]* **179**, 429–436.

Xia, S., Miyashita, T., Fu, T.F., Lin, W.Y., Wu, C.L., Pyzocha, L., Lin, I.R., Saitoe, M., Tully, T., and Chiang, A.S. (2005). NMDA receptors mediate olfactory learning and memory in *Drosophila*. *Curr. Biol.* **15**, 603–615.

Yarali, A., and Gerber, B. (2010). A neurogenetic dissociation between punishment-, reward-, and relief-learning in *Drosophila*. *Front. Behav. Neurosci.* **4**, 189.

Yarali, A., Krischke, M., Michels, B., Saumweber, T., Mueller, M.J., and Gerber, B. (2009). Genetic distortion of the balance between punishment and relief learning in *Drosophila*. *J. Neurogenet.* **23**, 235–247.

Young, J.M., and Armstrong, J.D. (2010a). Building the central complex in *Drosophila*: the generation and development of distinct neural subsets. *J. Comp. Neurol.* **518**, 1525–1541.

Young, J.M., and Armstrong, J.D. (2010b). Structure of the adult central complex in *Drosophila*: organization of distinct neuronal subsets. *J. Comp. Neurol.* **518**, 1500–1524.

Zars, T. (2000). Behavioral functions of the insect mushroom bodies. *Curr. Opin. Neurobiol.* **10**, 790–795.

Zars, T. (2001). Two thermosensors in *Drosophila* have different behavioral functions. *J. Comp. Physiol. [A]* **187**, 235–242.

Zars, T. (2009). Spatial orientation in *Drosophila*. *J. Neurogenet.* **23**, 104–110.

Zars, T. (2010a). Short-term memories in *Drosophila* are governed by general and specific genetic systems. *Learn. Mem.* **17**, 246–251.

Zars, T. (2010b). Visualizing PKA activity in a learning center. *Neuron* **65**, 442–444.

Zars, T. (2011). Flies race to a safe place. *Nature* **474**, 169–170.

Zars, M., and Zars, T. (2006). High and low temperatures have unequal reinforcing properties in *Drosophila* spatial learning. *J. Comp. Physiol. [A]* **192**, 727–735.

Zars, M., and Zars, T. (2009). Rapid matching in *Drosophila* place learning. *Naturwissenschaften* **96**, 927–931.

Zars, T., Fischer, M., Schulz, R., and Heisenberg, M. (2000a). Localization of a short-term memory in *Drosophila*. *Science* **288**, 672–675.

Zars, T., Wolf, R., Davis, R., and Heisenberg, M. (2000b). Tissue-specific expression of a type I adenylyl cyclase rescues the *rutabaga* mutant memory defect: In search of the engram. *Learn. Mem.* **7**, 18–31.

Zhang, K., Guo, J.Z., Peng, Y., Xi, W., and Guo, A. (2007). Dopamine-mushroom body circuit regulates saliency-based decision-making in *Drosophila*. *Science* **316**, 1901–1904.

Zhong, Y. (1995). Mediation of PACAP-like neuropeptide transmission by coactivation of Ras/Raf and cAMP signal transduction pathways in *Drosophila*. *Nature* **375**, 588–592.

STUDYING SENSORIMOTOR PROCESSING WITH PHYSIOLOGY IN BEHAVING *DROSOPHILA*

Johannes D. Seelig and Vivek Jayaraman

Howard Hughes Medical Institute, Janelia Farm Research Campus, 19700 Helix Drive, Ashburn, VA 20147, USA

The neural underpinnings of sensorimotor integration are best studied in the context of well-characterized behavior. A rich trove of *Drosophila* behavioral genetics research offers a variety of well-studied behaviors and candidate brain regions that can form the bases of such studies. The development of tools to perform *in vivo* physiology from the *Drosophila* brain has made it possible to monitor activity in defined neurons in response to sensory stimuli. More recently still, it has become possible to perform recordings from identified neurons in the brain of head-fixed flies during walking or flight behaviors. In this chapter, we discuss how experiments that simultaneously monitor behavior and physiology in *Drosophila* can be combined with other techniques to produce testable models of sensorimotor circuit function.

I. Motivation

Although this may not be the first thought in your mind when it happens, the humble fly that finds and settles on your artfully constructed fruit salad must perform

INTERNATIONAL REVIEW OF NEUROBIOLOGY, VOL. 99
DOI: 10.1016/B978-0-12-387003-2.00007-0

several sensorimotor feats in doing so. First, its olfactory receptors must detect the whiff of odors from the salad while it is in mid-flight (Budick and Dickinson, 2006). These receptors must then communicate the information to olfactory centers in the fly brain. Neural circuits in these centers must process the incoming information and quickly extract relevant features of the odor, such as its quality and intensity. Its brain must initiate actions to move toward or away from the source based on hard-wired impulses and past positive or negative associations with the odor. For the neural circuits involved in odor-guided orientation, this is an impressive accomplishment for several reasons. The odor was detected at a particular point in space and time by sensors located on the fly's antennae. Localizing the odor source based on information from moving sensors (the fly is flying and the position of the sensors relative to the odor source thus varies constantly) is nontrivial. If the odor is no longer present, the fly must search for it in its surroundings (Budick and Dickinson, 2006)—such active sampling requires sensorimotor circuits to incorporate positional information of the sensors, something that the brain controls through motor commands to the wings or antennae. Even the success or failure of a flight maneuver in producing the desired movement is something that requires visual and mechanosensory feedback to evaluate—for example, the same muscle output could produce a different change in the fly's flight trajectory if there were a breeze blowing. The difference between the predicted and actual position of the fly for the motor command issued must be combined with visual and olfactory information that indicates the presence of the food source, to issue new motor commands that eventually steer the fly toward the fruit salad (Frye, 2010). More generally, no matter what behavioral strategies flies use to locate food, maneuver around obstacles, or find a mate, sensorimotor integration by its little brain is what enables their execution.

A. Basic Questions and Challenges in Sensorimotor Processing

The basic computational problems underlying sensorimotor integration are the same across many species (Buschges and El Manira, 1998; Pouget and Snyder, 2000; Samuel and Sengupta, 2005; Sommer and Wurtz, 2008; Wolpert and Ghahramani, 2000), even if the solutions may be diverse. How does the nervous system extract behaviorally relevant features from sensory input to produce motor output (Lisberger, 2010)? How is information from different sensory organs integrated to resolve perceptual ambiguities that information from a single sensor may sometimes present (Frye, 2010; Green and Angelaki, 2010)? How is proprioceptive sensory information about current posture and limb position incorporated into sensorimotor processing to correctly hit motor targets (Laurent, 1991)? How can an animal avoid responding to sensory input that is generated from its own actions, for example, when singing (Crapse and Sommer, 2008; Poulet and Hedwig, 2007)? How is sensorimotor integration modulated by internal factors, such as hunger or behavioral state (Chiappe *et al.*, 2010; Maimon *et al.*, 2010), and external ones, such as the time of day (Pfeiffer and Homberg, 2007)? How are associations between sensory stimuli and motor outputs modulated by experience (Lalazar and Vaadia, 2008)?

Although elegant theoretical frameworks have been proposed that capture these facets of sensorimotor integration (Pouget and Snyder, 2000; Todorov, 2004; Wolpert and Ghahramani, 2000), it has been more difficult to identify the neural substrates and circuit mechanisms that underlie such computations. A major challenge is that while it is possible to record and interpret neural activity close to the sensory or motor periphery (Buschges *et al.*, 2008), where the inputs or outputs of a circuit are well understood, this is usually insufficient to understand the computations underlying sensorimotor integration. Centers that perform integrative functions often (Ritzmann and Buschges, 2007) have multiple inputs and complicated recurrent and feedback connections. It is difficult to conclusively establish such regions as being involved in a behavior and harder still to understand their computational roles.

A now traditional and powerful approach in such a situation is that of neuroethology (Bullock, 1999; Heiligenberg, 1991a). This approach takes the view that a circuit's computations are best understood in the context of an animal's natural behavior. Thus, neural circuits are assumed to extract and represent features of sensory input and motor output that enable the behavior. Indeed, neuroethology has led to insights into sensorimotor computations in systems as diverse as barn owls (Konishi, 2003) and weakly electric fish (Heiligenberg, 1991b). More generally, this history perhaps underscores the importance of carefully characterized behavior, whether natural or more artificial, to understand sensorimotor processing in neural circuits.

If the sensory input is well controlled, the behavior quantitatively characterized, and a specific circuit implicated in a particular sensorimotor operation, another major challenge remains: to reach a mechanistic understanding of the observed sensorimotor phenomenon. This is perhaps ideally achieved in a system in which it is possible to identify the important circuit components, map their interconnectivity, and selectively monitor and manipulate their activity in the context of a specific behavior. These are some of the key advantages of using a genetic model organism with a relatively small brain, such as *Drosophila melanogaster* (Olsen and Wilson, 2008; Simpson, 2009), in studies of sensorimotor integration.

B. Behaviors to Probe Sensorimotor Integration in *Drosophila*

Almost all behaviors that involve movement can be said to require sensorimotor integration. Even a seemingly simple reflex such as, for example, a fly's jump to escape a looming stimulus, involves more than a simple feed-forward process. The fly, in fact, makes a series of postural adjustments dependent on its own initial posture and the direction of approach of the looming object (Card and Dickinson, 2008).

Regardless of whether the specific behavior under investigation is natural (as is true for ethologically grounded studies) or more artificial, studies of sensorimotor integration require that the stimulus be well characterized, and the motor output carefully monitored. Quantitative descriptions of both can permit meaningful conclusions to be drawn about the underlying algorithms involved in the transformation.

In Table I, we list some well-explored *Drosophila* behaviors that seem well suited to enable an understanding of features of sensorimotor integration and have been shown to work in tethered preparations, making them ideal candidates for physiological recordings.

Table I
Tethered Fly Behaviors Suitable for Physiological Exploration.

Behaviors in tethered flies	Examples	Comments
Visually guided walking and flight behavior	Open- and closed-loop behavior, such as optomotor behavior in flying or walking flies (Borst *et al.*, 2010; Buchner, 1976; Gotz, 1968; Gotz and Wenking, 1973) or closed-loop stripe fixation in flying flies (Wehrhahn and Reichardt, 1973)	Tethered walking and flying preparations could be extended to study: gait changes (Strauss, 1995), influence of circadian rhythms (Rosato and Kyriacou, 2006), onset/offset of walking (Trimarchi and Schneiderman, 1995)/flight (Card and Dickinson, 2008; Fotowat *et al.*, 2009; Maimon *et al.*, 2008) (when flies escape or are startled for example)
Odor-guided walking and flight behavior	Osmotropotaxis	Flies walking on a ball react to changes in concentration gradients (Borst and Heisenberg, 1982)
	Tethered flying flies orienting to an odor source	This tracking behavior requires visual stimuli, permitting investigations of multisensory integration (Duistermars and Frye, 2008)
Motor sequences	Courtship	Wing extension and tracking behavior can be induced in tethered male flies walking on a ball (Kohatsu *et al.*, 2011)
	Proboscis extension	It is possible to perform physiology in the brain along with proboscis muscle recordings, which can provide a proxy for proboscis extension (Gordon and Scott, 2009; Haehnel *et al.*, 2009)
Learning and memory	Visual memory	Tethered flying flies can be conditioned to remember a visual pattern (Dill *et al.*, 1993)

This table is drawn with permission from a more exhaustive list in Simpson (2009).

C. Identifying Behaviorally Relevant Brain Regions for Physiological Experiments

Powerful techniques to selectively manipulate subpopulations of neurons are increasingly being applied in behavioral genetics screens to identify neurons and circuits involved in a variety of *Drosophila* behaviors (Simpson, 2009). With rapid improvements in the specificity of Gal4 drivers and in the efficacy of effectors (Pfeiffer *et al.*, 2010), studies using such techniques provide useful maps for physiologists interested in understanding the neural basis of a specific behavior. It should be noted that a Gal4 line implicated in such a screen rarely identifies the circuit computations responsible for a particular behavior. However, in cases where an implicated Gal4 line is specific, the neurons identified by the screen can provide a hook into an interesting circuit.

Older methods to identify neurons active during, and, by implication, possibly involved in, a particular behavior, such as (3H)-2-deoxyglucose labeling (Bausenwein *et al.*, 1994), have their caveats. Improved genetics-based successors to such techniques are not yet available in *Drosophila*. Immediate early gene expression has, however, been used with some success in rodents to identify regions activated during particular behaviors (Yassin *et al.*, 2010).

II. A Brief History of Physiology in the Adult *Drosophila* Brain

Physiologists have targeted the brains of larger flies, for example, *Calliphora*, for several decades now, but recording from the *Drosophila* brain presented problems. The techniques commonly used in larger insects, primarily sharp intracellular recordings and extracellular recordings with tungsten wires and silicon probes, were not successful in the smaller fly. The presence of an almost invisible—and thus hard-to-remove—sheath around the brain, and the small size of most of its neurons made such recordings difficult to perform in *Drosophila*. For similar reasons, loading cells with synthetic dyes for calcium imaging was also challenging.

The first breakthrough for physiologists seeking to work in *Drosophila* came with the advent of genetically encoded indicators. Being able to "load" such sensors directly into identified groups of neurons in the brain greatly reduced the barriers to *in vivo* imaging (Fiala *et al.*, 2002; Ng *et al.*, 2002; Wang *et al.*, 2003). A couple of years later, another technical advance made it possible to perform targeted *in vivo* electrophysiological recordings from identified neurons in the fly brain as well (Wilson *et al.*, 2004). The key here was to immobilize and desheath the brain, and to then use the whole-cell patch clamp technique to get stable, low-access-resistance recordings from the small *Drosophila* neurons.

Both techniques have their advantages. Calcium imaging does not require removing the sheath, which simplifies the dissection and makes it less invasive. Patch clamp recordings have subthreshold sensitivity and high temporal resolution. Calcium imaging gives access to signals in dendrites and neurons that are difficult to access with an electrode. Additionally imaging allows simultaneous recordings from multiple cells, and this can be used to perform more exploratory and correlational searches in the brain (Orger *et al.*, 2008). The two techniques are complementary, and although imaging is the easier technique to use, the data it produces are hard to interpret as a proxy of neural activity without calibration with electrical recordings (Jayaraman and Laurent, 2007; Reiff *et al.*, 2005). Early calcium indicators were found to not capture low levels of spiking, making neural activity seem sparser than it actually was (Hendel *et al.*, 2008; Jayaraman and Laurent, 2007; Reiff *et al.*, 2005). More recent indicators, however, have been promising (Mank *et al.*, 2008; Tian *et al.*, 2009), and their improving sensitivity and kinetics suggest a bright future for imaging.

III. Physiology in Behaving Flies

A. The Importance of Recording Neural Activity During Behavior

For studies of peripheral sensory regions, it is often convenient to perform physiological recordings in reduced and invasive preparations that permit easier optical or electrical access to neurons of interest. The possible assumption here is that such peripheral brain areas are likely to be driven largely by unimodal input (e.g., the antennal lobe by olfactory receptors) and lateral interactions rather than feedback from higher areas (although see, for example, Dacks *et al.* (2009)). They are thus likely to be less affected than deeper regions by the possibly compromised state of the brain/animal (although see Chiappe *et al.* (2010), Maimon *et al.* (2010)). The advantages that such reduced preparations confer are particularly compelling if the questions being addressed with them are explicitly related to neural processing at the earliest synapses of a pathway. The calculation is, however, dramatically different further away from the periphery and particularly in the context of sensorimotor processing.

First, as discussed in a previous section, activity in "deeper" circuits is perhaps best interpreted in the context of behavior. Although this does not require the behavior to be monitored simultaneously with physiology, there are obvious advantages in doing so. Simultaneous recordings of behavior and physiology offer experimenters the opportunity to directly link the two. Trial-to-trial variability (Lisberger, 2010) is often averaged over as noise, but in the event that physiological variability is predicted by changes in behavioral output or *vice versa*, the causes and consequences of such variability can be better understood and potentially assigned greater significance. Second, feedback is a prominent feature of most brains, and even early sensory responses can be affected by the animal's internal and behavioral state (Chiappe *et al.*, 2010; Maimon *et al.*, 2010). Third, the fact that a neuron responds to a stimulus does not automatically imply that the neuron is involved in enabling the behavior that the stimulus triggers in the animal. Monitoring the behavior simultaneously while recording and perturbing activity in a neuron or neural population can help establish such causal links. Fourth, a crucial aspect of sensorimotor integration is in understanding how the nervous system copes with sensory consequences of its own actions. For a freely walking or flying fly, any translation or rotation results in to retinal image shifts in the opposite direction. This can be simulated and manipulated in a tethered, head-fixed fly in an artificial closed-loop experiment. The fly's responses are determined by monitoring movements of an air-supported ball (Buchner, 1976) that the fly walks on (Bulthoff, 1982), or by the changes in flight detected by a wing beat analyzer (Gotz, 1987). These signals are then used as feedback for a visual display to close the loop for the fly.

There are numerous examples in larger insects of experiments featuring simultaneous physiology and behavior recordings (in both open- and closed-loop). Neural activity associated with sensory stimulation and locomotor activity has been studied in crickets (Bohm and Schildberger, 1992; Zorovic and Hedwig, 2011) and hearing flies (Ormia) (Mason *et al.*, 2001) during walking on a ball, and locust neurons have been recorded from during flight (Homberg, 1994).

B. Requirements for Recording in Behaving *Drosophila*

The small size of the fruit fly sets somewhat conflicting demands on a preparation that seeks to allow both physiology and behavior. On the one hand, the fly has to be able to move its legs or wings freely during recordings to perform walking and flight behavior. In addition, depending on the behavior under investigation, the eyes and antennae may need to be freely accessible for visual and olfactory stimulation. On the other hand, the brain has to be conveniently accessible for dissection with forceps and for recordings of neuronal activity with a water immersion objective and patch electrodes. These requirements are met in a similar fashion in the flying (Maimon *et al.*, 2010) and walking (Seelig *et al.*, 2010) preparations by separating a part of the head of the fly from the rest of the body with a membrane shaped appropriately for enabling walking or flight as schematically shown in Fig. 1A for flying and Fig. 1B for walking flies. The fly can perform flying

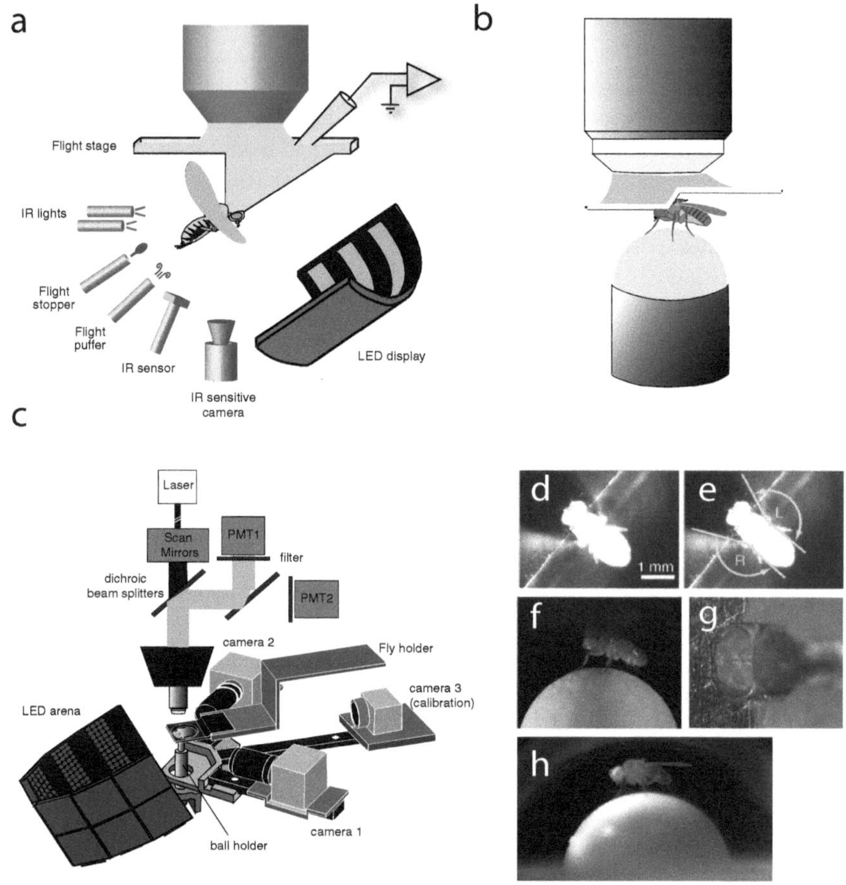

FIG. 1. Setup for physiological recordings in flying and walking flies. (A) Schematics of a fly inserted into a holder for patch clamp physiology. The setup includes an light-emitting diode (LED) display for visual stimulation, infrared (IR) lights for detecting the wing beat frequency and direction and the patch electrode. Flight can be initiated with an air puff (flight puffer) and interrupted with a piece of tissue (flight stopper). (B) Schematics of a fly mounted under the microscope and walking on an air-supported ball. (C) Schematics of the setup for two-photon calcium imaging in the walking fly, including an LED arena, two cameras for tracking the movement of the ball (camera 1 and 2), and a camera for positioning the fly on the ball and for calibrating the tracking system (camera 3). (D) A fly flying in the holder for patch clamp recordings. (E) Recording the wing beat at 100 Hz and integrating over several wing strokes allows the wings' edge to be detected. (F) A fly walking on an air-supported ball. (G) The dissected brain of the fly in the walking holder. (H) A fly walking freely on a ball is used as a guideline for positioning the fly in the walking setup. Flying fly figures (A, D, E) used with permission from (Maimon *et al.*, 2010). Walking fly figures (B, C, F, G) are used with permission from (Seelig *et al.*, 2010). (For color version of this figure, the reader is referred to the web version of this book.)

(Fig. 1D, E) or walking behavior (Fig. 1F). The side of the head that is shielded from the rest of the body by the membrane can then be immersed in saline and dissected for recordings (shown for a walking fly in Fig. 1G).

1. *Recording in Tethered Flying Flies*

Flight behavior requires space for wing movement around the head of the fly. Maimon *et al.* (2010) designed a pyramidal or wedge-shaped structure centered on the head of the fly that allows the fly to freely move its wings while at the same time allowing access to the brain (Fig. 1A) with an electrode. The fly holder is milled out of Delrin plastic using a computer-controlled mill. The fly is mounted in this holder after removing the proto- and mesothoracic legs or all six legs. The proboscis is fixed to prevent brain movement and the cuticle is removed to allow access to the brain.

A potential concern for any behaving animal preparation is the invasiveness of the dissection, which can interfere with natural behavior. This is particularly true for patch clamp recordings in fruit flies where the sheath that encloses the brain (which serves similar functions to the blood brain barrier in vertebrates) has to be removed. Both, mechanical damage to the brain surface as well as changes in the brain physiology could affect the behavior of the fly. Additionally, superficial cells such as the lobula plate tangential cells (LPTCs) can easily be damaged during the desheathing procedure. Maimon *et al.* (2010) developed a minimally invasive method for dissolving the sheath around the region of interest using locally applied collagenase combined with mechanical manipulations. This procedure removes the sheath only above the cells of interest and without causing visible damage to the underlying cells. Flies were able to fly and perform optomotor behavior in this setup, as shown in Fig. 2A, B.

Maimon *et al.* (2010) focused on comparing recordings from LPTCs during flight and nonflight behavior. They also developed a method to monitor a wing beat signal that provides data about the flies' turning response similar in quality to the data obtained with more traditional wing beat analyzers used in tethered flight behavior experiments (Maimon *et al.*, 2010). The wings are illuminated with infrared light and a camera records the wings of the fly from below at a frame rate of 100 Hz and an integration time of close to 10 ms (Fig. 1D), below the wing beat frequency of *Drosophila* (180–250 Hz), and therefore integrating more than one wing stroke per frame. The edge of the shape traced out by both wings during each frame is detected and is used as a parameter for quantifying the turning behavior of the fly (Fig. 1E). Changes in flight behavior induced with an either upward or downward moving grating could be detected in this way during patch clamp recordings. This optomotor response was found to be consistent with the behavior observed in tethered flies (Fig. 2A).

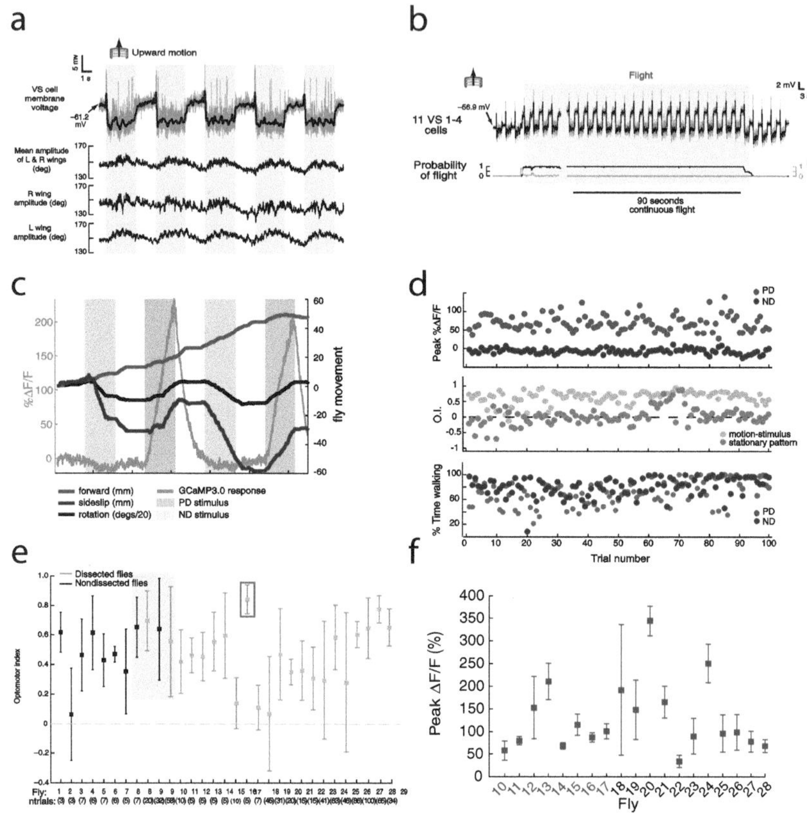

FIG. 2. Behavior during physiological recordings. (A) Simultaneously acquired wing stroke amplitude and whole-cell patch clamp recording from a fly during optomotor behavior. The black membrane voltage trace displays a low pass–filtered version of the gray trace (fourth-order Butterworth, 25-Hz cutoff). (B) Mean responses to repeated presentations of upward motion. The flies flew continuously without air puff stimulation for at least 90 s. (Downward glitches in the probability of flight trace during continuous flight are due to occasional errors in the flight detection algorithm.) (C) Horizontal system (HS)-neuron calcium changes in fluorescence and simultaneously recorded behavioral response (forward, rotational, and sideslip movement) during preferred direction (PD) stimulation, null direction (ND) stimulation, and no stimulation. (Rotation is shown in degs/20.) (D) Example of Fly 27 walking performance during 100 trials of PD (red) or ND (blue) stimulation. (a) Calcium response. (b) Optomotor index (O.I.); green: response during visual stimulation, orange: response during stationary pattern. (c) Percentage of walking per trial. (E) Optomotor indices (mean ± S.D.; *n* values are indicated in the figure) of flies nondissected and dissected flies walking in the holder. The gray box highlights optomotor indices of two flies before and after dissection. The trial highlighted in magenta is shown in Fig. 3. The dashed line corresponds to an optomotor index of 0, which indicates the absence of an optomotor response. (F) Difference between peaks of PD and ND relative fluorescence change for the flies performing optomotor behavior with corresponding numbers in (E). Figures A, B used with permission from (Maimon *et al.*, 2010). Figures C–F used with permission from (Seelig *et al.*, 2010). (For interpretation of the references to color in this figure legend, the reader is referred to the web version of this book.)

2. *Recording in Tethered Walking Flies*

Walking is less restrictive than flight with respect to the geometry of the fly holder as leg movement requires less space around the fly's head than wing movement. The considerations that entered the design of the fly holder (Fig. 1B) for walking (Seelig *et al.*, 2010) are ease of access to the brain (see Fig. 1G), minimal obstruction of the laser focus required for two-photon imaging, and the option to easily mount the head in different orientations, while leaving sufficient space for the legs to move freely.

The fly holder used in the walking preparation is cut from a stainless steel shim. The fly is positioned with respect to the shim using a micromanipulator. The mounting of the fly is monitored using cameras to make it possible to carefully control the positioning of the fly's head. The fly is fixed using high-viscosity UV glue applied locally on parts of the fly and the holder. The fly walks on an air-supported ball (Fig. 1C), a behavioral paradigm introduced for *Drosophila* by Buchner (1976), in a posture similar to that observed in flies freely walking on a ball (Fig. 1H). The rotation of the ball is monitored at high resolution and around all three axes of rotation using a custom-designed tracking system based on an optical mouse sensor. The setup allows the ball's rotation to be tracked at high spatial and temporal resolution (Fig. 1C).

There is a range of balls on which flies can perform walking behavior (Seelig *et al.*, 2010). Buchner originally reported using a Styrofoam ball of 9 mm diameter and a weight of about 10 mg (Buchner, 1976). We noticed that the flies would easily lift these low weight balls from its holder. We observed reliable optomotor behavior using 6 mm plastic balls with a weight of even 120 mg. We used a ball of 6 mm diameter and a weight of 40 mg, which increased the moment of inertia by about 80% compared with the ball used by Buchner, but resulted in smooth and reliable walking behavior both for walking tethered flies and flies mounted in the physiology holder.

Kohatsu *et al.* have more recently developed a walking preparation based on a polyethylene film that was glued to the fly. The head capsule was then cut open together with the film. The fly walked on an air-supported ball with a diameter of 5 mm (Kohatsu *et al.*, 2011).

C. Calcium Imaging and Electrophysiology During Behavior

Several factors make physiology in fruit flies challenging. Apart from the small dimension of the preparation and the small size of the neurons, the movement of the fly's brain complicates physiological recordings. Brain movement is a well-known problem for all imaging or patch clamp preparations in behaving animals, and reduces the quality, length, and stability of physiological recordings. Two-photon imaging requires even greater brain stability than patch clamp recordings because data are lost and artifacts introduced as soon as the structure of interest moves along the optical axis of the microscope. The mechanical stability provided by a high-resistance seal of a glass electrode on a cell makes whole-cell patch clamp electrophysiology somewhat less sensitive to brain movement.

Brain movement in *Drosophila* is caused mainly by the movement of muscles in its head, and by the fly extending its proboscis. Brain movement in behaving preparations can be minimized by fixing the proboscis and removing muscles as has been described previously (Jayaraman and Laurent, 2007). Additionally, depending on the brain regions of interest, air sacs, trachea, and fat may have to be removed for good optical and physiological access.

These steps did not prevent flies from performing optomotor flight behavior when stimulated with an upward or downward moving grating (Fig. 2A) or robust optomotor walking behavior when stimulated with a clockwise or counter-clockwise rotating grating. A single trial of a fly performing walking optomotor behavior is shown together with the calcium transients recorded from LPTCs in Fig. 2C. Walking flies perform this behavior for extended periods (Fig. 2D), and average data for multiple trials and multiple flies are shown in Fig. 2E (behavior) and Fig. 2F (simultaneously recorded changes in calcium signal) (Seelig *et al.*, 2010).

An alternative approach was used by Kohatsu *et al.* (2011) who covered the brain with transparent silicone gel. This has also been shown to be helpful in reducing brain movement in other preparations (Dombeck *et al.*, 2007).

For optical recordings during visual stimulation the fluorescence detectors have to be shielded from light emitted by the visual display. One method to achieve this is to spectrally separate the wavelengths of visual excitation and fluorescence by using appropriate filters. The visual sensitivity of flies is limited to the UV and green part of the spectrum, so that two-photon stimulation does not interfere with the presentation of visual patterns. Further, the intensity of visible light generated by two-photon-excited fluorescence did not affect walking behavior. An alternate technique to prevent visual cross talk is to temporally alternate visual stimulation and fluorescence detection (Reiff *et al.*, 2010).

Note that visible light has, however, been used with success for fluorescence imaging of responses in the male fly's brain to the tarsal touch of a female fly's abdomen or cuticular extract during walking behavior (Kohatsu *et al.*, 2011).

D. Comparison Between Tethered Behavior and Behavior During Physiological Recording

A good test of the quality of a behaving preparation is a comparison of the fly's behavior during stable, high-quality recordings with its tethered behavior (or with the equivalent behavior of freely walking or flying flies).

Flying flies have a rich repertoire of visually guided tethered behaviors that have been studied and characterized in depth. These behaviors provide a means to assess the quality of behavior elicited during physiological recordings.

Maimon *et al.* (2010) observed optomotor behavior during recordings similar to behavior in tethered flies. However, they observed some variability in the readiness of flies to fly. They enhanced flight activity by puffing air at the fly at a rate of about 0.2 Hz. Several flies also showed longer continuous flight bouts of several tens of seconds.

Detailed statistics for tethered walking flies are not as readily available as for flying flies. Behavior in flies walking in a tethered preparation has thus far focused mostly on basic visual paradigms, such as optomotor behavior. During two-photon calcium imaging experiments, flies could perform this behavior robustly for up to several hours. We compared the optomotor walking behavior of flies tethered in the holder before and after dissection and found no significant difference in their behavior as described with an optomotor index that quantifies the tendency of the fly to walk in the direction of motion of the visual stimulus (Seelig *et al.*, 2010). Further explorations of visually guided tethered walking behaviors are necessary to establish the quality of more complex walking behavior in the physiology configuration. It should be noted that the quality of behavior also varies substantially depending on the genotype of the fly.

E. Relating Tethered Behavior to Free Walking and Flight Behavior

Ultimately, one would like to relate the behavior in a physiology preparation to the behavior of freely walking or flying flies. Optomotor behaviors as observed in the tethered physiology preparations have also been observed in freely flying and walking flies (Katsov and Clandinin, 2008; Mronz and Lehmann, 2008) and the robustness of the tethered behavior suggest that the visual stimulus is relevant to the animal even in the highly artificial recording situation. Nevertheless, there are obvious differences between freely behaving flies and flies in the physiology preparations, for example, in the proprioceptive feedback or the absence of head movements that are prominent in freely behaving flies.

An approximation to the coupling between action and perception that occurs in freely behaving animals has been introduced for tethered flying flies through the closed-loop paradigm (Gotz, 1987; Wehrhahn and Reichardt, 1973) and it will be interesting to see whether this can be combined with physiological recordings.

It is currently not possible to relate the wing beat of a tethered flying fly to its trajectory or dynamics in free space and it would most likely require monitoring wing dynamics at high temporal and spatial resolution. The situation is different for tethered walking flies. Even though the fixed position of the fly relative to the surface on which it walks, and the ball's inertia and curvature, make free walking and tethered walking far from equivalent, a step on the ball can nevertheless be compared to a step in a virtual plane and the dynamics and the trajectory of a tethered fly can be related to the trajectories of freely walking flies. As shown the ball movement (Fig. 3A) is tracked around all three axes of rotation during an optomotor trial (Fig. 3B, C) and can therefore be translated into a virtual trajectory of a fly walking in a plane (Fig. 3C). This opens up several interesting opportunities for experiments that cannot be performed in tethered flying flies. For example, establishing closed-loop behavior in walking flies should make it possible to calibrate closed-loop parameters such as the gain of the feedback loop (which relates the angular displacement of the display to the angular displacement of the fly) and to relate these to equivalent parameters in freely walking flies. Further, converting the trajectories of a tethered fly to a two-dimensional path on a virtual surface should allow navigation in a virtual world (Harvey *et al.*, 2009) and exploration of complex behaviors such as spatial memory (Foucaud *et al.*, 2010; Ofstad *et al.*, 2011; Zars, 2009).

A recent example of a complex sequence of behaviors explored on the ball is courtship behavior, which includes proboscis extension, wing movement, and walking and tracking. Kohatsu *et al.* (2011) have shown that wing extension and tracking behavior can be induced in a tethered fly walking on a ball. They also recorded calcium signals from an important element of the circuit that is thought to underlie courtship behavior during tarsal stimulation. It is as yet unclear how much of the courtship process can be elicited during calcium imaging.

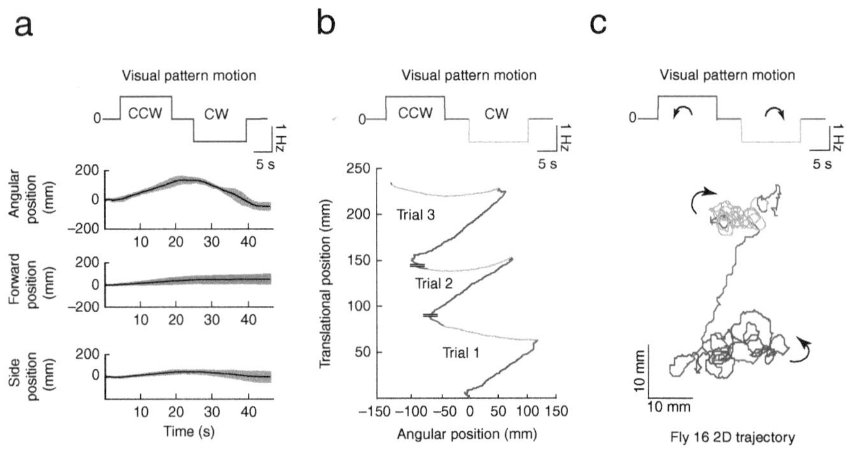

FIG. 3. Reconstructing walking trajectories of flies on a ball. (A) Rotational and translational velocity of a fly in response to clockwise (CW) and counterclockwise (CCW) visual stimuli (plotted as mean ± S. D. in gray ($n = 5$)). (B) Translational and rotational movement illustrates optomotor response of the fly. (C) The path of the fly in a virtual plane can be reconstructed as the full-rotational movement of the ball is recorded during each trial. Figure used with permission from (Seelig *et al.*, 2010). (For color version of this figure, the reader is referred to the web version of this book.)

IV. Results From Recording in Behaving *Drosophila*

Preparations that permit physiological recordings during behavior have only recently become possible. However, even the first studies using such preparations have been informative and indicate the potential of such approaches to elucidate important features of sensorimotor processing.

A. Neural Activity in Early Motion-Vision Neurons Is Modulated by Behavioral State

Flies use optic flow—retinal image shifts that are caused by the relative motion between their eyes and their visual surroundings—to stabilize their walking and flight paths, balance and gaze (Taylor and Krapp, 2007). Vertical system (VS) and horizontal system (HS) neurons of the lobula plate are important elements of the motion-vision pathway in flies. They are necessary for normal optic flow processing and the optomotor response (Hausen and Wehrhahn, 1990; Heisenberg *et al.*, 1978). LPTCs have been thoroughly characterized in flies in fixed configurations (i.e., in configurations that do not permit walking/flight behavior) and respond to motion in their preferred direction with membrane potential depolarizations (Joesch *et al.*, 2008; Schnell *et al.*, 2010).

Maimon *et al.* (2010) recorded from VS neurons in flies during flight (Fig. 4A), and found that these neurons showed an increase in their resting membrane potential, as well as an amplified motion response (Fig. 4B). Chiappe *et al.* (2010) performed two-photon imaging from HS-neurons (Fig. 4D) using GCaMP3.0 in walking *Drosophila*, and found similar boosting of responses to motion in the preferred direction (Fig. 4E). These state-dependent increases in sensitivity may be an energy conservation mechanism for the fly (Niven and Laughlin, 2008).

B. HS-Neurons Change Their Tuning to Speed of Image Motion During Walking

The overall directional tuning preferences of VS neurons were unchanged in flight (Maimon *et al.*, 2010), although both hyperpolarizing and depolarizing responses were amplified (Fig. 4C). HS neurons, however, showed a response to image motion that was correlated with walking speed (Fig. 4F). Further, they showed a change in their tuning to speed of image motion during walking (Chiappe *et al.*, 2010), significantly increasing their responses to higher motion speeds (higher temporal frequencies; Fig. 4G). The peaks of their temporal frequency tuning curves shifted toward higher image speeds during walking. This may represent a mechanism for the fly to only process faster image motion in situations where they are likely to be encountered (i.e., when the fly's own motion increases the retinal image shifts it is likely to encounter) (Warzecha and Egelhaaf, 1996).

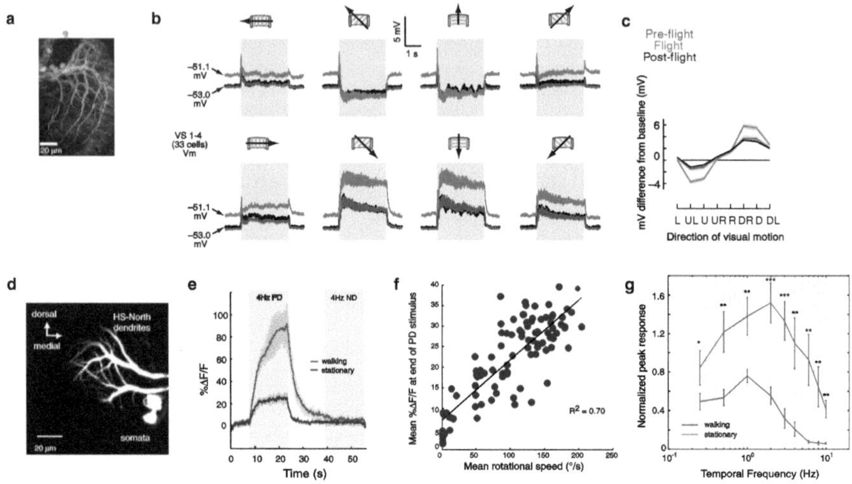

FIG. 4. Behavioral state dependent gain modulation and changes in tuning in *Drosophila* optic lobe neurons. (A) Whole-cell patch clamp recording from a VS1 neuron (*Gal4-3a* driving GFP (green)). Neuron is filled with biocytin (red) during recording. (B) VS1 shows an increase in resting membrane potential during flight and its visual responses are also boosted. (C) VS1 directional tuning curves under different behavioral conditions. Both depolarizing and hyperpolarizing responses of VS1 to motion in different directions are boosted, but overall directional preference remains unchanged. (D) Two-photon calcium imaging from dendrites of an HS-North neuron (*Gal4-R27B03* driving GCaMP3.0). (E) HS-North shows amplified responses to motion in its preferred direction during walking. (F) HS-North responses to the same preferred direction motion increase with increasing walking speed. (G) HS-North temporal frequency (motion speed) tuning curve. Tuning curve shows significant amplification and its peak shifts toward higher motion speeds during walking. Figures A–C used with permission from (Maimon *et al.*, 2010). Figures D–G used with permission from (Chiappe *et al.*, 2010). (For interpretation of the references to color in this figure legend, the reader is referred to the web version of this book.)

Taken together, these results of motor-state-dependent changes in responses and tuning of early visual system neurons suggest that sensorimotor integration begins fairly early in the "sensory" pathway. It is as yet unclear how widespread such gain modulation is in the *Drosophila* nervous system, and whether or not increased sensitivity is also common in other sensory modalities. In general, these results underscore the importance of measuring neural activity under conditions where flies are free to behave (although not yet free to roam).

V. Outlook for the Future

With a variety of different behaviors that involve innate and learned motor responses to different unimodal and multimodal sensory stimuli, a vast genetic toolkit with which to selectively and precisely perturb neurons in the brain, and the physiological tools with which to record from identified neurons in its brain during tethered flight and walking behavior, *D. melanogaster* is an excellent model system for researchers interested in understanding the circuit mechanisms underlying sensorimotor integration. Although only a few simple behaviors have as yet been explored "on the rig," it is likely that many more behaviors can be reproduced under tethered conditions, and then studied with simultaneous physiology. Importantly, with physiology now possible in behaving flies, neurons can be

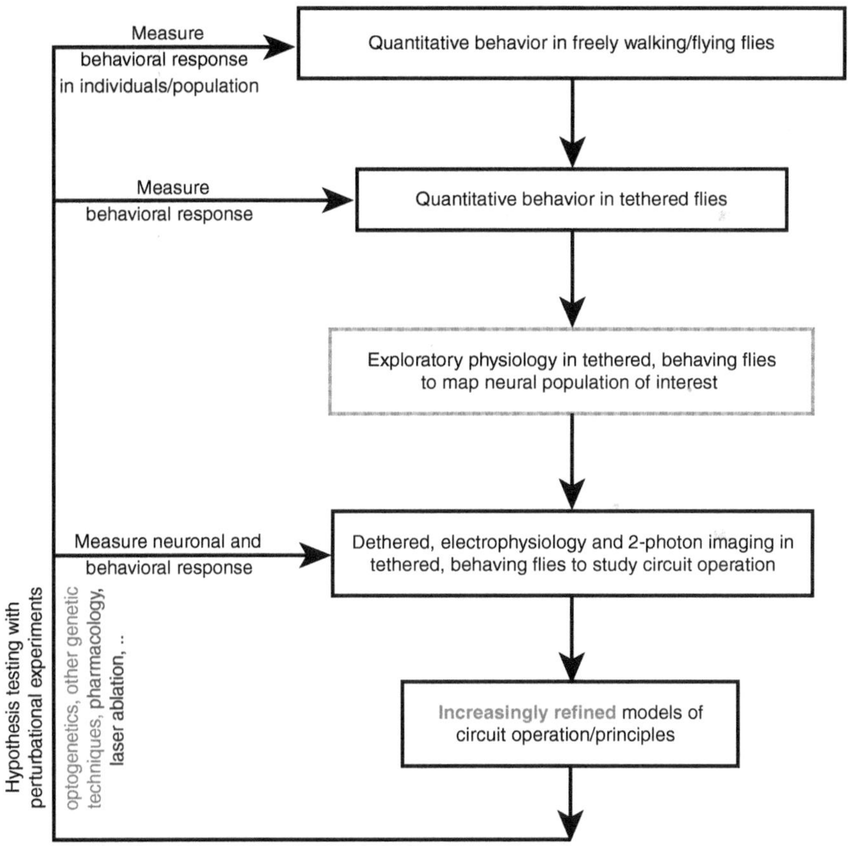

FIG. 5. Suggested experimental flowchart incorporating physiology during behavior. Behavioral findings from experiments in freely walking or flying flies are reproduced in a tethered "virtual reality" setup convenient for physiology. Findings from such experiments drive the creation of models of circuit function underlying the behavior. The models can then be tested and refined using the many genetic tools available for *Drosophila*. Tests can be performed at the level of both circuit function (tests on the rig) and behavior (on the rig/tethered/freely behaving). (For color version of this figure, the reader is referred to the web version of this book.)

manipulated in a situation where the impact of the perturbation, on both the circuit's operation and the behavior, can be monitored simultaneously. This should enable more powerful hypothesis testing than previously possible in *Drosophila*. Such advances should set the stage for rapid progress in understanding the different facets of sensorimotor integration, with researchers performing quantitative behavioral genetics studies to understand behaviors and identify brain regions involved, then moving the behaviors to tethered conditions and identifying possible computational roles for different neurons in the circuit, and finally building, testing and refining models of circuit function using combinations of the various genetic tools (see Fig. 5 for a possible experimental flowchart that incorporates physiology in behaving flies).

Acknowledgment

We thank Eugenia Chiappe for useful comments on the manuscript.

References

Bausenwein, B., Muller, N.R., and Heisenberg, M. (1994). Behavior-dependent activity labeling in the central complex of Drosophila during controlled visual stimulation. *J. Comp. Neurol.* **340**, 255–268.

Bohm, H., and Schildberger, K. (1992). Brain neurons involved in the control of walking in the cricket Gryllus-Bimaculatus. *J. Exp. Biol.* **166**, 113–130.

Borst, A., Haag, J., and Reiff, D.F. (2010). Fly motion vision. *Annu. Rev. Neurosci.* **33**, 49–70.

Borst, A., and Heisenberg, M. (1982). Osmotropotaxis in Drosophila melanogaster. *J. Comp. Physiol.* **147**, 479–484.

Buchner, E. (1976). Elementary movement detectors in an insect visual-system. *Biol. Cybern.* **24**, 85–101.

Budick, S.A., and Dickinson, M.H. (2006). Free-flight responses of Drosophila melanogaster to attractive odors. *J. Exp. Biol.* **209**, 3001–3017.

Bullock, T.H. (1999). Neuroethology has pregnant agendas. *J. Comp. Physiol. A* **185**, 291–295.

Bulthoff, H. (1982). Drosophila mutants disturbed in visual orientation. 2. Mutants affected in movement and position computation. *Biol. Cybern.* **45**, 71–77.

Buschges, A., Akay, T., Gabriel, J.P., and Schmidt, J. (2008). Organizing network action for locomotion: insights from studying insect walking. *Brain Res. Rev.* **57**, 162–171.

Buschges, A., and El Manira, A. (1998). Sensory pathways and their modulation in the control of locomotion. *Curr. Opin. Neurobiol.* **8**, 733–739.

Card, G., and Dickinson, M.H. (2008). Visually mediated motor planning in the escape response of Drosophila. *Curr. Biol.* **18**, 1300–1307.

Chiappe, M.E., Seelig, J.D., Reiser, M.B., and Jayaraman, V. (2010). Walking modulates speed sensitivity in Drosophila motion vision. *Curr. Biol.* **20**, 1470–1475.

Crapse, T.B., and Sommer, M.A. (2008). Corollary discharge across the animal kingdom. *Nat. Rev. Neurosci.* **9**, 587–600.

Dacks, A.M., Green, D.S., Root, C.M., Nighorn, A.J., and Wang, J.W. (2009). Serotonin modulates olfactory processing in the antennal lobe of Drosophila. *J. Neurogenet.* **23**, 366–377.

Dill, M., Wolf, R., and Heisenberg, M. (1993). Visual-pattern recognition in Drosophila involves retinotopic matching. *Nature* **365**, 751–753.

Dombeck, D.A., Khabbaz, A.N., Collman, F., Adelman, T.L., and Tank, D.W. (2007). Imaging large-scale neural activity with cellular resolution in awake, mobile mice. *Neuron* **56**, 43–57.

Duistermars, B.J., and Frye, M.A. (2008). Crossmodal visual input for odor tracking during fly flight. *Curr. Biol.* **18**, 270–275.

Fiala, A., Spall, T., Diegelmann, S., Eisermann, B., Sachse, S., Devaud, J.M., Buchner, E., and Galizia, C.G. (2002). Genetically expressed cameleon in Drosophila melanogaster is used to visualize olfactory information in projection neurons. *Curr. Biol.* **12**, 1877–1884.

Fotowat, H., Fayyazuddin, A., Bellen, H.J., and Gabbiani, F. (2009). A novel neuronal pathway for visually guided escape in Drosophila melanogaster. *J. Neurophysiol.* **102**, 875–885.

Foucaud, J., Burns, J.G., and Mery, F. (2010). Use of spatial information and search strategies in a water maze analog in Drosophila melanogaster. *PLoS One* **5**, e15231.

Frye, M.A. (2010). Multisensory systems integration for high-performance motor control in flies. *Curr. Opin. Neurobiol.* **20**, 347–352.

Gordon, M.D., and Scott, K. (2009). Motor control in a Drosophila taste circuit. *Neuron* **61**, 373–384.

Gotz, K.G. (1968). Flight control in Drosophila by visual perception of motion. *Kybernetik* **4**, 199–208.

Gotz, K.G. (1987). Course-control, metabolism and wing interference during ultralong tethered flight in Drosophila melanogaster. *J. Exp. Biol.* **128**, 35–46.

Gotz, K.G., and Wenking, H. (1973). Visual control of locomotion in the walking fruitfly Drosophila. *J. Comp. Physiol. A Neuroethol. Sens. Neural. Behav. Physiol.* **85**, 235–266.

Green, A.M., and Angelaki, D.E. (2010). Multisensory integration: resolving sensory ambiguities to build novel representations. *Curr. Opin. Neurobiol.* **20**, 353–360.

Haehnel, M., Froese, A., and Menzel, R. (2009). In vivo Ca2+ imaging of mushroom body neurons during olfactory learning in the honey bee. *J. Vis. Exp.* **30**, http://www.jove.com/details.php?id=1353, doi: 10.3791/1353.

Harvey, C.D., Collman, F., Dombeck, D.A., and Tank, D.W. (2009). Intracellular dynamics of hippocampal place cells during virtual navigation. *Nature* **461**, 941–946.

Hausen, K., and Wehrhahn, C. (1990). Neural circuits mediating visual flight control in flies. 2. Separation of 2 control-systems by microsurgical brain-lesions. *J. Neurosci.* **10**, 351–360.

Heiligenberg, W. (1991a). The neural basis of behavior: a neuroethological view. *Annu. Rev. Neurosci.* **14**, 247–267.

Heiligenberg, W. (1991b). Sensory control of behavior in electric fish. *Curr. Opin. Neurobiol.* **1**, 633–637.

Heisenberg, M., Wonneberger, R., and Wolf, R. (1978). Optomotor-BLIND–Drosophila mutant of lobula plate giant neurons. *J. Comp. Physiol.* **124**, 287–296.

Hendel, T., Mank, M., Schnell, B., Griesbeck, O., Borst, A., and Reiff, D.F. (2008). Fluorescence changes of genetic calcium indicators and OGB-1 correlated with neural activity and calcium in vivo and in vitro. *J. Neurosci.* **28**, 7399–7411.

Homberg, U. (1994). Flight-correlated activity changes in neurons of the lateral accessory lobes in the brain of the locust Schistocerca-Gregaria. *J. Comp. Physiol. A* **175**, 597–610.

Jayaraman, V., and Laurent, G. (2007). Evaluating a genetically encoded optical sensor of neural activity using electrophysiology in intact adult fruit flies. *Front. Neural Circuits* **1**, 3.

Joesch, M., Plett, J., Borst, A., and Reiff, D.F. (2008). Response properties of motion-sensitive visual interneurons in the lobula plate of Drosophila melanogaster. *Curr. Biol.* **18**, 368–374.

Katsov, A.Y., and Clandinin, T.R. (2008). Motion processing streams in Drosophila are behaviorally specialized. *Neuron* **59**, 322–335.

Kohatsu, S., Koganezawa, M., and Yamamoto, D. (2011). Female contact activates male-specific interneurons that trigger stereotypic courtship behavior in Drosophila. *Neuron* **69**, 498–508.

Konishi, M. (2003). Coding of auditory space. *Annu. Rev. Neurosci.* **26**, 31–55.

Lalazar, H., and Vaadia, E. (2008). Neural basis of sensorimotor learning: modifying internal models. *Curr. Opin. Neurobiol.* **18**, 573–581.

Laurent, G. (1991). Sensory control of locomotion in insects. *Curr. Opin. Neurobiol.* **1**, 601–604.

Lisberger, S.G. (2010). Visual guidance of smooth-pursuit eye movements: sensation, action, and what happens in between. *Neuron* **66**, 477–491.

Maimon, G., Straw, A.D., and Dickinson, M.H. (2008). A simple vision-based algorithm for decision making in flying Drosophila. *Curr. Biol.* **18**, 464–470.

Maimon, G., Straw, A.D., and Dickinson, M.H. (2010). Active flight increases the gain of visual motion processing in Drosophila. *Nat. Neurosci.* **13**, 393–399.

Mank, M., Santos, A.F., Direnberger, S., Mrsic-Flogel, T.D., Hofer, S.B., Stein, V., Hendel, T., Reiff, D.F., Levelt, C., and Borst, A et al., (2008). A genetically encoded calcium indicator for chronic in vivo two-photon imaging. *Nat. Methods* **5**, 805–811.

Mason, A.C., Oshinsky, M.L., and Hoy, R.R. (2001). Hyperacute directional hearing in a microscale auditory system. *Nature* **410**, 686–690.

Mronz, M., and Lehmann, F.O. (2008). The free-flight response of Drosophila to motion of the visual environment. *J. Exp. Biol.* **211**, 2026–2045.

Ng, M., Roorda, R.D., Lima, S.Q., Zemelman, B.V., Morcillo, P., and Miesenbock, G. (2002). Transmission of olfactory information between three populations of neurons in the antennal lobe of the fly. *Neuron* **36**, 463–474.

Niven, J.E., and Laughlin, S.B. (2008). Energy limitation as a selective pressure on the evolution of sensory systems. *J. Exp. Biol.* **211**, 1792–1804.

Ofstad, T.A., Zuker, C.S., and Reiser, M.B. (2011). *Nature* **474**, 204–207.

Olsen, S.R., and Wilson, R.I. (2008). Cracking neural circuits in a tiny brain: new approaches for understanding the neural circuitry of Drosophila. *Trends Neurosci.* **31**, 512–520.

Orger, M.B., Kampff, A.R., Severi, K.E., Bollmann, J.H., and Engert, F. (2008). Control of visually guided behavior by distinct populations of spinal projection neurons. *Nat. Neurosci.* **11**, 327–333.

Pfeiffer, B.D., Ngo, T.T., Hibbard, K.L., Murphy, C., Jenett, A., Truman, J.W., and Rubin, G.M. (2010). Refinement of tools for targeted gene expression in Drosophila. *Genetics* **186**, 735–755.

Pfeiffer, K., and Homberg, U. (2007). Coding of azimuthal directions via time-compensated combination of celestial compass cues. *Curr. Biol.* **17**, 960–965.

Pouget, A., and Snyder, L.H. (2000). Computational approaches to sensorimotor transformations. *Nat. Neurosci.* **3 Suppl.**, 1192–1198.

Poulet, J.F., and Hedwig, B. (2007). New insights into corollary discharges mediated by identified neural pathways. *Trends Neurosci.* **30**, 14–21.

Reiff, D.F., Ihring, A., Guerrero, G., Isacoff, E.Y., Joesch, M., Nakai, J., and Borst, A. (2005). In vivo performance of genetically encoded indicators of neural activity in flies. *J. Neurosci.* **25**, 4766–4778.

Reiff, D.F., Plett, J., Mank, M., Griesbeck, O., and Borst, A. (2010). Visualizing retinotopic half-wave rectified input to the motion detection circuitry of Drosophila. *Nat. Neurosci.* **13**, 973–978.

Ritzmann, R.E., and Buschges, A. (2007). Adaptive motor behavior in insects. *Curr. Opin. Neurobiol.* **17**, 629–636.

Rosato, E., and Kyriacou, C.P. (2006). Analysis of locomotor activity rhythms in Drosophila. *Nat. Protoc.* **1**, 559–568.

Samuel, A.D., and Sengupta, P. (2005). Sensorimotor integration: locating locomotion in neural circuits. *Curr. Biol.* **15**, R341–R343.

Schnell, B., Joesch, M., Foerstner, F., Raghu, S.V., Otsuna, H., Ito, K., Borst, A., and Reiff, D.F. (2010). Processing of horizontal optic flow in three visual interneurons of the Drosophila brain. *J. Neurophysiol.* **103**, 1646–1657.

Seelig, J.D., Chiappe, M.E., Lott, G.K., Dutta, A., Osborne, J.E., Reiser, M.B., and Jayaraman, V. (2010). Two-photon calcium imaging from head-fixed Drosophila during optomotor walking behavior. *Nat. Methods* **7**, 535–540.

Simpson, J.H. (2009). Mapping and manipulating neural circuits in the fly brain. *Adv. Genet.* **65**, 79–143.

Sommer, M.A., and Wurtz, R.H. (2008). Brain circuits for the internal monitoring of movements. *Annu. Rev. Neurosci.* **31**, 317–338.

Strauss, R. (1995). A screen for EMS-induced X-linked locomotor mutants in Drosophila melanogaster. *J. Neurogenet.* **10**, 53–54.

Taylor, G.K., and Krapp, H.G. (2007). Sensory systems and flight stability: What do insects measure and why? *Adv. Insect. Physiol.* **34**, 231–316.

Tian, L., Hires, S.A., Mao, T., Huber, D., Chiappe, M.E., Chalasani, S.H., Petreanu, L., Akerboom, J., McKinney, S.A., and Schreiter, E.R et al., (2009). Imaging neural activity in worms, flies and mice with improved GCaMP calcium indicators. *Nat. Methods* **6**, 875–881.

Todorov, E. (2004). Optimality principles in sensorimotor control. *Nat. Neurosci.* **7**, 907–915.

Trimarchi, J.R., and Schneiderman, A.M. (1995). Different neural pathways coordinate Drosophila flight initiations evoked by visual and olfactory stimuli. *J. Exp. Biol.* **198**, 1099–1104.

Wang, J.W., Wong, A.M., Flores, J., Vosshall, L.B., and Axel, R. (2003). Two-photon calcium imaging reveals an odor-evoked map of activity in the fly brain. *Cell* **112**, 271–282.

Warzecha, A.K., and Egelhaaf, M. (1996). Intrinsic properties of biological motion detectors prevent the optomotor control system from getting unstable. *Philos. Trans. R. Soc. B* **351**, 1579–1591.

Wehrhahn, C., and Reichardt, W. (1973). Visual orientation of the fly Musca domestica towards a horizontal stripe. *Naturwissenschaften* **60**, 203–204.

Wilson, R.I., Turner, G.C., and Laurent, G. (2004). Transformation of olfactory representations in the Drosophila antennal lobe. *Science* **303**, 366–370.

Wolpert, D.M., and Ghahramani, Z. (2000). Computational principles of movement neuroscience. *Nat. Neurosci.* **3 Suppl.**, 1212–1217.

Yassin, L., Benedetti, B.L., Jouhanneau, J.S., Wen, J.A., Poulet, J.F., and Barth, A.L. (2010). An embedded subnetwork of highly active neurons in the neocortex. *Neuron* **68**, 1043–1050.

Zars, T. (2009). Spatial orientation in Drosophila. *J. Neurogenet.* **23**, 104–110.

Zorovic, M., and Hedwig, B. (2011). Processing of species-specific auditory patterns in the cricket brain by ascending, local and descending neurons during standing and walking. *J. Neurophysiol.*

MODELING HUMAN TRINUCLEOTIDE REPEAT DISEASES IN *DROSOPHILA*

Zhenming Yu[1] and Nancy M. Bonini[1,2]

[1]Department of Biology, 415 S University Ave., University of Pennsylvania
[2]Howard Hughes Medical Institute, Philadelphia, PA 19104, USA

Drosophila is a powerful model system to study human trinucleotide repeat diseases. Findings in *Drosophila* models highlighted importance of host proteins, chaperons, and protein clearance pathways in polyglutamine diseases as well as that of RNA-binding proteins in noncoding repeat RNA toxicity diseases. Recent novel aspects revealed in *Drosophila* models include pleiotropic Ataxin 2 interactions, antisense transcription in trinucleotide repeat diseases, contribution of CAG RNA in polyglutamine diseases, and the role of RNA foci in CUG expansion diseases. *Drosophila* models have been also used for repeat stability studies.

I. Introduction

More than 20 neurological diseases are caused by trinucleotide repeat expansions in the human genome (La Spada and Taylor, 2010; McMurray, 2010). *Drosophila* has proved to be a powerful model system to study the mechanisms of such diseases (Bilen and Bonini, 2005; Cauchi and van den Heuvel, 2006; Marsh and Thompson, 2006; Zoghbi and Botas, 2002). Here, we will discuss fly models of CAG/polyglutamine expansion diseases (Fernandez-Funez *et al.*, 2000; Jackson *et al.*, 1998; Takeyama *et al.*, 2002; Warrick *et al.*, 1998) and noncoding repeat expansion RNA toxicity diseases (de Haro *et al.*, 2006; Garcia-Lopez *et al.*, 2008; Houseley *et al.*, 2005; Jin *et al.*, 2003; Mutsuddi *et al.*, 2004; Yu *et al.*, 2011a). We will review additional findings made in the *Drosophila* system of possible contributions of

INTERNATIONAL REVIEW OF
NEUROBIOLOGY, VOL. 99
DOI: 10.1016/B978-0-12-387003-2.00008-2

0074-7742/10 $35.00

expanded CAG RNA in polyglutamine repeat diseases (Li *et al.*, 2008), as well as of antisense transcription in trinucleotide repeat expansion diseases (Yu *et al.*, 2011a). We will also discuss recent efforts to approach repeat instability in *Drosophila* models (Jung and Bonini, 2007; Jung *et al.*, 2011). Other reviews of the fly as a model for human neurodegenerative disease include Ambegaokar *et al.* (2010), Bilen and Bonini (2005), Marsh and Thompson (2006), Shulman *et al.* (2003), Whitworth and Pallanck (2008), and Zoghbi and Botas (2002).

II. Results

A. Human Diseases Caused by Expanded Trinucleotide Repeats

A remarkable feature of the human genome is the wide prevalence of microsatellite sequences including trinucleotide tandem repeats (Lander *et al.*, 2001). A number of the microsatellite sequences, with trinucleotide repeats being the most common, can undergo dynamic expansion and lead to human neurological diseases (Mirkin, 2007; Pearson *et al.*, 2005). In 1991, the dynamic expansion of the CGG trinucleotide repeat in fragile X mental retardation 1 gene (*FMR1*) was found to underlie fragile X syndrome (FRAXA) and expansion of a CAG trinucleotide repeat in androgen receptor gene (*AR*) was found to underlie spinal bulbar muscular atrophy (SBMA) (La Spada *et al.*, 1991; Verkerk *et al.*, 1991). Since then, more than 20 human neurological, neurodegenerative, and neuromuscular diseases have been found associated with expansion of trinucleotide repeats (La Spada and Taylor, 2010; McMurray, 2010). These repeats can reside in the 5′ untranslated region (5′UTR), coding sequence (CDS), 3′ untranslated region (3′UTR), or introns.

At least three molecular mechanisms are now known for pathogenesis of trinucleotide repeat diseases (Table I; see also La Spada and Taylor (2010), Orr and Zoghbi (2007)). Huntington disease (HD), spinocerebellar ataxia (SCA) 1, 2, 3, 6, 7 and 17, SBMA, dentatorubral-pallidoluysian atrophy (DRPLA) belong to a category of diseases known as the polyglutamine diseases, and are due to a CAG repeat expansion encoding polyglutamine within the coding region of the genes. This is typically considered to confer a gain-of-function effect on the host protein. Myotonic dystrophy (DM1) and fragile X-associated tremor/ataxia syndrome (FXTAS) are stereotypical diseases due to repeat expansions within the noncoding region of their respective host genes. In this case, the disease is thought to be due to gain-of-function effect of the RNA of the gene, bearing the expanded repeat sequence. FRAXA, fragile XE syndrome (FRAXE), and Friedreich ataxia (FRDA) constitute a third class, namely due to repeat expansions in the 5′UTR or intron of the respective genes, though to confer silencing of gene expression. Notably, these mechanisms

Table I
SUMMARY OF TRINUCLEOTIDE REPEAT DISEASES.

Disease	Repeat (Gene)	Repeat Location	Normal Repeat Length	Disease Repeat Length	Main Clinical Symptoms
Polyglutamine diseases					
HD	CAG (Htt)	CDS	6–35	36–121	Motor disability, chorea, memory/cognitive decline
SCA1	CAG (ATXN1)	CDS	6–44	39–82	Ataxia, dysarthria, spasticity, ophthalmoplegia
SCA2	CAG (ATXN2)	CDS	15–31	36–63	Ataxia, dysarthria, slow eye movement, hyporeflexia, supranuclear ophthalmoplegia, parkinsonism
SCA3	CAG (ATXN3)	CDS	12–40	55–84	Ataxia, spasticity, and ocular movement abnormalities
SCA6	CAG (CACNA1A)	CDS	4–18	21–33	Slow progressive or episodic ataxia
SCA7	CAG (ATXN7)	CDS	4–35	37–306	Ataxia, dysarthria, cone–rod dystrophy retinal disease
SCA17	CAG (TBP)	CDS	25–44	45–63	Ataxia, dementia, chorea, seizures, dystonia
SBMA	CAG (AR)	CDS	9–36	38–62	Proximal muscle atrophy, lower motor neuron disease
DRPLA	CAG (ATN1)	CDS	6–35	49–88	Seizures, choreoathetosis, ataxia, dementia, cognitive decline
Expanded repeat RNA toxicity					
DM1	CTG (DMPK)	3′UTR	5–37	50–>1000	Muscle weakness, myotonia, cardiac arrhythmia, mental retardation
FXTAS	CGG (FMR1)	5′UTR	6–53	55–200	Ataxia, intention tremor, cognitive decline, parkinsonism

(Continued)

Table I (*continued*)

Disease	Repeat (Gene)	Repeat Location	Normal Repeat Length	Disease Repeat Length	Main Clinical Symptoms
Loss of function of the host gene expression					
FRAXA	CGG (FMR1)	5′UTR	6–53	>200	Mental retardation, facial dysmorphism, macroorchidism, autism
FRAXE	GCC (FMR2)	5′UTR	6–35	61–200 (pre) >200 (full)	Mental retardation, hyperactivity and autistic behavior
FDRA	GAA (FXN)	Intron	5–30	70–>1000	Ataxia, muscle weakness, dysarthria, sensory loss
Polyglutamine protein and expanded repeat RNA toxicity					
SCA8	CTG/CAG (ATXN8OS)	Noncoding RNA/CDS	15–50	71–1300	Ataxia, dysarthria, nystagmus, spasticity, decreased vibration sense
HDL2	CTG/CAG (JPH3)	3′UTR/CDS?	6–28	>41	Similar to HD
Unknown mechanism					
SCA12	CAG (PPP2R2B)	Promoter/ 5′UTR?	7–32	51–78	Tremor, ataxia, spasticity, dementia

DM, dystrophy myotonic; DRPLA, dentatorubral-pallidoluysian atrophy; FDRA, Friedreich's ataxia; FRAXA, fragile X syndrome; FRAXE, fragile XE syndrome; FXTAS, fragile X-associated tremor/ataxia syndrome; HD, Huntington disease; HDL2, Huntington disease-like 2; SBMA, spinal bulbar muscular atrophy; SCA, spinocerebellar ataxia.

are not mutually exclusive: for example, in the case of SCA8, both polyglutamine protein toxicity and CUG RNA gain-of-function toxicity may contribute to the disease pathology (Daughters *et al.*, 2009; Ikeda *et al.*, 2008; Moseley *et al.*, 2006).

B. *Drosophila* as a Model to Study Human Disease

Drosophila, the fruit fly, has proved to be a powerful model system for the study of human neurodegenerative diseases including trinucleotide repeat diseases. Fundamental biological pathways are highly conserved between human and flies, with counterparts of the same genes in the analogous pathways, yet typically the fly will be a simpler, less complex situation. Thus, findings in fly models are physiologically relevant and can be powerfully extended to the human disease.

Notably, numerous genetic techniques and tools are readily available to manipulate gene expression in flies. For example, the GAL4/UAS system to direct transgene expression in a temporally and spatially controlled manner allows expression of toxic gene products, and modeling of neural features of disease (Brand and Perrimon, 1993). Many chromosomal deletion collections, RNAi collections, and transposon insertion collections are available and maintained in public *Drosophila* stock centers (Bellen *et al.*, 2004; Dietzl *et al.*, 2007; Parks *et al.*, 2004; Ryder *et al.*, 2007; Thibault *et al.*, 2004). These greatly facilitate study of altered activity of the function of various genes and pathways, to understand disease gene mechanism. The short generation time, small size, and large number of progeny make it possible to perform large-scale genetic modifier screens and pharmacological compound screens with relatively low cost and in a relatively short time frame. To date, *Drosophila* models have been established for polyglutamine protein diseases as well as trinucleotide repeat RNA toxicity diseases. Here we focus on the common themes reflected in these models, and novel insights obtained from studying such disease models in a system like *Drosophila*.

C. *Drosophila* Models of Polyglutamine Diseases

Drosophila models of SCA3 were initially generated by introducing the C-terminally truncated domain of SCA3, with 27 glutamines (SCA3trQ27, control) or 78 glutamines (SCA3trQ78, disease range) into flies (Warrick *et al.*, 1998). Expression of SCA3trQ78, but not SCA3trQ27, in the eye causes loss of pigmentation, as well as collapse of the retinal structure. Moreover, expression of the pathogenic SCA3trQ78 led to formation of nuclear inclusions of the disease protein, reminiscent of SCA3 human disease (Paulson *et al.*, 1997; Warrick *et al.*, 1998). A model for SCA3 using the full-length protein was subsequently generated

(Warrick *et al.*, 2005). Comparing the toxicity of the full-length protein with the truncated protein, the full-length protein showed a milder effect that was more selective to neurons. Further study indicated that the SCA3 protein has biological activity *in vivo* to mitigate toxicity of the pathogenic protein (Warrick *et al.*, 2005). This remarkable finding speaks to the importance of host protein context for the expanded repeat, and how the function of the protein interplays in disease.

To model HD in flies, exon 1 of the Htt protein bearing 2, 75, or 120 glutamine residues was directed to the fly eye (Jackson *et al.*, 1998). Expression of Htt with 75 and 120 glutamines causes repeat length-dependent degeneration of photoreceptor neurons. Moreover, these flies were born with normal retinal morphology, but the pathogenic proteins caused a progressive and late-onset degeneration. In this model, the HD protein was noted to accumulate in the nucleus, however, without the formation of nuclear inclusions (Jackson *et al.*, 1998), thus linking nuclear localization of the protein to pathogenic activities. Gunawardena *et al.* (2003) showed that expression of an exon 1 Htt protein bearing a pathogenic polyglutamine expansion causes axonal transport defects and cell death, thus providing insight into additional biological features of the toxicity (Gunawardena *et al.*, 2003). Expression of an Htt protein fragment that includes and extends well beyond just exon 1 also revealed cytoplasmic accumulations, as well as axonal transport blockage (Lee *et al.*, 2004). However, in a full-length HD fly, axon transportation defects were not observed. Rather, increased neurotransmission seemed to be a root cause of neurodegeneration in the early stage of toxic pathogenic effects (Sofola *et al.*, 2007).

Fernandez-Funez *et al.* (2000) generated a *Drosophila* model for SCA1 using the complete full-length protein. Flies expressing a pathogenic version of SCA1 with 82 glutamines (SCA1Q82) showed progressive neurodegeneration. Interestingly, flies expressing the wild-type control of SCA1 with a normal-length glutamine domain of 30 residues also showed a rough eye effect when the protein was expressed at high levels. This raised the possibility that a component of SCA1 toxicity may be inherent in deleterious effects of the SCA1 protein when simply expressed at levels that are too high. Such an effect was confirmed in mice with the control protein (Fernandez-Funez *et al.*, 2000). A genetic screen using SCA1Q82 to define other genes that may modulate the toxicity led to the identification of genes in diverse pathways, such as protein folding, protein clearance, RNA processing, transcriptional regulation, and cellular detoxification (Fernandez-Funez *et al.*, 2000).

A *Drosophila* model for SBMA was created by introducing the human androgen receptor (hAR) protein carrying 52 glutamine residues (hARQ52) into flies. Strikingly, these animals failed to reveal an effect unless the animals were given a ligand for the androgen receptor. In that case, the protein translocated to the nucleus, for onset of neurodegeneration (Takeyama *et al.*, 2002). This example is the most striking case of host protein context dramatically influencing toxicity (see below). Expression of naked, raw polyglutamine domains causes toxicity in flies, highlighting the intrinsic and fundamental deleterious effects of just a polyglutamine expansion (Kazemi-Esfarjani and Benzer, 2000; Marsh *et al.*, 2000). Interesting, a genetic modifier screen using such a model led to the identification of two suppressors of toxicity, both of which contain a chaperone-related J domain, suggestive of protein folding and clearance being critical (Kazemi-Esfarjani and Benzer, 2000).

1. *The Host Protein in Which Polyglutamine Tract Resides Plays a Crucial Role in the Pathogenic Features*

A puzzling observation in the field of polyglutamine disease research is that, regardless the fundamental cytotoxicity of polyglutamine tracks and overlapping expression patterns of the different host proteins, different polyglutamine proteins selectively target specific groups of neurons and thus lead to distinct clinical manifestations. This suggests that the polyglutamine host protein may play a critical role in disease pathogenesis, beyond simply serving as a vector of a toxic polyglutamine expansion. Indeed, findings in animal models, including the fly, corroborate this notion. Most strikingly, as noted, toxicity of the AR protein in the fly SBMA model is dependent on the presence of the ligand. Thus, feeding flies such a ligand induces neurodegeneration, highlighting the key contribution of the host protein function in this disease (Takeyama *et al.*, 2002). Further study suggests that ligand-mediated nuclear translocation of AR is necessary, but not sufficient for deleterious effects. Rather, both the DNA-binding domain and the coregulator interaction domain AF2 are required for hARQ52-mediated disruption of the fly eye (Nedelsky *et al.*, 2010). These results raise the possibility that the polyglutamine expansion in mutant hAR mediates deleterious effects in SBMA through altered interactions with normal protein partners.

In addition, the truncated Ataxin 3 protein generally shows much stronger toxicity than the full-length Ataxin 3 protein in fly models (Warrick *et al.*, 2005). Further, the normal Ataxin 3 protein can mitigate toxicity of a pathogenic Ataxin 3 protein, as well as pathogenicity of other polyglutamine proteins. Ubiquitin-associated activities of the Ataxin 3 are likely to underlie these findings. This further highlights the importance of host proteins in polyglutamine toxicity. In SCA1, phosphorylation of the protein by the Akt kinase and interaction with the 14-3-3 protein modulate toxicity in mammalian cell culture and mouse models. A genetic interaction between SCA1 and Akt was re-capitulated in flies as well (Chen *et al.*, 2003). Moreover, differential effects on axonal transport were observed between HD fly models generated with only the N-terminal Htt fragment (exon 1), compared with the full-length Htt protein (Gunawardena *et al.*, 2003; Lee *et al.*, 2004; Romero *et al.*, 2008). This indicates that protein context may dictate pathogenic outcome at least in some situations.

2. *Chaperones and Protein-Folding Pathways Are Strong Modifiers of Polyglutamine Toxicity*

Targeted expression of heat-shock protein 70 (Hsp70), one of the key molecular chaperones that can help refold other proteins, in both SCA3 and HD fly models strongly mitigates the pathogenic protein toxicity (Chan *et al.*, 2000; Warrick *et al.*, 1999). Moreover, nonbiased genome-wide screens of the raw polyglutamine fly model, the SCA1 fly model and the SCA3 model all led to identification of select chaperones and co-chaperones as suppressors of protein toxicity (see Fig. 1 for examples) (Bilen and Bonini, 2007; Kazemi-Esfarjani and Benzer, 2000). These results strongly suggest that chaperones and protein-folding pathways play central roles in polyglutamine disease. Further studies show that Hsp70 increases the solubility of SCA3 polyglutamine protein in flies (Chan *et al.*, 2000). These results echo the findings that the heat-shock protein HDJ-2/HDJs (Hsp40) colocalizes with ubquitinated nuclear inclusions in SCA1 postmortem patient tissues and their overexpression affects SCA1 aggregates in cell culture (Cummings *et al.*, 1998). Consistent with the protective effect in fly models of polyglutamine diseases, upregulation of Hsp70 suppresses neuropathology in SCA1 mice (Cummings *et al.*, 2001). Moreover, expression of a dominant-negative Hsp70 form in the fly can cause a strong degeneration-like effect (Bilen and Bonini, 2007). This finding suggests that intact chaperone activity is key to maintain proper protein homeostasis and normal function in cells. Together, these data

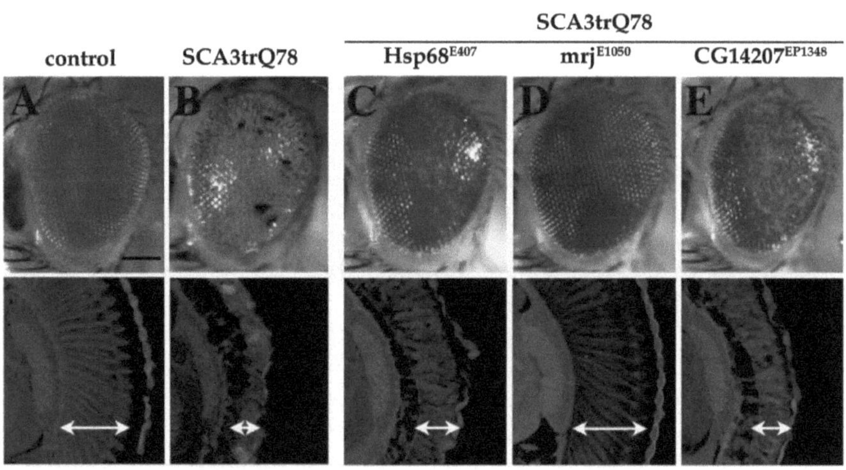

FIG. 1. Chaperones modulate polyglutamine toxicity. Several chaperones were identified as strong suppressors of SCA3trQ78-induced neurodegeneration in *Drosophila.* External eye (top) and internal retinal sections of 1-day-old flies. (A) Control with normal eye and retinal thickness (white arrow). (B) Expression of strong SCA3trQ78 causes external and internal degeneration, with pigment loss and severely reduced retinal thickness. (C–E) Chaperone suppressors Hsp68, mrj, and CG14207 rescue pigmentation and internal retinal structure. Image from Bilen and Bonini (2007) (For color version of this figure, the reader is referred to the web version of this book.)

indicate that insufficient chaperone activity may contribute to pathogenesis in polyglutamine diseases, and thus protein-folding pathways may be ideal therapeutic targets for these disorders.

3. *Cellular Protein Degradation Machinery in Polyglutamine Disease*

Ubiquitin-positive nuclear inclusions were found in multiple polyglutamine diseases including HD, SCA1, and SCA3 (Davies *et al.*, 1997; DiFiglia *et al.*, 1997; Paulson *et al.*, 1997; Skinner *et al.*, 1997). Interestingly, nuclear inclusions were found to colocalize with proteasome components in SCA1 patients and transgenic mice (Cummings *et al.*, 1998). Thus, the ubiqutin/proteasome system (UPS) in affected neurons might be altered. It is now generally thought that the UPS may be impaired by the polyglutamine expansion mutation in diseases such as HD and SCAs (Bence *et al.*, 2001; Bennett *et al.*, 2005; Duenas *et al.*, 2006). However, controversies remain on this issue, as some groups find no change in proteasome activity or even increased proteasome activity in polyglutamine diseases (Bett *et al.*, 2006; Bowman *et al.*, 2005; Diaz-Hernandez *et al.*, 2003; Ding *et al.*, 2002). It could be that this is a pathway that plays different functional roles in disease at different times in the process.

Modifiers obtained in a forward genetic screen using the fly model of SCA1 suggest that compromising the UPS system enhances SCA1 toxicity (Fernandez-Funez *et al.*, 2000). In addition, components in the UPS have been identified as modifiers in a genome-wide screen for SCA3trQ78 modifiers (Bilen and Bonini, 2007). Therefore, data in the fly support the hypothesis that UPS activity may be compromised in the pathological situation, and certainly that enhancement of these processes may be beneficial. Interestingly, Ataxin 3 itself has a ubiquitin protease domain and ubiquitin-interacting motifs, and the normal protein can mitigate polyglutamine toxicity in a proteasome activity-dependent manner in flies (Warrick *et al.*, 2005). Intriguingly, the normal Ataxin 3 protein is found in nuclear inclusions of other polyglutamine proteins, including SCA1 and SCA2 in cell-based studies and postmortem patient brains (Perez *et al.*, 1998; Uchihara *et al.*, 2001). It is tempting to hypothesize that the normal function of Ataxin 3 may be protective against polyglutamine toxicity not only in SCA3 but also in other polyglutamine diseases.

Accumulation of autophagic vacuoles has been observed in affected neurons in a number of different neurodegenerative diseases including Parkinson's, Alzheimer's, Creutzfeldt–Jakob, and many of the polyglutamine diseases (Anglade *et al.*, 1997; Nixon *et al.*, 2005; Sapp *et al.*, 1997; Sikorska *et al.*, 2004). Studies in animal models including the fly suggest that autophagy plays a critical role in maintaining neuronal integrity and, like the ubiquitin proteasome system, is protective in the pathogenic protein situation. In a fly model for SBMA, genetic inhibition of autophagy greatly enhances degeneration caused by polyglutamine protein, and pharmacological augmentation of autophagy protects animals from toxicity (Pandey *et al.*, 2007). Collectively, these results highlight the critical roles of cellular protein degradation pathways in polyglutamine disease and raise the possibility that these pathways, too, may be key therapeutic targets.

4. *Genetic Interactions of Ataxin 2 in Flies*

Reducing activity of the *Drosophila* homologue of Ataxin 2 (dAtx2) by mutation or increasing its activity by transgenic upregulation led to pleiotropic deleterious effects in flies including female sterility, tissue degeneration, and lethality (Satterfield *et al.*, 2002). This suggests that Ataxin 2 may be a key protein generally for cellular maintenance. Moreover, dAtx2 upregulation was found to enhance toxicity caused by SCA1 and SCA3 in genetic modifier screens in *Drosophila* models (Fig. 2). Knocking down endogenous *dAtx2* in *Drosophila* blocked neurodegeneration in both SCA1 and SCA3 flies, suggesting that Ataxin 2 is a key mediator of toxicity induced by pathogenic Ataxin 1 and Ataxin 3 (Al-Ramahi *et al.*, 2007; Lessing and Bonini, 2008). Further, recent studies in yeast and flies led to the identification of Ataxin 2 as a central modifier of toxicity caused by TAR DNA-binding protein 43 (TDP-43), a protein involved in amyotrophic lateral sclerosis

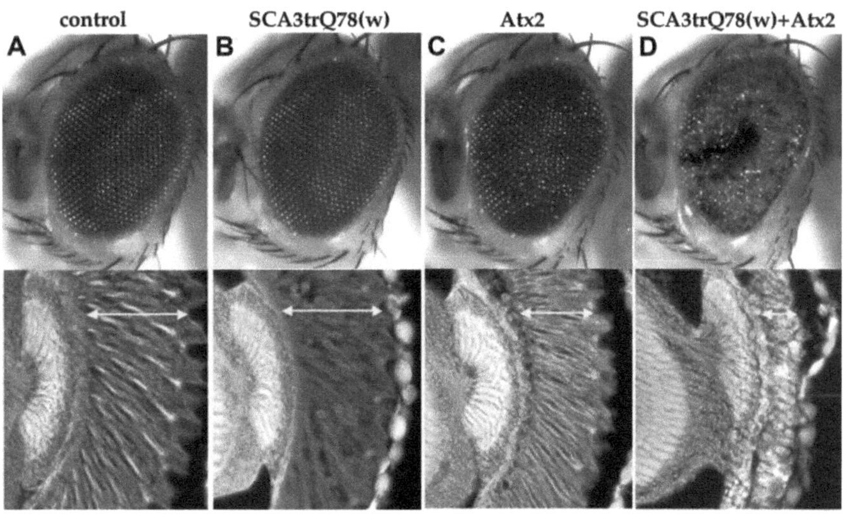

FIG. 2. dAtaxin 2 enhances Ataxin 3-dependent neurodegeneration. Eye (top) and internal retinal sections (bottom) of flies. (A) Control with normal eye and retinal thickness. (B) Flies with weak expression of a pathogenic SCA3trQ78 protein have a normal eye, with slightly disorganized structures.(C) Flies expressing dAtaxin 2 have a mildly rough external eye. (D) Expression of dAtaxin 2 with SCA3trQ78 results in severe degeneration, with loss of pigmentation and collapse of the retina. Image from Lessing and Bonini (2008). (For color version of this figure, the reader is referred to the web version of this book.)

(ALS) and frontotemporal lobar degeneration (FTLD) (Neumann *et al.*, 2006). Further studies suggest that Ataxin 2 and TDP-43 form a complex in a RNA-dependent manner and reveal mis-localization of Ataxin 2 in ALS and TDP-43 in SCA2. More importantly, this study established Ataxin 2 with intermediate polyglutamine repeat expansions, higher than normal, but below the threshold for SCA2, as a major risk factor in ALS (Elden *et al.*, 2010). Interestingly, Ataxin 2 with an expanded polyglutamine domain can contribute to at least three clinically distinct diseases, depending on length and sequence of the polyglutamine region (Yu *et al.*, 2011b). Collectively, these results suggest that Ataxin 2 plays a pivotal role in mediating interactions of multiple components to maintain integrity and normal function of the nervous system.

D. *Drosophila* Models of Noncoding Trinucleotide Diseases With RNA-Based Toxicity

Gain of function effects of repeat expansions within the mRNA for the gene causes or contributes to neurological diseases including DM1, FXTAS, and SCA8. *Drosophila* models have been generated to study these diseases, as well. To model FXTAS, Jin *et al* introduced a (CGG)90 expansion in the 5′UTR of a control GFP reporter gene (Jin *et al.*, 2003). These flies displayed repeat length-dependent degeneration; thus, an expansion in a noncoding part of an RNA can cause effects. Interestingly, ubiquitin-positive inclusions were also detected in these flies, and upregulation of Hsc70 suppressed CGG toxicity, indicating a link between CGG RNA-based toxicity and protein-folding and degradation pathways (Jin *et al.*, 2003).

A *Drosophila* SCA8 model was generated by expressing the human SCA8 noncoding RNA in flies (Mutsuddi *et al.*, 2004). An SCA8 RNA carrying a (CTG)8 repeat or a (CTG)112 repeat both caused late-onset neurodegeneration. A targeted genetic screen of RNA-binding proteins for modifiers of the SCA8 (CTG)112 identified of four modifiers including the Staufen and Muscleblind proteins. Although SCA8 carrying (CTG)8 and (CTG)112 show a similar effects, Mutsuddi *et al.* (2004) did observe differential interactions with select RNA-binding proteins, suggesting that the CTG expansion confers specificity to the SCA8 pathogenic mechanism. Houseley *et al.* (2005) generated a fly model of DM1 by incorporating an uninterrupted (CTG)162 in the 3′UTR of the GFP reporter gene. Despite the presence of robust ribonuclear accumulations of the RNA in select tissues including muscle, these flies did not show deleterious effects of the RNA (Houseley *et al.*, 2005). This may be due to the fact that a (CTG)162 expansion is below the length threshold of the repeat to induce disruptive effects in flies (Yu *et al.*, 2011a). DM1 models created using interrupted 480 CTG expansions indeed show repeat length-dependent RNA-based toxicity. Importantly, Muscleblind like 1 (MBNL1) and CUG-binding protein 1 (CUG-BP1), two RNA-binding proteins thought to be critical in DM1 pathogenesis (Ranum and Cooper, 2006), modify the toxicity associated with the expanded CTG repeat in these models (de Haro *et al.*, 2006; Garcia-Lopez *et al.*, 2008). A DM1 model generated by expressing an uninterrupted CTG repeat expansion of various lengths ranging from 19 to 270 in the 3′UTR of the DsRed reporter shows features reminiscent of DM1, such as repeat length-dependent toxicity, formation of ribonuclear foci in muscles and affected alternative splicing (Fig. 3; Yu *et al.* (2011a)).

1. *Reevaluating the Role of Ribonuclear Foci in CUG Repeat RNA Toxicity*

Ribonuclear foci formation is a hallmark for DM1 and DM2 diseases (Ranum and Cooper, 2006). It is generally believed that these foci contribute to pathogenesis by sequestrating RNA-binding proteins such as MBNL1, which subsequently affects the role of MBNL1 in the alternative splicing of genes such as chloride channel 1 (*Clc-1*) and insulin receptor (*IR*), underlying DM1 symptoms such as myotonia and insulin resistance (Day and Ranum, 2005; Osborne and Thornton, 2006; Ranum and Cooper, 2006). Recent data, including those in fly models, however, raise the possibility that foci formation is neither sufficient nor required for RNA toxicity. In fly models of CTG repeat toxicity, ribonuclear foci formation was observed in select tissues including muscles even without any appreciable

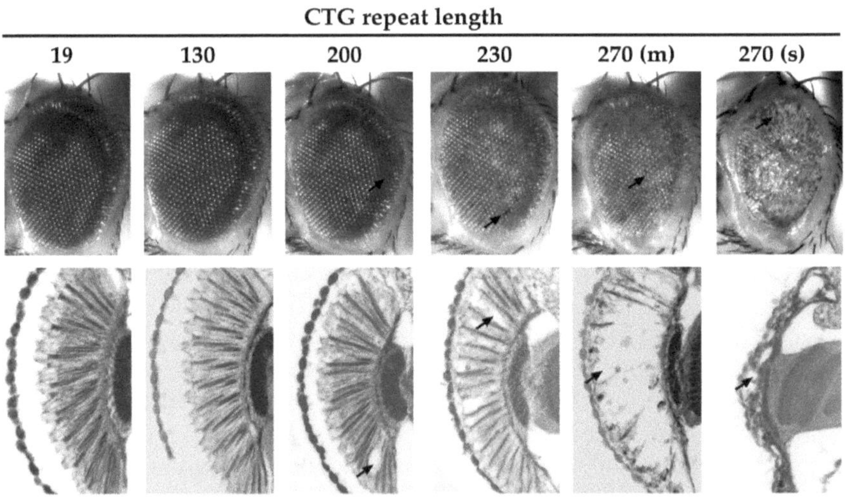

FIG. 3. Noncoding CTG repeat transcripts cause repeat length-dependent toxicity in *Drosophila*. External eye (top) and internal retinal sections (bottom) of 14-day-old flies expressing distinct length CTG repeat expansions. CTG repeat length is noted. The effect of a (CTG)270 was variable; shown are of mild (m) and severe (s). Arrows highlight necrotic patches on external eyes and loss of retinal tissue internally. Figure from Yu *et al.* (2011a). (For color version of this figure, the reader is referred to the web version of this book.)

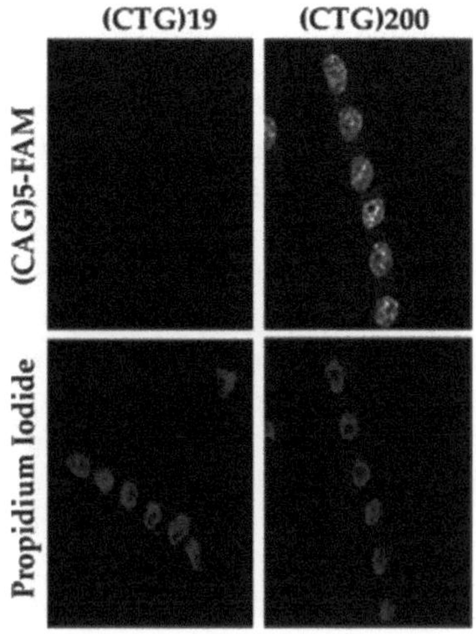

FIG. 4. Expression of (CTG)200 expansion in *Drosophila* muscles leads to ribonuclear foci formation but without appreciable toxicity. Ribonuclear foci are detectable in muscles expressing a (CTG)200 RNA, but not in those expressing (CTG)19, yet expressing (CTG)200 in muscles does not show any appreciable toxicity. Image from Yu *et al.* (2011a). (For color version of this figure, the reader is referred to the web version of this book.)

toxicity at the organismal level (Fig. 4; also Houseley *et al.* (2005)). In mammalian cell culture, interrupted (CAG)480 forms ribonuclear foci that colabel for MBNL1, just as (CTG)480 does. However, repeat-associated splicing changes associated with altered MBNL activity were only associated with (CTG)480, and not (CAG) 480 (Ho *et al.*, 2005). These data suggest that foci formation is not sufficient for repeats to affect alternative splicing and confer toxicity. In addition, Mahadevan *et al.* (2006) showed that expression of the *DMPK* 3′UTR carrying as short as five CTG repeats in transgenic mice causes cardinal features characteristic of DM1 including myotonia, cardiac conduction abnormalities, and abnormal RNA splicing, in the absence of detectable nuclear foci formation, suggesting that foci are not required for CUG repeat toxicity (Mahadevan *et al.*, 2006). Together, these data suggest that foci formation, although a hallmark feature characteristic of CUG repeat RNAs, may be neither necessary nor sufficient for toxicity, at least in animal and cell culture models. However, we caution that these data cannot not rule out the possibility that ribonuclear foci contribute to some aspects of CTG toxicity. Further studies are needed to confirm their potential significance.

2. *Critical Roles of RNA-Binding Proteins in RNA Repeat Diseases*

RNA-binding proteins such as MBNL1 and CUGBP1 are known to play critical roles in CUG toxicity diseases such as DM1 and SCA8 (Ranum and Cooper, 2006). As noted above, studies in fly models further highlight the importance of these RNA-binding proteins in noncoding repeat RNA toxicity at the organismal level (de Haro *et al.*, 2006; Garcia-Lopez *et al.*, 2008). Similarly, two CGG RNA repeat binding proteins, Pur α and hnRNP A2/B1, were found to suppress CGG repeat RNA toxicity in a fly model of FXTAS (Jin *et al.*, 2007; Sofola *et al.*, 2007). This finding suggests that sequestration of CGG-binding proteins may contribute to pathogenesis in FXTAS disease. Interestingly, CUGBP1 was found to also interact with CGG repeat expansion through hnRNPA2/B1 in the *Drosophila* model of FXTAS (Sofola *et al.*, 2007). Moreover, MBNL1 was identified as one of the CGG-binding proteins in the postmortem brain tissue of FXTAS patients (Iwahashi *et al.*, 2006). These results raise the possibility that RNA-binding proteins such as MBNL1 and CUGBP1 may have multiple activities in mediating pathogenesis of distinct RNA expansion diseases, and thus underlie some of the commonality among these disorders. Together these data suggest that RNA-binding proteins play important roles in mediating repeat RNA toxicity, and this rationalizes various approaches to studying RNA-binding proteins in such RNA-associated diseases.

3. *CAG Repeat RNA Toxicity Contributes to SCA3 Pathogenesis*

Expanded CAG RNAs can form similar hairpin structures as CUG expansion RNAs (Kiliszek *et al.*, 2009, 2010). Moreover, some of the RNA-binding proteins, such as MBNL1, are able to interact with both CUG and CAG repeat expansions (Ho *et al.*, 2005; Yuan *et al.*, 2007). In a genetic screen for modifiers of SCA3 toxicity, *Drosophila* Mbl was identified as an enhancer of SCA3 toxicity (Li *et al.*, 2008). This promoted the question of whether the RNA in polyglutamine diseases could have a toxicity, independent of the protein. Changing the polyglutamine-coding sequences from a pure CAG expansion to an interrupted CAA/CAG expansion (which encodes the identical polyglutamine protein, but would no longer make a hairpin structure) shifted the toxicity. Further, expression of a noncoding CAG expansion in the 3′UTR region of a DsRed reporter gene also leads to neuronal toxicity in a repeat length-dependent manner (Li *et al.*, 2008). Non-coding CAG repeat expansions have been found to cause deleterious effects in muscle and sperm in transgenic mice, indicating that a CAG toxicity mechanism also exists in mammalian systems (Hsu *et al.*, 2011). These results raise the possibility that RNA toxicity may constitute a pathogenic component in polyglutamine diseases such as SCA3.

4. *Antisense Transcripts in Trinucleotide Repeat Diseases*

Many the genomic loci show evidence of transcription from both sense and antisense strands in the mammalian genome and these include most of the trinucleotide repeat disease genes (Dion and Wilson, 2009; Ge *et al.*, 2006; He *et al.*, 2008; Katayama *et al.*, 2005; Yelin *et al.*, 2003). To date, antisense transcription spanning the repeat region has been extensively studied in SCA8 and DM1 (Cho *et al.*, 2005; Moseley *et al.*, 2006; Zu *et al.*, 2011). Although long exogenous double-stranded RNAs elicit the interferon response in mammalian cells (Ui-Tei *et al.*, 2000), studies indicate that endogenous convergent transcripts, and presumably resultant endogenous double-stranded RNAs, may be processed into ~21–23nt small RNAs in human cells (Kawaji *et al.*, 2008; Yang and Kazazian, 2006). Indeed, data in mammalian cell culture suggest that, in DM1, the sense and antisense transcript spanning the CTG expansion region in 3′UTR of *DMPK* interact to produce small RNAs with local effects on chromatin silencing (Cho *et al.*, 2005). An interesting question remains whether repeat expansions in the sense and antisense transcripts in trinucleotide diseases such as DM1 can interact through the RNA interference pathway. If so, then trinucleotide repeat-derived small RNAs may have widespread effects, given the abundance of genes in the human genome that contain short CAG or CTG tracts. Coexpression of a noncoding CAG RNA expansion together with a noncoding CUG RNA expansion indeed leads to dramatically enhanced toxicity in flies (Yu *et al.*, 2011a). This toxicity is *dicer-2* and *ago-2*-dependent, and is associated with ~21nt repeat-derived siRNAs. Downregulation of CAG-containing genes by siRNAs derived from CUG expansions is at least partially responsible for the toxicity (Yu *et al.*, 2011a). Although it remains unknown whether similar mechanisms operate in the physiological disease situation, these results raised the possibility that antisense transcripts spanning repeat region, which may occur frequently in trinucleotide repeat diseases, may contribute to pathogenesis by interacting with the sense transcripts through the RNA interference pathway.

E. Instability of CAG Repeats—Insights From the *Drosophila* Model

Most of the trinucleotide repeat diseases share a remarkable clinical feature called genetic anticipation. The diseases become progressively worse and with earlier onset from generation to generation (Cummings and Zoghbi, 2000; Ohara, 2001; Orr and Zoghbi, 2007). It is thought that instability of the repeat in the germline cells underlies this process, such that the repeat expands from generation to generation. Somatic instability of the repeats is also observed (Chong *et al.*, 1995; Ito *et al.*, 1998; Telenius *et al.*, 1994; Ueno *et al.*, 1995) and may influence disease onset (Swami *et al.*, 2009). As repeat instability is the cause of the trinucleotide repeat diseases, the mechanisms of instability may provide insight into the fundamental aspects of genome integrity that underlie these many disorders.

Drosophila does show repeat instability, in the germline, in a transcription-dependent manner, meaning if the repeat is expressed, the repeat will undergo changes from generation to generation (Jung and Bonini, 2007). The instability correlates with repeat length and is biased for expansion, which is similar to that observed in mammals (Ashley and Warren, 1995; Fortune *et al.*, 2000; Wheeler *et al.*, 1999). The integrity of transcription coupled repair and/or nuclear excision repair seems important for repeat instability, as loss of function of *mus201*, a gene critical for both pathways, leads to greatly reduced instability (Jung and Bonini, 2007). Animals heterozygous for CREB (cAMP response element binding protein)-binding protein (CBP), a histone acetyltransferase (HAT) protein that plays critical role in DNA repair and is implicated in the toxicity of polyglutamine diseases (Steffan *et al.*, 2001), cause increased repeat instability. Further, pharmacological inhibition of histone deacetylase (HDAC) activity decreases CAG repeat instability, suggesting that the balance between the activities of HAT and HDAC in cells may be important in determining trinucleotide repeat stability (Jung and Bonini, 2007). As CBP HAT activity is thought to be inhibited by polyglutamine protein in diseases (Bodai *et al.*, 2003; Steffan *et al.*, 2001), these results raise the possibility that polyglutamine toxicity may also contribute to disease pathogenesis by increasing repeat stability and repeat length in cells, thereby forming a positive feedback loop. In support of this notion, coexpression of an expanded polyglutamine was found to enhance instability of a noncoding (CAG)270 expanded repeat in flies (Jung *et al.*, 2011). Given that little is known about the mechanisms of repeat instability and most of the studies so far have taken candidate gene approaches, unbiased genetic screens for modifiers of repeat instability in *Drosophila* models may reveal insight into the fundamental etiology of repeat expansion diseases possibly shared by all or many unstable repeat diseases (Jung *et al.*, 2011).

III. Conclusion

Drosophila is a powerful system for studying human trinucleotide repeat diseases, including polyglutamine diseases and RNA-based toxicity diseases. Genetic modifiers identified using forward genetic screens or candidate gene approaches have provided valuable insights into pathogenic mechanisms. With the rapid advance of fly molecular genetic tools, it is likely that additional screens will result in additional modifiers, leading to greater understanding of pathogenic mechanisms. This understanding should provide the foundation for therapeutic targets. In addition, *Drosophila* disease models can be used for chemical compound screens to identify possible therapeutic drugs (Chang *et al.*, 2008; Garcia-Lopez *et al.*, 2008). Thus, application of the fly system to human diseases, including the trinucleotide repeat diseases, will continue to prove a fruitful approach.

References

Al-Ramahi, I., Perez, A.M., Lim, J., Zhang, M., Sorensen, R., de Haro, M., Branco, J., Pulst, S.M., Zoghbi, H.Y., and Botas, J. (2007). dAtaxin-2 mediates expanded Ataxin-1-induced neurodegeneration in a Drosophila model of SCA1. *PLoS Genet.* **3**, e234.

Ambegaokar, S.S., Roy, B., and Jackson, G.R. (2010). Neurodegenerative models in Drosophila: polyglutamine disorders, Parkinson disease, and amyotrophic lateral sclerosis. *Neurobiol. Dis.* **40**, 29–39.

Anglade, P., Vyas, S., Javoy-Agid, F., Herrero, M.T., Michel, P.P., Marquez, J., Mouatt-Prigent, A., Ruberg, M., Hirsch, E.C., and Agid, Y. (1997). Apoptosis and autophagy in nigral neurons of patients with Parkinson's disease. *Histol. Histopathol.* **12**, 25–31.

Ashley Jr., C.T., and Warren, S.T. (1995). Trinucleotide repeat expansion and human disease. *Annu. Rev. Genet.* **29**, 703–728.

Bellen, H.J., Levis, R.W., Liao, G., He, Y., Carlson, J.W., Tsang, G., Evans-Holm, M., Hiesinger, P.R., Schulze, K.L., and Rubin, G.M et al., (2004). The BDGP gene disruption project: single transposon insertions associated with 40% of Drosophila genes. *Genetics* **167**, 761–781.

Bence, N.F., Sampat, R.M., and Kopito, R.R. (2001). Impairment of the ubiquitin-proteasome system by protein aggregation. *Science (New York, NY)* **292**, 1552–1555.

Bennett, E.J., Bence, N.F., Jayakumar, R., and Kopito, R.R. (2005). Global impairment of the ubiquitin-proteasome system by nuclear or cytoplasmic protein aggregates precedes inclusion body formation. *Mol. Cell* **17**, 351–365.

Bett, J.S., Goellner, G.M., Woodman, B., Pratt, G., Rechsteiner, M., and Bates, G.P. (2006). Proteasome impairment does not contribute to pathogenesis in R6/2 Huntington's disease mice: exclusion of proteasome activator REGgamma as a therapeutic target. *Hum. Mol Genet.* **15**, 33–44.

Bilen, J., and Bonini, N.M. (2005). Drosophila as a model for human neurodegenerative disease. *Annu. Rev. Genet.* **39**, 153–171.

Bilen, J., and Bonini, N.M. (2007). Genome-wide screen for modifiers of ataxin-3 neurodegeneration in Drosophila. *PLoS Genet.* **3**, 1950–1964.

Bodai, L., Pallos, J., Thompson, L.M., and Marsh, J.L. (2003). Altered protein acetylation in polyglutamine diseases. *Curr. Med. Chem.* **10**, 2577–2587.

Bowman, A.B., Yoo, S.Y., Dantuma, N.P., and Zoghbi, H.Y. (2005). Neuronal dysfunction in a polyglutamine disease model occurs in the absence of ubiquitin-proteasome system impairment and inversely correlates with the degree of nuclear inclusion formation. *Hum. Mol. Genet.* **14**, 679–691.

Brand, A.H., and Perrimon, N. (1993). Targeted gene expression as a means of altering cell fates and generating dominant phenotypes. *Development (Cambridge, England)* **118**, 401–415.

Cauchi, R.J., and van den Heuvel, M. (2006). The fly as a model for neurodegenerative diseases: is it worth the jump? *Neurodegener. Dis.* **3**, 338–356.

Chan, H.Y., Warrick, J.M., Gray-Board, G.L., Paulson, H.L., and Bonini, N.M. (2000). Mechanisms of chaperone suppression of polyglutamine disease: selectivity, synergy and modulation of protein solubility in Drosophila. *Hum. Mol. Genet.* **9**, 2811–2820.

Chang, S., Bray, S.M., Li, Z., Zarnescu, D.C., He, C., Jin, P., and Warren, S.T. (2008). Identification of small molecules rescuing fragile X syndrome phenotypes in Drosophila. *Nat. Chem. Biol.* **4**, 256–263.

Chen, H.K., Fernandez-Funez, P., Acevedo, S.F., Lam, Y.C., Kaytor, M.D., Fernandez, M.H., Aitken, A., Skoulakis, E.M., Orr, H.T., and Botas, J et al., (2003). Interaction of Akt-phosphorylated ataxin-1 with 14-3-3 mediates neurodegeneration in spinocerebellar ataxia type 1. *Cell* **113**, 457–468.

Cho, D.H., Thienes, C.P., Mahoney, S.E., Analau, E., Filippova, G.N., and Tapscott, S.J. (2005). Antisense transcription and heterochromatin at the DM1 CTG repeats are constrained by CTCF. *Mol. Cell* **20**, 483–489.

Chong, S.S., McCall, A.E., Cota, J., Subramony, S.H., Orr, H.T., Hughes, M.R., and Zoghbi, H.Y. (1995). Gametic and somatic tissue-specific heterogeneity of the expanded SCA1 CAG repeat in spinocerebellar ataxia type 1. *Nat. Genet.* **10**, 344–350.

Cummings, C.J., Mancini, M.A., Antalffy, B., DeFranco, D.B., Orr, H.T., and Zoghbi, H.Y. (1998). Chaperone suppression of aggregation and altered subcellular proteasome localization imply protein misfolding in SCA1. *Nat. Genet.* **19**, 148–154.

Cummings, C.J., Sun, Y., Opal, P., Antalffy, B., Mestril, R., Orr, H.T., Dillmann, W.H., and Zoghbi, H.Y. (2001). Over-expression of inducible HSP70 chaperone suppresses neuropathology and improves motor function in SCA1 mice. *Hum. Mol. Genet.* **10**, 1511–1518.

Cummings, C.J., and Zoghbi, H.Y. (2000). Fourteen and counting: unraveling trinucleotide repeat diseases. *Hum. Mol. Genet.* **9**, 909–916.

Daughters, R.S., Tuttle, D.L., Gao, W., Ikeda, Y., Moseley, M.L., Ebner, T.J., Swanson, M.S., and Ranum, L.P. (2009). RNA gain-of-function in spinocerebellar ataxia type 8. *PLoS Genet.* **5**, e1000600.

Davies, S.W., Turmaine, M., Cozens, B.A., DiFiglia, M., Sharp, A.H., Ross, C.A., Scherzinger, E., Wanker, E.E., Mangiarini, L., and Bates, G.P. (1997). Formation of neuronal intranuclear inclusions underlies the neurological dysfunction in mice transgenic for the HD mutation. *Cell* **90**, 537–548.

Day, J.W., and Ranum, L.P. (2005). RNA pathogenesis of the myotonic dystrophies. *Neuromuscul. Disord.* **15**, 5–16.

de Haro, M., Al-Ramahi, I., De Gouyon, B., Ukani, L., Rosa, A., Faustino, N.A., Ashizawa, T., Cooper, T.A., and Botas, J. (2006). MBNL1 and CUGBP1 modify expanded CUG-induced toxicity in a Drosophila model of myotonic dystrophy type 1. *Hum. Mol. Genet.* **15**, 2138–2145.

Diaz-Hernandez, M., Hernandez, F., Martin-Aparicio, E., Gomez-Ramos, P., Moran, M.A., Castano, J.G., Ferrer, I., Avila, J., and Lucas, J.J. (2003). Neuronal induction of the immunoproteasome in Huntington's disease. *J. Neurosci.* **23**, 11653–11661.

Dietzl, G., Chen, D., Schnorrer, F., Su, K.C., Barinova, Y., Fellner, M., Gasser, B., Kinsey, K., Oppel, S., and Scheiblauer, S et al., (2007). A genome-wide transgenic RNAi library for conditional gene inactivation in Drosophila. *Nature* **448**, 151–156.

DiFiglia, M., Sapp, E., Chase, K.O., Davies, S.W., Bates, G.P., Vonsattel, J.P., and Aronin, N. (1997). Aggregation of huntingtin in neuronal intranuclear inclusions and dystrophic neurites in brain. *Science (New York, NY)* **277**, 1990–1993.

Ding, Q., Lewis, J.J., Strum, K.M., Dimayuga, E., Bruce-Keller, A.J., Dunn, J.C., and Keller, J.N. (2002). Polyglutamine expansion, protein aggregation, proteasome activity, and neural survival. *J. Biol. Chem.* **277**, 13935–13942.

Dion, V., and Wilson, J.H. (2009). Instability and chromatin structure of expanded trinucleotide repeats. *Trends Genet.* **25**, 288–297.

Duenas, A.M., Goold, R., and Giunti, P. (2006). Molecular pathogenesis of spinocerebellar ataxias. *Brain* **129**, 1357–1370.

Elden, A.C., Kim, H.J., Hart, M.P., Chen-Plotkin, A.S., Johnson, B.S., Fang, X., Armakola, M., Geser, F., Greene, R., and Lu, M.M et al., (2010). Ataxin-2 intermediate-length polyglutamine expansions are associated with increased risk for ALS. *Nature* **466**, 1069–1075.

Fernandez-Funez, P., Nino-Rosales, M.L., de Gouyon, B., She, W.C., Luchak, J.M., Martinez, P., Turiegano, E., Benito, J., Capovilla, M., and Skinner, P.J et al., (2000). Identification of genes that modify ataxin-1-induced neurodegeneration. *Nature* **408**, 101–106.

Fortune, M.T., Vassilopoulos, C., Coolbaugh, M.I., Siciliano, M.J., and Monckton, D.G. (2000). Dramatic, expansion-biased, age-dependent, tissue-specific somatic mosaicism in a transgenic mouse model of triplet repeat instability. *Hum. Mol. Genet.* **9**, 439–445.

Garcia-Lopez, A., Monferrer, L., Garcia-Alcover, I., Vicente-Crespo, M., Alvarez-Abril, M.C., and Artero, R.D. (2008). Genetic and chemical modifiers of a CUG toxicity model in Drosophila. *PloS One* **3**, e1595.

Ge, X., Wu, Q., Jung, Y.-C., Chen, J., and Wang, S.M. (2006). A large quantity of novel human antisense transcripts detected by LongSAGE. *Bioinformatics* **22**, 2475–2479.

Gunawardena, S., Her, L.S., Brusch, R.G., Laymon, R.A., Niesman, I.R., Gordesky-Gold, B., Sintasath, L., Bonini, N.M., and Goldstein, L.S. (2003). Disruption of axonal transport by loss of huntingtin or expression of pathogenic polyQ proteins in Drosophila. *Neuron* **40**, 25–40.

He, Y., Vogelstein, B., Velculescu, V.E., Papadopoulos, N., and Kinzler, K.W. (2008). The antisense transcriptomes of human cells. *Science (New York, NY)* **322**, 1855–1857.

Ho, T.H., Savkur, R.S., Poulos, M.G., Mancini, M.A., Swanson, M.S., and Cooper, T.A. (2005). Colocalization of muscleblind with RNA foci is separable from mis-regulation of alternative splicing in myotonic dystrophy. *J. Cell Sci.* **118**, 2923–2933.

Houseley, J.M., Wang, Z., Brock, G.J., Soloway, J., Artero, R., Perez-Alonso, M., O'Dell, K.M., and Monckton, D.G. (2005). Myotonic dystrophy associated expanded CUG repeat muscleblind positive ribonuclear foci are not toxic to Drosophila. *Hum. Mol. Genet.* **14**, 873–883.

Hsu, R.J., Hsiao, K.M., Lin, M.J., Li, C.Y., Wang, L.C., Chen, L.K., and Pan, H. (2011). Long tract of untranslated CAG repeats is deleterious in transgenic mice. *PloS One* **6**, e16417.

Ikeda, Y., Daughters, R.S., and Ranum, L.P. (2008). Bidirectional expression of the SCA8 expansion mutation: one mutation, two genes. *Cerebellum (London, England)* **7**, 150–158.

Ito, Y., Tanaka, F., Yamamoto, M., Doyu, M., Nagamatsu, M., Riku, S., Mitsuma, T., and Sobue, G. (1998). Somatic mosaicism of the expanded CAG trinucleotide repeat in mRNAs for the responsible gene of Machado-Joseph disease (MJD), dentatorubral-pallidoluysian atrophy (DRPLA), and spinal and bulbar muscular atrophy (SBMA). *Neurochem. Res.* **23**, 25–32.

Iwahashi, C.K., Yasui, D.H., An, H.J., Greco, C.M., Tassone, F., Nannen, K., Babineau, B., Lebrilla, C.B., Hagerman, R.J., and Hagerman, P.J. (2006). Protein composition of the intranuclear inclusions of FXTAS. *Brain* **129**, 256–271.

Jackson, G.R., Salecker, I., Dong, X., Yao, X., Arnheim, N., Faber, P.W., MacDonald, M.E., and Zipursky, S.L. (1998). Polyglutamine-expanded human huntingtin transgenes induce degeneration of Drosophila photoreceptor neurons. *Neuron* **21**, 633–642.

Jin, P., Duan, R., Qurashi, A., Qin, Y., Tian, D., Rosser, T.C., Liu, H., Feng, Y., and Warren, S.T. (2007). Pur alpha binds to rCGG repeats and modulates repeat-mediated neurodegeneration in a Drosophila model of fragile X tremor/ataxia syndrome. *Neuron* **55**, 556–564.

Jin, P., Zarnescu, D.C., Zhang, F., Pearson, C.E., Lucchesi, J.C., Moses, K., and Warren, S.T. (2003). RNA-mediated neurodegeneration caused by the fragile X premutation rCGG repeats in Drosophila. *Neuron* **39**, 739–747.

Jung, J., and Bonini, N. (2007). CREB-binding protein modulates repeat instability in a Drosophila model for polyQ disease. *Science (New York, NY)* **315**, 1857–1859.

Jung, J., van Jaarsveld, M.T., Shieh, S.Y., Xu, K., and Bonini, N.M. (2011). Defining genetic factors that modulate intergenerational CAG repeat instability in Drosophila melanogaster. *Genetics* **187**, 61–71.

Katayama, S., Tomaru, Y., Kasukawa, T., Waki, K., Nakanishi, M., Nakamura, M., Nishida, H., Yap, C.C., Suzuki, M., and Kawai, J et al., (2005). Antisense transcription in the mammalian transcriptome. *Science (New York, NY)* **309**, 1564–1566.

Kawaji, H., Nakamura, M., Takahashi, Y., Sandelin, A., Katayama, S., Fukuda, S., Daub, C.O., Kai, C., Kawai, J., and Yasuda, J et al., (2008). Hidden layers of human small RNAs. *BMC Genomics* **9**, 157.

Kazemi-Esfarjani, P., and Benzer, S. (2000). Genetic suppression of polyglutamine toxicity in Drosophila. *Science (New York, NY)* **287**, 1837–1840.

Kiliszek, A., Kierzek, R., Krzyzosiak, W.J., and Rypniewski, W. (2009). Structural insights into CUG repeats containing the 'stretched U-U wobble': implications for myotonic dystrophy. *Nucleic Acids Res.* **37**, 4149–4156.

Kiliszek, A., Kierzek, R., Krzyzosiak, W.J., and Rypniewski, W. (2010). Atomic resolution structure of CAG RNA repeats: structural insights and implications for the trinucleotide repeat expansion diseases. *Nucleic Acids Res.* **38**, 8370–8376.

La Spada, A.R., and Taylor, J.P. (2010). Repeat expansion disease: progress and puzzles in disease pathogenesis. *Nat. Rev.* **11**, 247–258.

La Spada, A.R., Wilson, E.M., Lubahn, D.B., Harding, A.E., and Fischbeck, K.H. (1991). Androgen receptor gene mutations in X-linked spinal and bulbar muscular atrophy. *Nature* **352**, 77–79.

Lander, E.S., Linton, L.M., Birren, B., Nusbaum, C., Zody, M.C., Baldwin, J., Devon, K., Dewar, K., Doyle, M., and FitzHugh, W et al., (2001). Initial sequencing and analysis of the human genome. *Nature* **409**, 860–921.

Lee, W.C., Yoshihara, M., and Littleton, J.T. (2004). Cytoplasmic aggregates trap polyglutamine-containing proteins and block axonal transport in a Drosophila model of Huntington's disease. *Proc. Natl. Acad. Sci. U S A* **101**, 3224–3229.

Lessing, D., and Bonini, N.M. (2008). Polyglutamine genes interact to modulate the severity and progression of neurodegeneration in Drosophila. *PLoS Biol.* **6**, e29.

Li, L.B., Yu, Z., Teng, X., and Bonini, N.M. (2008). RNA toxicity is a component of ataxin-3 degeneration in Drosophila. *Nature* **453**, 1107–1111.

Mahadevan, M.S., Yadava, R.S., Yu, Q., Balijepalli, S., Frenzel-McCardell, C.D., Bourne, T.D., and Phillips, L.H. (2006). Reversible model of RNA toxicity and cardiac conduction defects in myotonic dystrophy. *Nat. Genet.* **38**, 1066–1070.

Marsh, J.L., and Thompson, L.M. (2006). Drosophila in the study of neurodegenerative disease. *Neuron* **52**, 169–178.

Marsh, J.L., Walker, H., Theisen, H., Zhu, Y.Z., Fielder, T., Purcell, J., and Thompson, L.M. (2000). Expanded polyglutamine peptides alone are intrinsically cytotoxic and cause neurodegeneration in Drosophila. *Hum. Mol. Genet.* **9**, 13–25.

McMurray, C.T. (2010). Mechanisms of trinucleotide repeat instability during human development. *Nat. Rev.* **11**, 786–799.

Mirkin, S.M. (2007). Expandable DNA repeats and human disease. *Nature* **447**, 932–940.

Moseley, M.L., Zu, T., Ikeda, Y., Gao, W., Mosemiller, A.K., Daughters, R.S., Chen, G., Weatherspoon, M.R., Clark, H.B., and Ebner, T.J et al., (2006). Bidirectional expression of CUG and CAG expansion transcripts and intranuclear polyglutamine inclusions in spinocerebellar ataxia type 8. *Nat. Genet.* **38**, 758–769.

Mutsuddi, M., Marshall, C.M., Benzow, K.A., Koob, M.D., and Rebay, I. (2004). The spinocerebellar ataxia 8 noncoding RNA causes neurodegeneration and associates with staufen in Drosophila. *Curr. Biol.* **14**, 302–308.

Nedelsky, N.B., Pennuto, M., Smith, R.B., Palazzolo, I., Moore, J., Nie, Z., Neale, G., and Taylor, J.P. (2010). Native functions of the androgen receptor are essential to pathogenesis in a Drosophila model of spinobulbar muscular atrophy. *Neuron* **67**, 936–952.

Neumann, M., Sampathu, D.M., Kwong, L.K., Truax, A.C., Micsenyi, M.C., Chou, T.T., Bruce, J., Schuck, T., Grossman, M., and Clark, C.M et al., (2006). Ubiquitinated TDP-43 in frontotemporal lobar degeneration and amyotrophic lateral sclerosis. *Science (New York, NY)* **314**, 130–133.

Nixon, R.A., Wegiel, J., Kumar, A., Yu, W.H., Peterhoff, C., Cataldo, A., and Cuervo, A.M. (2005). Extensive involvement of autophagy in Alzheimer disease: an immuno-electron microscopy study. *J. Neuropathol. Exp. Neurol.* **64**, 113–122.

Ohara, K. (2001). Anticipation, imprinting, trinucleotide repeat expansions and psychoses. *Prog. Neuropsychopharmacol. Biol Psychiatry* **25**, 167–192.

Orr, H.T., and Zoghbi, H.Y. (2007). Trinucleotide repeat disorders. *Annu. Rev. Neurosci.* **30**, 575–621.

Osborne, R.J., and Thornton, C.A. (2006). RNA-dominant diseases. *Hum. Mol. Genet.* **2**, R162–R169.

Pandey, U.B., Nie, Z., Batlevi, Y., McCray, B.A., Ritson, G.P., Nedelsky, N.B., Schwartz, S.L., DiProspero, N.A., Knight, M.A., and Schuldiner, O et al., (2007). HDAC6 rescues neurodegeneration and provides an essential link between autophagy and the UPS. *Nature* **447**, 859–863.

Parks, A.L., Cook, K.R., Belvin, M., Dompe, N.A., Fawcett, R., Huppert, K., Tan, L.R., Winter, C.G., Bogart, K.P., and Deal, J.E et al., (2004). Systematic generation of high-resolution deletion coverage of the Drosophila melanogaster genome. *Nat. Genet.* **36**, 288–292.

Paulson, H.L., Perez, M.K., Trottier, Y., Trojanowski, J.Q., Subramony, S.H., Das, S.S., Vig, P., Mandel, J.L., Fischbeck, K.H., and Pittman, R.N. (1997). Intranuclear inclusions of expanded polyglutamine protein in spinocerebellar ataxia type 3. *Neuron* **19**, 333–344.

Pearson, C.E., Nichol Edamura, K., and Cleary, J.D. (2005). Repeat instability: mechanisms of dynamic mutations. *Nat. Rev.* **6**, 729–742.

Perez, M.K., Paulson, H.L., Pendse, S.J., Saionz, S.J., Bonini, N.M., and Pittman, R.N. (1998). Recruitment and the role of nuclear localization in polyglutamine-mediated aggregation. *J. Cell Biol.* **143**, 1457–1470.

Ranum, L.P., and Cooper, T.A. (2006). RNA-mediated neuromuscular disorders. *Annu. Rev. Neurosci.*

Romero, E., Cha, G.H., Verstreken, P., Ly, C.V., Hughes, R.E., Bellen, H.J., and Botas, J. (2008). Suppression of neurodegeneration and increased neurotransmission caused by expanded full-length huntingtin accumulating in the cytoplasm. *Neuron* **57**, 27–40.

Ryder, E., Ashburner, M., Bautista-Llacer, R., Drummond, J., Webster, J., Johnson, G., Morley, T., Chan, Y.S., Blows, F., and Coulson, D et al., (2007). The DrosDel deletion collection: a Drosophila genomewide chromosomal deficiency resource. *Genetics* **177**, 615–629.

Sapp, E., Schwarz, C., Chase, K., Bhide, P.G., Young, A.B., Penney, J., Vonsattel, J.P., Aronin, N., and DiFiglia, M. (1997). Huntingtin localization in brains of normal and Huntington's disease patients. *Ann. Neurol.* **42**, 604–612.

Satterfield, T.F., Jackson, S.M., and Pallanck, L.J. (2002). A Drosophila homolog of the polyglutamine disease gene SCA2 is a dosage-sensitive regulator of actin filament formation. *Genetics* **162**, 1687–1702.

Shulman, J.M., Shulman, L.M., Weiner, W.J., and Feany, M.B. (2003). From fruit fly to bedside: translating lessons from Drosophila models of neurodegenerative disease. *Curr. Opin. Neurol.* **16**, 443–449.

Sikorska, B., Liberski, P.P., Giraud, P., Kopp, N., and Brown, P. (2004). Autophagy is a part of ultrastructural synaptic pathology in Creutzfeldt-Jakob disease: a brain biopsy study. *Int. J. Biochem. Cell Biol.* **36**, 2563–2573.

Skinner, P.J., Koshy, B.T., Cummings, C.J., Klement, I.A., Helin, K., Servadio, A., Zoghbi, H.Y., and Orr, H.T. (1997). Ataxin-1 with an expanded glutamine tract alters nuclear matrix-associated structures. *Nature* **389**, 971–974.

Sofola, O.A., Jin, P., Qin, Y., Duan, R., Liu, H., de Haro, M., Nelson, D.L., and Botas, J. (2007). RNA-binding proteins hnRNP A2/B1 and CUGBP1 suppress fragile X CGG premutation repeat-induced neurodegeneration in a Drosophila model of FXTAS. *Neuron* **55**, 565–571.

Steffan, J.S., Bodai, L., Pallos, J., Poelman, M., McCampbell, A., Apostol, B.L., Kazantsev, A., Schmidt, E., Zhu, Y.Z., and Greenwald, M et al., (2001). Histone deacetylase inhibitors arrest polyglutamine-dependent neurodegeneration in Drosophila. *Nature* **413**, 739–743.

Swami, M., Hendricks, A.E., Gillis, T., Massood, T., Mysore, J., Myers, R.H., and Wheeler, V.C. (2009). Somatic expansion of the Huntington's disease CAG repeat in the brain is associated with an earlier age of disease onset. *Hum. Mol. Genet.* **18**, 3039–3047.

Takeyama, K., Ito, S., Yamamoto, A., Tanimoto, H., Furutani, T., Kanuka, H., Miura, M., Tabata, T., and Kato, S. (2002). Androgen-dependent neurodegeneration by polyglutamine-expanded human androgen receptor in Drosophila. *Neuron* **35**, 855–864.

Telenius, H., Kremer, B., Goldberg, Y.P., Theilmann, J., Andrew, S.E., Zeisler, J., Adam, S., Greenberg, C., Ives, E.J., and Clarke, L.A et al., (1994). Somatic and gonadal mosaicism of the Huntington disease gene CAG repeat in brain and sperm. *Nat. Genet.* **6**, 409–414.

Thibault, S.T., Singer, M.A., Miyazaki, W.Y., Milash, B., Dompe, N.A., Singh, C.M., Buchholz, R., Demsky, M., Fawcett, R., and Francis-Lang, H.L et al., (2004). A complementary transposon tool kit for Drosophila melanogaster using P and piggyBac. *Nat. Genet.* **36**, 283–287.

Uchihara, T., Fujigasaki, H., Koyano, S., Nakamura, A., Yagishita, S., and Iwabuchi, K. (2001). Non-expanded polyglutamine proteins in intranuclear inclusions of hereditary ataxias—triple-labeling immunofluorescence study. *Acta Neuropathol.* **102**, 149–152.

Ueno, S., Kondoh, K., Kotani, Y., Komure, O., Kuno, S., Kawai, J., Hazama, F., and Sano, A. (1995). Somatic mosaicism of CAG repeat in dentatorubral-pallidoluysian atrophy (DRPLA). *Hum. Mol. Genet.* **4**, 663–666.

Ui-Tei, K., Zenno, S., Miyata, Y., and Saigo, K. (2000). Sensitive assay of RNA interference in Drosophila and Chinese hamster cultured cells using firefly luciferase gene as target. *FEBS Lett.* **479**, 79–82.

Verkerk, A.J., Pieretti, M., Sutcliffe, J.S., Fu, Y.H., Kuhl, D.P., Pizzuti, A., Reiner, O., Richards, S., Victoria, M.F., and Zhang, F.P et al., (1991). Identification of a gene (FMR-1) containing a CGG repeat coincident with a breakpoint cluster region exhibiting length variation in fragile X syndrome. *Cell* **65**, 905–914.

Warrick, J.M., Chan, H.Y., Gray-Board, G.L., Chai, Y., Paulson, H.L., and Bonini, N.M. (1999). Suppression of polyglutamine-mediated neurodegeneration in Drosophila by the molecular chaperone HSP70. *Nat. Genet.* **23**, 425–428.

Warrick, J.M., Morabito, L.M., Bilen, J., Gordesky-Gold, B., Faust, L.Z., Paulson, H.L., and Bonini, N. M. (2005). Ataxin-3 suppresses polyglutamine neurodegeneration in Drosophila by a ubiquitin-associated mechanism. *Mol. Cell* **18**, 37–48.

Warrick, J.M., Paulson, H.L., Gray-Board, G.L., Bui, Q.T., Fischbeck, K.H., Pittman, R.N., and Bonini, N.M. (1998). Expanded polyglutamine protein forms nuclear inclusions and causes neural degeneration in Drosophila. *Cell* **93**, 939–949.

Wheeler, V.C., Auerbach, W., White, J.K., Srinidhi, J., Auerbach, A., Ryan, A., Duyao, M.P., Vrbanac, V., Weaver, M., and Gusella, J.F et al., (1999). Length-dependent gametic CAG repeat instability in the Huntington's disease knock-in mouse. *Hum. Mol. Genet.* **8**, 115–122.

Whitworth, A.J., and Pallanck, L.J. (2008). Genetic models of Parkinson's disease: mechanisms and therapies. *SEB Exp. Biol. Ser.* **60**, 93–113.

Yang, N., and Kazazian Jr., H.H. (2006). L1 retrotransposition is suppressed by endogenously encoded small interfering RNAs in human cultured cells. *Nat. Struct. Mol. Biol.* **13**, 763–771.

Yelin, R., Dahary, D., Sorek, R., Levanon, E.Y., Goldstein, O., Shoshan, A., Diber, A., Biton, S., Tamir, Y., and Khosravi, R et al., (2003). Widespread occurrence of antisense transcription in the human genome. *Nat. Biotechnol.* **21**, 379–386.

Yu, Z., Teng, X., and Bonini, N.M. (2011 a) Triplet repeat-derived siRNAs enhance RNA-mediated toxicity in a Drosophila model for myotonic dystrophy. *PLoS Genet.* **7**(3); e1001340.

Yu, Z., Zhu, Y., Chen-Plotkin, A.S., Clay-Falcone, D., McCluskey, L., Elman, L., Kalb, R., Trojanowski, J.Q., Lee, V.M., and Van Deerlin, V.M et al., (2011b). PolyQ repeat expansions in ATXN2 associated with ALS are CAA interrupted repeats. *PloS One* **6**(3); 17951.

Yuan, Y., Compton, S.A., Sobczak, K., Stenberg, M.G., Thornton, C.A., Griffith, J.D., and Swanson, M.S. (2007). Muscleblind-like 1 interacts with RNA hairpins in splicing target and pathogenic RNAs. *Nucleic Acids Res.* **35**, 5474–5486.

Zoghbi, H.Y., and Botas, J. (2002). Mouse and fly models of neurodegeneration. *Trends Genet.* **18**, 463–471.

Zu, T., Gibbens, B., Doty, N.S., Gomes-Pereira, M., Huguet, A., Stone, M.D., Margolis, J., Peterson, M., Markowski, T.W., and Ingram, M.A et al., (2011). Non-ATG-initiated translation directed by microsatellite expansions. *Proc. Natl. Acad. Sci. U S A* **108**, 260–265.

FROM GENETICS TO STRUCTURE TO FUNCTION: EXPLORING SLEEP IN *DROSOPHILA*

Daniel Bushey and Chiara Cirelli

Department of Psychiatry, University of Wisconsin, 6001 Research Park Blvd.
Madison, 53719, WI, USA

Sleep consists of quiescent periods with reduced responsiveness to external stimuli. Despite being maladaptive in that when asleep, animals are less able to respond to dangerous stimuli; sleep behavior is conserved in all animal species studied to date. Thus, sleep must be performing at least one fundamental, conserved function that is necessary, and/or whose benefits outweigh its maladaptive consequences. Currently, there is no consensus on what that function might be. Over the last 10 years, multiple groups have started to characterize the molecular mechanisms and brain structures necessary for normal sleep in *Drosophila melanogaster*. These researchers are exploiting genetic tools developed in *Drosophila* over the past century to identify and manipulate gene expression. Forward genetic screens can identify molecular components in complex biological systems and once identified, these genes can be manipulated within specific brain areas to determine which neuronal groups are important to initiate and maintain sleep. Screening for mutations and brain regions necessary for normal sleep has revealed that several genes that affect sleep are involved in synaptic plasticity and have preferential expression in the mushroom bodies (MBs). Moreover, altering MB neuronal activity alters sleep. Previous genetic screens found that the same genes enriched in MB are necessary for learning and memory. Increasing evidence in mammals, including humans, points to a beneficial role for sleep in synaptic plasticity, learning and memory. Thus, results from both flies and mammals suggest a strong link between sleep need and wake plasticity.

INTERNATIONAL REVIEW OF
NEUROBIOLOGY, VOL. 99
DOI: 10.1016/B978-0-12-387003-2.00009-4

I. Introduction

Sleep is conserved across all the animal species that have been carefully studied so far (Cirelli and Tononi, 2008), and is necessary to maintain cognitive function and performance (Killgore, 2010). Yet, why and how sleep benefits the brain remains unclear. This surprising deficit arises from a simple fact, the brain is extremely complicated and the methods to directly assess the effects of sleep on brain functioning with high (cellular) resolution are limited and still suffer from many technical limitations. For instance, repeated *in vivo* two-photon imaging has recently been used in zebrafish (Appelbaum *et al.*, 2010) and mice (Maret *et al.*, 2011) to study how axonal terminals and single dendritic spines are affected by sleep and wake, but the analysis remains so far confined to few superficial areas of the brain. Most sleep studies have been conducted in mammals, especially rats and mice, and, to a much less extent, in birds (Lesku *et al.*, 2011). Sleep in mammals and birds shows electroencephalographic (EEG) patterns similar to those observed in human sleep: during nonrapid eye movement (NREM) sleep, which accounts for most of sleep, large slow waves predominate, whereas REM sleep is characterized by an "activated" high-frequency low-voltage pattern similar to that of wake. The rodent brain, however, is still very complex, and genetic molecular techniques have only recently been developed to probe the cellular mechanisms underlying sleep functions. This is why the use of simpler model organisms, including the zebrafish, *Danio reiro*, and the nematode, *Caenorhabditis elegans*, has been of great help (Cirelli and Tononi, 2008; Crocker and Sehgal, 2010; Harbison *et al.*, 2009; Zimmerman *et al.*, 2008).

Here we focus on *Drosophila melanogaster* as a simple system to investigate sleep. Work over the last 30 years has demonstrated that fruit flies are a practical system to explore complex behavior, including circadian rhythms. By using a combination of forward and reverse genetics to isolate fly mutants with abnormal circadian behavior, neurogeneticists were able to characterize the complex transcriptional feedback system that forms the basis of the circadian molecular clock, and to identify mammalian orthologues (Peschel and Helfrich-Forster, 2011). Current studies of fly sleep mainly rely on the same infrared-based technology originally designed to study circadian behavior, but with higher temporal resolution (seconds/minutes rather than hours/days). Sleep/wake are defined based on measures of locomotor activity: sleep is defined as any period of immobility >5 min because flies quiescent for >5 min have a reduced arousal threshold, which is the essential feature that distinguishes sleep from quiet wake (Bushey *et al.*, 2007, 2009; Cirelli *et al.*, 2005a).

The first articles on sleep/rest in *Drosophila* confirmed that the rest behavior observed in fruit flies shares most of the features of sleep in humans (Hendricks *et al.*, 2000; Shaw *et al.*, 2000). Quiescent periods in flies were entrained by the 24-h circadian cycle, occurring primarily at night. Interfering with these quiescent periods resulted in a homeostatic response with increased rest the following day. Sleep deprivation also resulted in decreased performance. Hypnotics and some stimulants (i.e., caffeine and modafinil) produced similar affects on sleep in *Drosophila* as they did in mammals. Moreover, quiescent periods were more abundant in young flies than in older flies. Together these results suggested that sleep behavior is present in flies, and may involve at least some of the same biochemical pathways known to affect sleep in mammals (reviewed in Cirelli (2003), Ho and Sehgal (2005), Shaw (2003)). Based on these encouraging results, several laboratories started genetic screens, using both reverse and forward approaches, searching for fly mutants with either reduced sleep time or alterations in the homeostatic response to sleep deprivation.

II. Successful (and Less Successful) Screenings of Sleep Phenotypes in Flies

As we will discuss below, the sleep screens are primarily concentrating on total sleep time over the 24-h period (Cirelli *et al.*, 2005a; Koh *et al.*, 2008; Wu *et al.*, 2008). Our group has also tried to identify mutations that specifically affect the homeostatic regulation of sleep by studying the response to 24 h of sleep loss. Flies were allowed to recover starting at light onset, that is, in the morning when they are normally awake. This is because flies sleep a lot during the night even in baseline conditions, and thus a further increase in sleep after sleep deprivation may not occur due to a ceiling effect. Almost all mutant lines tested so far showed an increase in sleep duration and a decrease in sleep fragmentation after 24 h of sleep deprivation. As in wild-type flies, the sleep rebound was most pronounced during the first 4–6 h immediately after the end of the sleep deprivation period, and in most cases did not persist the second day after sleep loss. Similarly to wild-type flies, most mutant lines only recovered a small fraction (10–40%) of the sleep lost. We have over the past several years identified several putative "sleep deprivation" mutants, but in all cases this phenotype was not confirmed after repeated testing. Most likely, this is because even in wild-type flies sleep rebound is extremely variable. Genetic screens work best when variation falls within a very narrow range and thus outliers can be readily identified. When a phenotype is variable and outliers are not easily discernable, false positives cannot be tested quickly enough to efficiently identify variations due to mutations. So thus far, screening specifically for homeostatic mutants has not been successful (but mutants that are both short sleeping and show abnormal response to sleep loss have been identified, see below).

Another likely reason why screening for sleep deprivation mutants has failed is due to the fact that in flies it is easier to assess sleep quantity than sleep quality (intensity). Even in mammals, the strongest and most consistent effects of sleep loss are not on total sleep duration (which may or may not increase, depending on the duration of sleep deprivation and on the time of day when animals are allowed to recover sleep), but on sleep depth: after sleep deprivation, sleep is more consolidated (with less fragmentation due to brief awakenings), and more "intense." In birds and mammals, this intensity is best measured using slow wave activity (SWA), which is defined as the EEG power spectrum in the 0.5–4 Hz range during NREM sleep: the longer and/or more "intense" is wake, the higher is SWA at sleep onset (Achermann and Borbely, 2003). Moreover, SWA changes are larger in the brain areas more directly affected by the wake experience. For example, birds watching for the first time a movie with only one eye open show increased SWA in the visual brain contralateral to the open eye (Lesku *et al.*, 2011). Also, in humans and rats, learning a motor task increases SWA specifically in the trained cortical area, and does so more than performing a previously learned task (Huber *et al.*, 2004a; Hanlon *et al.*, 2009). Interestingly, social enrichment increases sleep time also in *Drosophila*, demonstrating that experience in flies also contributes to sleep need (Ganguly-Fitzgerald *et al.*, 2006). In flies, however, we can measure some aspects of sleep depth (e.g., brief awakenings) (Huber *et al.*, 2004b), but sleep/wake electrophysiological recordings are invasive and difficult (Nitz *et al.*, 2002; Van Swinderen *et al.*, 2004), and no attempt has been done so far to implement "high-density" recordings as it is done in mammals.

A likely source of variability when studying the sleep deprivation phenotype also stems from the complex interaction between circadian and homeostatic factors, the influence of which can vary to a variable extent across individual flies. When allowing flies to sleep rebound in the morning, the homeostatic process will promote sleep, whereas the circadian process will oppose it. The influence of the latter can be removed genetically by using mutations affecting key canonical circadian genes. Some of these mutant flies (tim^{01}, per^{01}, Clk^{jrk}), which are arrhythmic (do not consolidate sleep during the subjective night in constant darkness), show prominent (100%, as compared to 30–40% in wild-type canton-S flies) rebound after 12–24 h of sleep deprivation in dark only conditions (Shaw *et al.*, 2002). cyc^{01} mutants are unique, in that they show an exaggerated sleep rebound after as little as 3 h of sleep deprivation, and die after longer periods of sleep loss (Shaw *et al.*, 2002). cyc^{01} mutants have reduced heat-shock protein expression and mutations affecting *hsp83*, a heat-shock chaperone, also result in an exaggerated homeostatic response and death after sleep deprivation, demonstrating an important role for stress response genes in protecting against the lethal effects of sleep loss (Shaw *et al.*, 2002).

Sleep can be extremely variable because it is a complex behavior dependent on both genetic factors and experience in both mammals and *Drosophila*. Genetic background can be controlled by testing flies in a common genetic background. Over successive generations mutants can be backcrossed into a specific background and compared to wild-type flies from the same background, or comparisons can be made between siblings that inherit a common background. However, it is much harder to control for "quality of wake," that is, individual experiences. Experience in the small glass tubes where flies are normally housed for sleep recordings (one fly/tube) is theoretically uniform, but the visual/mechanical/acoustic stimuli used during sleep deprivation may introduce a new level of variation. On the contrary, the glass tubes represent a very impoverished environment, where flies can only move a few centimeters, and cannot fly. As we will see below, compelling evidence in mammals and flies now shows that sleep is necessary to process/mitigate the effects of experience on neuronal function (e.g., (Hanlon *et al.*, 2009; Huber *et al.*, 2004a; Ganguly-Fitzgerald *et al.*, 2006). If so, then testing in a uniform, impoverished environment may reduce our ability to assess the effects of sleep on brain functions. In line with this, learning mutants do not show increases in sleep time after social experience (Ganguly-Fitzgerald *et al.*, 2006). Unfortunately, it is difficult to design an automated fly tracking system that can also monitor multiple flies in complex environment, but an effort in this direction should definitively be made.

A. Reverse Genetic Screen Demonstrates the Role of the cAMP/PKA/CREB Pathway in Sleep Regulation

In flies a role in sleep regulation was first demonstrated for the cAMP/PKA/CREB pathway using reverse genetics (Hendricks *et al.*, 2001). In reverse genetics, a candidate gene is mutated first, and then the effects of the mutation on sleep are assessed (from genotype to phenotype). Candidate genes are chosen based on their known function, which makes them likely to be relevant for sleep. The cAMP/PKA/CREB pathway was a good candidate because it was known that cAMP levels and CREB activity are important for learning and memory (discussed below), and that CREB expression in the fly brain shows a circadian pattern. Furthermore, CREB responsive transcription has a diurnal cycle. Hendricks *et al.* (2001) tested previously identified mutations and found that increasing cAMP or CREB activity decreased both sleep time and sleep rebound after sleep deprivation, while inhibiting cAMP or CREB had the opposite effect (Table I). Furthermore, they found that sleep deprivation increased CREB-driven reporter expression. A later study in mice found that mutations knocking down 2 of the 3 CREB mouse isoforms increased sleep time (Graves *et al.*, 2003). Finally, a more recent study in mice found that a few hours of sleep deprivation selectivity impair long-term synaptic potentiation (LTP) in the hippocampus, and do so specifically for the LTP forms that require cAMP and PKA signaling (Vecsey *et al.*, 2009). The

Table I
Mutations Affecting *Drosophila* Sleep, and Their Effects on Learning and Memory.

Table 1	Total Sleep Time	Arousal Threshold	Response to Sleep Deprivation (Homeostatic Response)	References (Effects on Sleep)
Potassium channels and genes affecting Shaker currents				
Shaker	Reduced sleep (LOF)	Normal	Normal	(Cirelli *et al.*, 2005a)
Quiver/sleepless	Reduced sleep (LOF)		No response to sleep deprivation	(Koh *et al.*, 2008)
Hyperkinetic	Reduced sleep (LOF)	Normal	Normal	(Bushey *et al.*, 2007)
Neurotransmitters				
Dopamine transporter (DAT)	Reduced sleep (fumin)	Reduced at low stimulation	Normal	(Kume *et al.*, 2005) (Wu *et al.*, 2008)
Octopamine biosynthesis: *Tyrosine decarboxylase 2 (Tdc2)/Tyramine* β *hydroxylase (Tbh)*	LOF mutations increase sleep; feeding flies octopamine decreases sleep	Increased		(Crocker and Sehgal, 2008)
d5-HT1A serotonin receptor	Reduced sleep time Increasing serotonin levels increases sleep			(Yuan *et al.*, 2006)
cAMP\PKA\CREB pathway				
Dunce (3′,5′-cyclic-AMP phosphodiesterase activity)	Reduced sleep (LOF)			(Hendricks *et al.*, 2001)
Rutabaga (adenylate cyclase activity)	Increased sleep (LOF); LOF mutations prevent experience dependent increases in sleep			(Hendricks *et al.*, 2001; Donlea *et al.*, 2009)
Cyclic-AMP response element binding protein 2 (Creb2)	Increased/decreased sleep		Increased sleep rebound after sleep loss	(Hendricks *et al.*, 2001)
cAMP-dependent protein kinase 1 (PKA-C1)	Expression of a constitutively active form reduces sleep			(Joiner *et al.*, 2006)
Epidermal growth factor				

Learning/memory	References (Effects on Learning/Memory)	Brain Areas With High Expression	Reference	Demonstrated Ability to Rescue Memory Deficits	References
Impaired courtship conditioning; Impaired aversive olfactory learning; Impaired short-term memory	(Bushey *et al.*, 2007) (Cowan and Siegel, 1984) (Cowan and Siegel, 1986)	MB	(Rogero *et al.*, 1997) (Schwarz *et al.*, 1990)		
		MB	(Wu *et al.*, 2010)		
Impaired short-term memory	(Bushey *et al.*, 2007)				
Poor memory retention	(Zhang *et al.*, 2008) (Seugnet *et al.*, 2008) (Waddell, 2010) (Schwaerzel *et al.*, 2003)			D1 receptor expression	(Seugnet *et al.*, 2008)
Impaired appetitive olfactory memory	(Schwaerzel *et al.*, 2003)	Octopamine receptor enriched in MB	(Han *et al.*, 1998)		
Impaired aversive olfactory memory	(Dudai *et al.*, 1976) (Qiu and Davis, 1993)	MB	(Qiu and Davis, 1993)		
Impaired memory	(Schwaerzel *et al.*, 2003) (Blum *et al.*, 2009)	MB	(Han *et al.*, 1992)	MB, PDF	(Schwaerzel *et al.*, 2003) (Blum *et al.*, 2009; Donlea *et al.*, 2009) blist
Impaired (OE enhances long-term memory?)	(Yin *et al.*, 1995) (Perazzona *et al.*, 2004) (Sakai *et al.*, 2004)	Enriched in cell bodies, not in neuropil	(Yin *et al.*, 1995)		
Impaired learning; Heterozygotes have normal memory and are resistant to age-related memory impairment	(Skoulakis *et al.*, 1993) (Yamazaki *et al.*, 2007)	MB	(Skoulakis *et al.*, 1993)		

(*Continued*)

Table I (*Continued*)

Table 1	Total Sleep Time	Arousal Threshold	Response to Sleep Deprivation (Homeostatic Response)	References (Effects on Sleep)
Epidermal growth factor receptor (EGFR)	LOF mutations prevent experience dependent increases in sleep			(Donlea *et al.*, 2009)
rhomboid (Rho)	OE increases sleep (in combination with star); LOF mutations reduce sleep		Dominant negative mutants recover less sleep after sleep deprivation	(Foltenyi *et al.*, 2007)
Star	OE increases sleep (in combination with rho)			(Foltenyi *et al.*, 2007)
s-Spitz	OE mutations increase sleep			(Foltenyi *et al.*, 2007)
Steroids				
Ecdysone receptor (EcR)	LOF mutations reduce sleep		LOF mutations reduce change in sleep bout duration after sleep loss	(Ishimoto and Kitamoto, 2010)
Molting defective (mld, DTS-3)	LOF mutations reduce sleep		LOF mutations reduce change in sleep bout duration after sleep loss	(Ishimoto and Kitamoto, 2010)
mRNA transport and control				
dFmr1	LOF mutations increase sleep; OE decreases sleep	LOF or OE increase arousal threshold	LOF or OE reduce sleep rebound	(Bushey *et al.*, 2009)
Heat-shock proteins/chaperones				
Binding immunoglobulin protein (BiP)	No effect		OE increases sleep rebound, LOF mutations reduce sleep rebound	(Naidoo *et al.*, 2007)
hsp83	No effect		LOF mutations increase sleep rebound	(Shaw *et al.*, 2002)
Circadian genes/genes expressed in circadian neurons				
Cycle	LOF mutations increases sleep during starvation		LOF mutations show high mortality rate after sleep deprivation and increased sleep rebound	(Shaw *et al.*, 2002) (Keene *et al.*, 2010)
Period			Normal	

Learning/memory	References (Effects on Learning/Memory)	Brain Areas With High Expression	Reference	Demonstrated Ability to Rescue Memory Deficits	References
		Pars Intercerebralis	(Foltenyi *et al.*, 2007)		
Impaired courtship	(Ishimoto *et al.*, 2009)			OE in MB increases sleep; LOF in MB impairs long-term memory	(Ishimoto *et al.*, 2009)
Olfactory learning and memory impaired, Impaired courtship	(Bolduc *et al.*, 2008) (Mcbride *et al.*, 2005)	MB	(Schenck *et al.*, 2002)	Greatest effect on sleep is in MB	(Bushey *et al.*, 2009)
	(Sakai *et al.*, 2004)				

(*Continued*)

Table I (*Continued*)

Table 1	Total Sleep Time	Arousal Threshold	Response to Sleep Deprivation (Homeostatic Response)	References (Effects on Sleep)
	Necessary for experience dependent increases in sleep			(Donlea *et al.*, 2009) (Shaw *et al.*, 2002) (Thimgan *et al.*, 2010)
Clock	LOF mutations increases sleep during starvation			(Keene *et al.*, 2010)
BLISTER (*bs*, serum response factor)	LOF mutations show no increase in sleep after social enrichment			(Donlea *et al.*, 2009)
Pigment-dispersing factor (PDF)	LOF mutations increase sleep			(Parisky *et al.*, 2008)
Resistant to dieldrin (Rdl)	Reduced sleep (LOF); increased sleep (GOF)			(Parisky *et al.*, 2008) (Agosto *et al.*, 2008)
Chromatin structure/microtubule dynamics				
Elongator Protein 3	LOF mutations reduce sleep			(Singh *et al.*, 2010)
Triglyceride storage				
Lipid Storage Droplet 2 (Lsd-2)	LOF mutations reduce sleep		LOF mutations show no sleep rebound	(Thimgan *et al.*, 2010)
Brummer	No effect		LOF mutations show increased sleep rebound	(Thimgan *et al.*, 2010)

GOF, gain of function; LOF, loss of function; OE, overexpression.

Learning/memory	References (Effects on Learning/Memory)	Brain Areas With High Expression	Reference	Demonstrated Ability to Rescue Memory Deficits	References
Impaired long-term memory and courtship (OE enhances it)					
		Expressed throughout the brain but rescue in sLNvs	(Donlea *et al.*, 2009)	Expressed throughout brain but rescue in sLNvs	(Donlea *et al* 2009)
(no studies on me mory) Increases bouton number at NMJ	(Singh *et al.*, 2010)				
		Fat Body	(Gronke *et al.*, 2003)		
		Fat Body	(Gronke *et al.*, 2005)		

study also demonstrated that the LTP impairment due to sleep loss resulted from increased nucleotide phosphodiesterase (PDE) activity (which breaks down cAMP), because selective PDE inhibition was able to rescue the LTP deficits, as well as the deficits in a hippocampus-dependent memory task.

B. Forward Genetic Screens Identify Shaker and Sleepless

A major barrier to studying sleep is the ignorance concerning the specific molecular events occurring during this behavioral state as compared to wake. Forward genetics is an unbiased approach in which prior knowledge of the "important" genes is not required, and novel genes can thus be discovered (Cirelli, 2009). The starting point is the phenotype, and a significant part of the work involves going back to the genotype to identify the responsible gene (from phenotype to genotype). Forward genetic methods include quantitative trait loci (QTLs) analysis and mutagenesis screening. In the latter, random small mutations are induced over the entire genome, and hundreds/thousands of mutated individuals are screened for the phenotype of interest. Insertional mutagenesis uses transposable elements (in flies) to induce mutations, whereas chemical mutagenesis uses ethylmethane sulfonate (EMS, in flies) or *N*-ethyl *N*-nitrosourea (ENU, in mice).

In our screen, mutant flies are continuously recorded for 1 week, including 2–3 baseline days, 24 h of sleep deprivation, and 1–3 days of recovery after sleep deprivation. Ten to 16 flies (4–7-day-old at the beginning of the experiment) are tested for each line. This relatively high number of flies is needed because sleep pattern and sleep amount, although consistent across different days in each individual adult fly, may vary among different flies. The analysis of thousand of lines (>15,000) has confirmed a significant difference between male and female flies: females sleep almost exclusively during the night, whereas males also show a long period of sleep in the middle of the day. The daily amount of sleep in the mutant lines tested so far shows a normal distribution, with female flies for most lines sleeping between 400 and 800 min/day, with a mean of ~600 min, similar to that of wild-type flies.

Our first EMS screen looked for mutations on the X chromosome. Screening hemizygous males (only inherit the mutagnized X-chromosome) for a short sleeping phenotype isolated X-chromosomes with a mutation producing a sleep phenotype. In genetic screens, the likelihood of identifying a mutation depends on the number of individual mutagenized chromosomes that can be tested. This is why *Drosophila* is so amply used in genetic screens, as hundreds of individuals containing a given mutagenized chromosome can be quickly generated and tested. The first X-chromosome EMS screening identified a short sleeping line carrying *minisleep*, a mutation in *Shaker (Sh)*, a gene coding the α subunit of a voltage-dependent potassium channel (Cirelli *et al.*, 2005a) (Table I). The screen used restrictive criteria to select short sleeping mutants, with only mutant lines whose sleep time was less than two standard deviations from the mean of all tested lines were studied. Thus, the screen did not identify other genes on the X chromosome, such as *dunce* and *Hyperkinetic (Hk)*, whose loss also results in a short sleeping phenotype (Bushey *et al.*, 2007; Hendricks *et al.*, 2001), but not as pronounced as that seen in *Shaker* mutants. Of note, *Hk* encodes the β (regulatory) subunit of the Shaker channel, and null *Hk* mutations decrease but do not abolish the Shaker current (since the α subunit is still functioning). Both *Sh* and *Hk* loss of function (null/hypomorphic) mutants that have a short sleeping phenotype also have short-term memory deficits in the heat box paradigm, as well as decreased life span (Bushey *et al.*, 2010). Further studies also found that short sleeping *Sh* mutants were also more resistant to volatile anesthetics (Tinklenberg *et al.*, 1991; Weber *et al.*, 2009).

Channels homologous to Sh in vertebrates have similar properties and, in both mammals and flies, the Sh current plays a major role in the control of membrane repolarization and transmitter release. Consistent with the results in flies, knocking out the *Sh* orthologue in mice, *Kcna2*, results in reduced NREM sleep (Douglas *et al.*, 2007). However, the *Kcna2* short sleeping phenotype is far from being as dramatic as in *Shaker* flies, perhaps because of redundancy—there is one *Shaker* gene in *Drosophila*, but at least 16 genes code for α subunits of voltage-dependent potassium channels in mammals (Misonou and Trimmer, 2004; Yuan and Chen, 2006). Finally in mice, the injection of an antibody against the kv1.2 potassium channel into the central medial thalamus induces arousal from anesthesia (Alkire *et al.*, 2009).

After the identification of *minisleep*, an independent genetic screen using insertional mutagenesis identified *Sleepless*, a mutation in the *quiver* (*qvr*) locus, which also shows very significantly reduced sleep time (to only ~2 h a day, 85% less than controls) (Koh *et al.*, 2008). *qvr* codes a ly-6/neurotoxin family member that has been shown to interact directly with the Shaker channel, and its loss reduces Shaker localization, kinetics, and current density (Wu *et al.*, 2010). Thus, two independent genetic screens identified a major role for the Shaker current in sleep regulation. There are nevertheless some interesting differences between *Shaker* and *Sleepless* flies, most notably that only the latter show a reduced homeostatic response, as indicated by no changes in sleep duration after sleep deprivation.

III. Identified Sleep Mutants Suggest a Link Between Sleep and Brain Plasticity, Especially in the Mushroom Bodies

Table I lists most of the sleep mutants identified in either forward or reverse genetic screens, and the distribution of most of the corresponding genes is shown in Fig. 1. The isolated genes include those coding for the voltage-dependent potassium channel Shaker, several neurotransmitters, molecules that are part of the cAMP pathway, steroids, heat-shock proteins/chaperones, circadian proteins, and proteins involved in mRNA transport (Fmr1) and chromatin structure. In most cases, previous research had found that these "sleep" genes are also important for

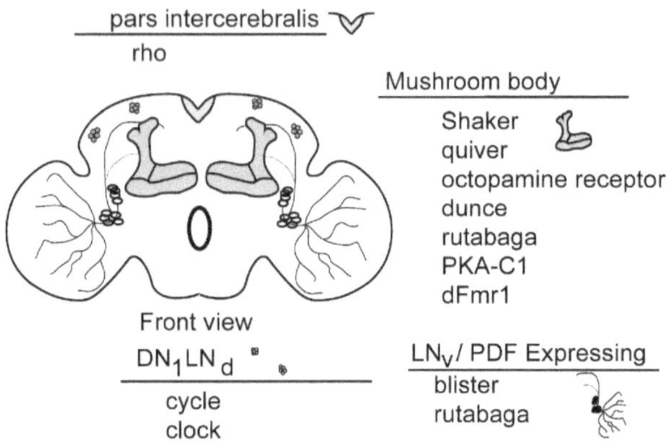

FIG. 1. "Sleep" genes and their distribution in the fly brain.

learning and/or memory, and many are enriched in the mushroom bodies (MBs), a brain area crucial for olfactory learning and memory. A genetic screen that restricted *UAS-shi*^{TS1} expression within specific neuronal groups, to block synaptic transmission in specific brain regions, found that silencing activity in the MB produced the greatest reduction in sleep time (Pitman *et al.*, 2006). Another group independently demonstrated the importance of the MB in sleep regulation after screening GAL4 lines in combination with the *UAS-mc** transgene, which expresses a constitutively active PKAc isoform (Joiner *et al.*, 2006). Finally, blocking neurotransmission within the MB protects against the impairment in learning caused by sleep deprivation (Li *et al.*, 2009).

For most of these mutations, transgenic rescue experiments confirm that the gene products are indeed necessary within the MB to restore normal learning and memory (Table I). The dual role in sleep and learning of two neurotransmitters, octopamine and dopamine, and their effects on the MB through activation of the PKA pathway, has been especially well characterized. Octopamine and dopamine are necessary for normal learning in appetitive and aversive learning, respectively (Schwaerzel *et al.*, 2003), and PKA may interact with Shaker and quiver to modulate excitability in the MB (Yao and Wu, 2001). The fragile x-mental retardation (*Fmr1*) gene product is also enriched in the MB (Schenck *et al.*, 2002). *Fmr1* protein product, FMRP, is present in dendritic spines (Feng *et al.*, 1997) and a hallmark of loss of function *Fmr1* mutations, in both flies and mammals, is the failure to remove immature synapses (Comery *et al.*, 1997; Hinton *et al.*, 1991; Irwin *et al.*, 2002; Pan *et al.*, 2004; Restivo *et al.*, 2005). *Fmr1* overexpression in flies results in the opposite phenotype, with dendritic and axonal underbranching and loss of synapse differentiation (Pan and Broadie, 2007; Pan *et al.*, 2004). In a previous study we found that sleep duration increases when *Fmr1* function is lost and decreases when *Fmr1* is overexpressed, even when overexpression is confined to the MB of adult flies (Bushey *et al.*, 2009).

MB are also closely involved in the temporal and spatial reorganization of cellular memory "traces," as identified using calcium imaging (Berry *et al.*, 2008). Aversive olfactory conditioning results in immediate memory traces (3–6 min) in the antennal lobe, a region necessary for short-term memory (Yu *et al.*, 2004), whereas intermediate memory traces occur in the DPM neuron that innervates the MB 30 min–1 h after training (Yu *et al.*, 2005). After aversive olfactory conditioning designed to trigger long-term memory, calcium influx occurs in the α/β lobes of the MB within 3 and 9 h after training, and persists for 24 h (Yu *et al.*, 2006). Of note, neurotransmission from the α/β lobes is also necessary for memory retrieval (Krashes *et al.*, 2007). The fact that these memory traces can persist over a 24-h period suggests that sleep may have a functional role, perhaps to favor the reorganization of memories across different brain circuits, similar to the "reactivation and redistribution of hippocampus-dependent memories to neocortical sites" proposed to occur during NREM sleep, at least in mammals (Diekelmann and Born, 2010; Rattenborg *et al.*, 2010). Unfortunately, calcium-imaging experiments are invasive, and not conducive to analysis of physiological behavioral states. Thus, it remains unknown whether sleep is important for the occurrence/transfer of these memory traces.

A recently characterized general mechanism that acts systemically to control memory formation and sleep need is that of the steroid hormone ecdysone (Table I). Ecdysone coordinates postembryonic development and is necessary for neuronal rewiring (Hewes, 2008). Administering ecdysone to adult flies increases sleep time in a dose-dependent manner, whereas blocking ecdysone synthesis (DTS-3) or the ecdysone receptor (EcR) reduces sleep time (Ishimoto and Kitamoto, 2010). Adding ecdysone during training also enhances courtship long-term memories, whereas adding ecdysone after training impairs memory (Ishimoto and Kitamoto, 2010). Both experience in the courtship conditioning assay and sleep deprivation increase ecdysone levels, while overexpressing the ecdysone receptor within the MB increases sleep time, and targeted knockdown using RNAi impairs long-term memory. Finally, treatment with ecdysone enhances CREB-dependent expression (Ishimoto and Kitamoto, 2010). Overall, these results show that the ecdysone system may represent a systemic mechanism capable not only of controlling brain plasticity in response to environmental stimulation, especially in the MB, but also of directly affecting sleep.

IV. Distinct Circadian Neurons Promote Either Wake or Sleep

Like humans, flies are more active during the day, and show the longest and most consolidated sleep bouts at night, even in constant darkness. Treating flies with a GABA antagonist results in increased sleep latency and decreased total sleep time, whereas mutations in the *Drosophila gamma-Aminobutyric acid (GABA)$_A$receptor* (*Rdl*) that increase the time the channel remains open result in reduced sleep latency and longer sleep (Agosto *et al.*, 2008) (Table I). The Rdl $GABA_A$ receptors are expressed in the ventral lateral clock neurons (LNvs) (Parisky *et al.*, 2008), which express the peptide pigment dispersing factor (PDF). PDF immunoreactivity peaks at the beginning of the light period (Park *et al.*, 2000), consistent with a role for PDF in initiating and/or maintaining arousal. In line with this, reducing PDF expression and enhancing neuronal excitability decrease sleep time (Parisky *et al.*, 2008). It has been suggested that PDF may play in *Drosophila* a role similar to that played in mammals by the arousal-promoting peptide orexin/hypocretin (Sakurai, 2007).

Drosophila clock neurons have been shown to regulate sleep time in a complex way that is more than a simple reflection of the time of day. For instance, social enrichment in older flies results in longer sleep, but only if the expression of *Rutabaga* and *period* in PDF expressing neurons is normal (Donlea *et al.*, 2009). Also, starvation increases wakefulness in *Drosophila* as it does in mammals (Keene *et al.*, 2010; Thimgan *et al.*, 2010), but the clock neurons that do not express PDF (DN1s or LNds neurons) promote sleep during starvation and their loss enhances sleep loss during starvation (Keene *et al.*, 2010). Of note, sleep loss due to starvation is not followed by a sleep rebound when flies are placed back on their normal diet (Thimgan *et al.*, 2010). Also, in contrast to sleep deprivation due to mechanical or other kinds of stimulation, starvation does not impair performance in an aversive phototaxis memory test (Thimgan *et al.*, 2010). Moreover, *lipid storage droplet 2 (Lsd-2)* mutants that show reduced triglycerides storage sleep less on normal media and are not significantly impaired in the aversive phototaxis memory test after sleep deprivation (performance in *Lsd2* mutant flies is slightly lower than wild-type controls, but still in the range of wild-type flies) (Thimgan *et al.*, 2010). However, starvation in these experiments never lasted more than 12 h, leaving open the possibility that longer fasting may cause sleep rebound and/or memory impairment. Altogether, these results show that different groups of circadian neurons in *Drosophila* can promote either wake or sleep, in addition to affect neuronal plasticity in response to social enrichment.

V. Fly Sleep and Hypotheses on Sleep Functions

Sleep is perhaps the only major behavior still in search of a function. Although the entire body certainly benefits from sleep (Knutson *et al.*, 2007), most researchers agree that sleep may be especially important for the brain and supply something not provided by quiet wake. There is great uncertainty, however, when it comes to which chemical or molecular pathway in the brain may be depleted during wake and restored during sleep or, alternatively, about which toxic substance might accumulate during wake and dissipate during sleep. For instance, it was suggested that sleep favors the replenishment of glycogen in glial stores (Benington and Heller, 1995), but recent evidence show that this may be the case only in a few brain regions, and not in all strains of mice (Franken *et al.*, 2003, 2006). The demonstration that flies sleep, and the subsequent identification of several fly sleep mutants (Table I), have already contributed to the ongoing debate concerning sleep functions, as will be discussed below.

As mentioned above, growing evidence in mammals including humans, birds, and flies points to a link between sleep need and neuronal plasticity, but why and how sleep may benefit the brain by modifying synapses remains unclear. There are three main hypotheses on how sleep could do so, which suggest a role for sleep in promoting (1) synaptic homeostasis (Tononi and Cirelli, 2006), (2) macromolecule synthesis (Mackiewicz *et al.*, 2009), and (3) memory consolidation (Diekelmann and Born, 2010). Importantly, these hypotheses are not mutually exclusive. For instance, there is compelling evidence that sleep benefits several forms of memory (Diekelmann and Born, 2010), but the underlying mechanisms remain unclear, and could involve the specific strengthening of a few synapses already potentiated during wake, and/or, as hypothesized by the synaptic homeostasis hypothesis, a more generalized downregulation of synaptic strength during sleep, with subsequent increase in the signal to noise ratio (details in Tononi and Cirelli (2006)).

Some early clues about the functions of sleep were obtained by considering the extensive changes in brain gene expression that occur between sleep and wake or after sleep deprivation (Cirelli and Tononi, 2000b; Cirelli *et al.*, 2004, 2005c, 2006; Jones *et al.*, 2008; Maret *et al.*, 2007; Mackiewicz *et al.*, 2007; Mackiewicz *et al.*, 2009; Nikonova *et al.*, 2010; Terao *et al.*, 2003a, 2003b, 2006; Zimmerman *et al.*, 2006). In all species studied so far (flies, mice, rats, hamsters, and sparrows) wake leads to the upregulation of transcripts involved in the response to cellular stress and in activity-dependent processes of synaptic potentiation. By contrast, transcripts expressed at higher levels during sleep are involved in synaptic depression, in the synthesis/maintenance of membranes and in lipid metabolism, including the synthesis and transport of cholesterol (Cirelli *et al.*, 2004; Mackiewicz *et al.*, 2007, 2009). Lesions of the locus coeruleus that deplete the cerebral cortex of noradrenaline, which promotes arousal, attention, and the response to novelty (Sara, 2009), blunt the homeostatic response to sleep deprivation and abolish the upregulation of a few wake-related genes, mostly plasticity-related genes (Cirelli *et al.*, 1996, 2005b; Cirelli and Tononi, 2004). Of note, adrenalectomy has almost opposite effects: sleep homeostasis is not affected, but about two-thirds of all wake-related transcripts are no longer upregulated by sleep deprivation, with the notable exception of some stress response genes including the endoplasmic reticulum chaperone BiP (also called glucose-regulated protein 78 and, in *Drosophila*, heat-shock protein cognate 3) and plasticity-related genes (Mongrain *et al.*, 2010). Although these results may at first appear disparate, they may also reflect a coherent set of functional changes at the cellular level. One way to make sense of these findings is in terms of plastic processes. Specifically, in the synaptic homeostasis hypothesis we have suggested that during wake there is a net increase in synaptic strength in many brain areas; these plastic changes are a major determinant of sleep need and sleep would be needed to renormalize such changes (Tononi and Cirelli, 2003, 2006). Why would wake result in a net increase in synaptic strength? Because the awake brain is always "learning," whether it is performing a learning task or simply adapting to an ever-changing environment, and novelty exposure, enrichment and learning mostly occur through synaptic potentiation, not synaptic depression (Feldman, 2009; Nithianantharajah and Hannan, 2006). Moreover, wake is associated with high levels of acetylcholine and noradrenaline (Jones, 2005), which together favor synaptic potentiation (Cirelli *et al.*, 1996; Cirelli and Tononi, 2000a; Seol *et al.*, 2007).

Why would then sleep be needed to revert the net increase of synaptic strength at the end of wake? Because such increase would result in higher energy consumption (Attwell and Laughlin, 2001; Rothman *et al.*, 2003), in larger synapses that take up precious space (Chklovskii *et al.*, 2002), and in the saturation of the capacity to learn. Also, a net strengthening of synapses likely represents a major source of cellular stress, due to the need to synthesize and deliver cellular constituents ranging from mitochondria to synaptic vesicles to various proteins and lipids (Mackiewicz *et al.*, 2009). In this view, then, sleep would be necessary to downregulate synapses to a baseline level that is sustainable and ensures cellular homeostasis. Importantly, downregulation would have to occur off-line, that is, during sleep, because it should affect most synapses, whether or not they are engaged in behavior. How would sleep bring about a net decrease in synaptic strength? One obvious cellular mechanism is the reduced level of neuronal excitability that characterizes most neurons during most of sleep. During NREM sleep, which accounts for 70–80% of all sleep in mammals, neurons are relatively hyperpolarized and fire less as compared to wake (Steriade, 2003), including neurons that release noradrenaline, serotonin, hypocretin, and histamine (Jones, 2005; Saper *et al.*, 2010). Synaptic downregulation may be favored by periods of neuronal silence (Birtoli and Ulrich, 2004; Czarnecki *et al.*, 2007; Kemp and Bashir, 2001; Lubenov and Siapas, 2008; Rosanova and Ulrich, 2005; Werk and Chapman, 2003; Werk *et al.*, 2006), as well as by the low levels of noradrenaline, serotonin, hypocretin, and histamine (Harley, 1991; Seol *et al.*, 2007). These electrophysiological and biochemical changes across the sleep/wake cycle are well characterized in mammals, but recent evidence shows that they also occur in flies (Crocker and Sehgal, 2010). For instance, *Drosophila* neurons in the medial brain, as well as LNvs neurons, which release the wake-promoting neuropeptide PDF, are more active in wake than in sleep (Nitz *et al.*, 2002; Sheeba *et al.*, 2008). Moreover, noradrenaline is wake-promoting in mammals as octopamine, its equivalent in insects, is wake-promoting in flies, an effect mediated in both species by the cAMP/PKA/CREB-signaling pathway (Crocker and Sehgal, 2008;

Crocker *et al.*, 2010; Graves *et al.*, 2003; Hendricks *et al.*, 2001). Specific molecular mechanisms that may mediate synaptic renormalization are also conserved between mammals and flies, including the gene *Fmr1* (fragile X mental retardation 1), for which extensive evidence is available (Comery *et al.*, 1997; Hinton *et al.*, 1991; Irwin *et al.*, 2002; Pan *et al.*, 2004; Restivo *et al.*, 2005).

The evidence supporting the synaptic homeostasis hypothesis comes from mammals as well as flies. In rats, a recent study examined molecular and electrophysiological markers of synaptic function during sleep and wake (Vyazovskiy *et al.*, 2008). It was found that the levels of AMPA receptors in cortical synaptoneurosomes decrease by ~40% after several hours of sleep (Vyazovskiy *et al.*, 2008). Phosphorylation changes of AMPA receptors, and of the enzymes CamKII and GSK3β, were also in line with a net decrease in synaptic strength during sleep. Electrophysiologically, it was shown using cortical electrical stimulation and local field potential recordings that both slope and amplitude of cortical evoked responses (classical *in vivo* measures of synaptic strength) also decrease after sleep (Vyazovskiy *et al.*, 2008). Similar observations have been made in humans using transcranial magnetic stimulation and high-density electroencephalogram (EEG) analysis (Bellina *et al.*, 2008). Direct evidence for a net decrease in synaptic strength after sleep comes from a recent study in which miniature excitatory postsynaptic currents (mEPSCs) were recorded from frontal cortex slices of mice and rats. Changes in mEPSCs frequency are thought to result from modification of the presynaptic component of synaptic transmission, whereas amplitude changes indicate alterations in the postsynaptic component (e.g., Ungless *et al.* (2001)). It was found that, in both rats and mice, the frequency and amplitude of mEPSCs increase after wake and decrease after sleep (Liu *et al.*, 2010). Recovery sleep after sleep deprivation also decreases mEPSCs, suggesting that sleep brings about a net decrease in synaptic strength (Liu *et al.*, 2010). Finally, mean firing rates in the rat cerebral cortex increase after periods of wake and decrease after periods of sleep, consistent with a net change in synaptic strength (Vyazovskiy *et al.*, 2009). In line with this observation, the levels of glutamate in the rat cortical extrasynaptic space also increase progressively during wake and decrease during NREM sleep (Dash *et al.*, 2009). In line with these mammalian studies, we recently found in flies that overall protein levels of both pre- and postsynaptic components of central synapses are high after wake and low after sleep (Gilestro *et al.*, 2009). These changes are related to behavioral state rather than time of day and occur in all major areas of the *Drosophila* brain. Moreover, the decrease of synaptic markers during sleep is progressive (Gilestro *et al.*, 2009), consistent with the synaptic homeostasis hypothesis. Another study also found morphological evidence, in a specific neural circuit (large LNvs), for synaptic growth in flies that were sleep deprived for 48 h after chronic social enrichment, compared to flies that were left undisturbed (Donlea *et al.*, 2009). Furthermore, new data from our laboratory show that that number/size of synapses in three different *Drosophila* neural circuits is higher after wake and lower after sleep, and that synaptic renormalization can only occur if flies are allowed to sleep, but not if they are sleep deprived (Bushey *et al.*, 2011). Finally, a recent study in zebrafish larvae (Appelbaum *et al.*, 2010) found that presynaptic terminals of hypocretin neurons projecting to the pineal gland undergo both circadian and sleep–wake-dependent structural changes, the latter consistent with sleep-dependent downregulation. Finally, the findings from fly mutagenesis screens that genes necessary for brain plasticity are also necessary for sleep are in line with the synaptic homeostasis hypothesis.

The results from fly mutagenesis screens, on the contrary, so far do not provide direct evidence that sleep is necessary for the synthesis of macromolecules such as brain lipids (Mackiewicz *et al.*, 2009). This, however, may be due to these screens focusing on short sleeping mutants, while loss of function mutations decreasing lipid synthesis should produce long sleepers. Finally, the finding that in both mammals and flies prolonged wake is associated with increased expression of heat-shock proteins and chaperones such as BiP may suggest that sleep loss causes the accumulation of unfolded proteins, which then need to be processed and eliminated during sleep. The fact that *Drosophila* mutations that affect the expression of BiP and hsp83 result in an exaggerated sleep rebound (Table I) is consistent with this idea. However, there are no data showing that the overexpression of heat-shock proteins reduces sleep time in normal physiological conditions (Naidoo *et al.*, 2007; Shaw *et al.*, 2002), suggesting perhaps that abnormal accumulation of unfolded proteins occurs only when wake is prolonged beyond its physiological duration. Yet, in both flies and mammals BiP expression also increases during spontaneous wake, and in sleep-deprived mice BiP is still induced after adrenalectomy (Mongrain *et al.*, 2010), suggesting that BiP induction may not be simply a sign of cellular stress. Long-term sensitization training in *Aplysia* induces BiP (Kuhl *et al.*, 1992), and BiP may be involved in the trafficking of glutamatergic AMPA receptors (Rubio and Wenthold, 1999). Moreover, the induction of BiP and of the unfolded protein response promotes the surface expression of GluR1-containing AMPA receptors (Vandenberghe *et al.*, 2005). Thus, BiP induction in the context of the unfolded protein response during wake/sleep deprivation may be the result of a physiological increase in glutamatergic signaling and long-term potentiation.

VI. Closing Remarks

A decade of screening and studying sleep in *Drosophila* has provided insight into the molecular pathways involved in both sleep and wake. Consistent with data in mammals including humans, studies in flies show that sleep need is a function of brain plasticity, as genes necessary for neuronal plasticity and brain regions where this plasticity occurs are directly involved in sleep homeostasis. Thus, any hypothesis about sleep functions must explain the strong link between sleep need and plastic changes.

Acknowledgments

This study was supported by NIGMS (R01 GM075315 to C.C.), Army Research Office (DURIP Award W911NF-08-1-0169 to C.C.), and Canadian Institutes of Health Research (to D.B.).

References

Achermann, P. and Borbely, A.A. (2003). Mathematical models of sleep regulation. *Front. Biosci.* **8**, s683–s693. Available from http://www.ncbi.nlm.nih.gov/entrez/query.fcgi?cmd=Retrieve&db=PubMed&dopt=Citation&list_uids=12700054.

Agosto, J., Choi, J.C., Parisky, K.M., Stilwell, G., Rosbash, M. and Griffith, L.C. (2008). Modulation of GABAa receptor desensitization uncouples sleep onset and maintenance in *Drosophila*. *Nat. Neurosci.* **11**(3); 354–359. Available from http://www.ncbi.nlm.nih.gov/entrez/query.fcgi?cmd=Retrieve&db=PubMed&dopt=Citation&list_uids=18223647.

Alkire, M.T., Asher, C.D., Franciscus, A.M. and Hahn, E.L. (2009). Thalamic microinfusion of antibody to a voltage-gated potassium channel restores consciousness during anesthesia. *Anesthesiology* **110**(4); 766–773.

Appelbaum, L., Wang, G., Yokogawa, T., Skariah, G.M., Smith, S.J., Mourrain, P. and Mignot, E. (2010). Circadian and homeostatic regulation of structural synaptic plasticity in hypocretin neurons. *Neuron* **68**(1); 87–98. Available from http://www.ncbi.nlm.nih.gov/entrez/query.fcgi?cmd=Retrieve&db=PubMed&dopt=Citation&list_uids=20920793.

Attwell, D. and Laughlin, S.B. (2001). An energy budget for signaling in the grey matter of the brain. *J. Cereb. Blood Flow Metab.* **21**(10); 1133–1145.

Bellina, V., Huber, R., Rosanova, M., Mariotti, M., Tononi, G. and Massimini, M. (2008). Cortical excitability and sleep homeostasis in humans: a tms/hd-eeg study. *J. Sleep Res.* **17**(Suppl. 1); 39.

Benington, J.H. and Heller, H.C. (1995). Restoration of brain energy metabolism as the function of sleep. *Prog. Neurobiol.* **45**, 347–360.

Berry, J., Krause, W.C. and Davis, R.L. (2008). Olfactory memory traces in *Drosophila*. *Prog. Brain Res.* **169**, 293–304. Available from http://www.ncbi.nlm.nih.gov/entrez/query.fcgi?cmd=Retrieve&db=PubMed&dopt=Citation&list_uids=18394482.

Birtoli, B. and Ulrich, D. (2004). Firing mode-dependent synaptic plasticity in rat neocortical pyramidal neurons. *J. Neurosci.* **24**(21); 4935–4940. Available from http://www.ncbi.nlm.nih.gov/entrez/query.fcgi?cmd=Retrieve&db=PubMed&dopt=Citation&list_uids=15163685.

Blum, A.L., Li, W., Cressy, M. and Dubnau, J. (2009). Short- and long-term memory in *Drosophila* require camp signaling in distinct neuron types. *Curr. Biol.* **19**(16); 1341–1350. Available from http://www.ncbi.nlm.nih.gov/entrez/query.fcgi?cmd=Retrieve&db=PubMed&dopt=Citation&list_uids=19646879.

Bolduc, F.V., Bell, K., Cox, H., Broadie, K.S. and Tully, T. (2008). Excess protein synthesis in *Drosophila* fragile x mutants impairs long-term memory. *Nat. Neurosci.* **11**(10); 1143–1155. Available from http://www.ncbi.nlm.nih.gov/entrez/query.fcgi?cmd=Retrieve&db=PubMed&dopt=Citation&list_uids=18776892.

Bushey, D., Huber, R., Tononi, G. and Cirelli, C. (2007). *Drosophila* hyperkinetic mutants have reduced sleep and impaired memory. *J. Neurosci.* **27**(20); 5384–5393. Available from http://www.ncbi.nlm.nih.gov/entrez/query.fcgi?cmd=Retrieve&db=PubMed&dopt=Citation&list_uids=17507560.

Bushey, D., Tononi, G. and Cirelli, C. (2009). The *Drosophila* fragile x mental retardation gene regulates sleep need. *J. Neurosci.* **29**(7); 1948–1961. Available from http://www.ncbi.nlm.nih.gov/entrez/query.fcgi?cmd=Retrieve&db=PubMed&dopt=Citation&list_uids=19228950.

Bushey, D., Hughes, K.A., Tononi, G. and Cirelli, C. (2010). Sleep duration affects lifespan in Drosophila. *BMC Neurosci.* **11**, 56.

Bushey, D., Tononi, G. and Cirelli, C. (2011). Increased pre-synaptic size and post-synaptic complexity during wake as compared to sleep in D. melanogaster. *Sleep* **34**(Suppl.).

Chklovskii, D.B., Schikorski, T. and Stevens, C.F. (2002). Wiring optimization in cortical circuits. *Neuron* **34** (3); 341–347. Available from http://www.ncbi.nlm.nih.gov/entrez/query.fcgi?cmd=Retrieve&db=PubMed&dopt=Citation&list_uids=11988166.

Cirelli, C. (2003). Searching for sleep mutants of *Drosophila melanogaster*. *Bioessays* **25**(10); 940–949. Available from http://www.ncbi.nlm.nih.gov/entrez/query.fcgi?cmd=Retrieve&db=PubMed&dopt=Citation&list_uids=14505361.

Cirelli, C. (2009). The genetic and molecular regulation of sleep: from fruit flies to humans. *Nat. Rev. Neurosci.* **10**(8); 549–560. Available from http://www.ncbi.nlm.nih.gov/entrez/query.fcgi?cmd=Retrieve&db=PubMed&dopt=Citation&list_uids=19617891.

Cirelli, C., Bushey, D., Hill, S., Huber, R., Kreber, R., Ganetzky, B. and Tononi, G. (2005a). Reduced sleep in Drosophila shaker mutants. *Nature* **434**(7037); 1087–1092. Available from http://www.ncbi.nlm.nih.gov/entrez/query.fcgi?cmd=Retrieve&db=PubMed&dopt=Citation&list_uids=15858564.

Cirelli, C., Faraguna, U. and Tononi, G. (2006). Changes in brain gene expression after long-term sleep deprivation. *J. Neurochem.* **98**(5); 1632–1645. Available from http://www.ncbi.nlm.nih.gov/entrez/query.fcgi?cmd=Retrieve&db=PubMed&dopt=Citation&list_uids=16923172.

Cirelli, C., Gutierrez, C.M. and Tononi, G. (2004). Extensive and divergent effects of sleep and wakefulness on brain gene expression. *Neuron* **41**(1); 35–43. Available from http://www.ncbi.nlm.nih.gov/entrez/query.fcgi?cmd=Retrieve&db=PubMed&dopt=Citation&list_uids=14715133.

Cirelli, C., Huber, R., Gopalakrishnan, A., Southard, T.L. and Tononi, G. (2005b). Locus ceruleus control of slow-wave homeostasis. *J. Neurosci.* **25**(18); 4503–4511. Available from http://www.ncbi.nlm.nih.gov/entrez/query.fcgi?cmd=Retrieve&db=PubMed&dopt=Citation&list_uids=15872097.

Cirelli, C., Lavaute, T.M. and Tononi, G. (2005c). Sleep and wakefulness modulate gene expression in *Drosophila*. *J. Neurochem.* **94**(5); 1411–1419. Available from http://www.ncbi.nlm.nih.gov/entrez/query.fcgi?cmd=Retrieve&db=PubMed&dopt=Citation&list_uids=16001966.

Cirelli, C., Pompeiano, M. and Tononi, G. (1996). Neuronal gene expression in the waking state: a role for the locus coeruleus. *Science* **274**(5290); 1211–1215. Available from http://proquest.umi.com/pqdweb?SK=3&ScQ=000028298|000000000556484|*&StPt=51&FC=25&RQT=317&SrtM=2.

Cirelli, C. and Tononi, G. (2000a). Differential expression of plasticity-related genes in waking and sleep and their regulation by the noradrenergic system. *J. Neurosci.* **20**(24); 9187–9194. Available from http://www.ncbi.nlm.nih.gov/cgi-bin/Entrez/referer?http://www.jneurosci.org/cgi/content/abstract/20/24/9187.

Cirelli, C. and Tononi, G. (2000b). Gene expression in the brain across the sleep-waking cycle. *Brain Res.* **885**(2); 303–321. Available from http://www.sciencedirect.com/science?_ob=IssueURL&_tockey=%23TOC%234841%232000%23991149997%23219523%23FLA%23Volume_885,_Issue_2,_Pages_143-324_(8_December_2000)&_auth=y&_acct=C000020958&_version=1&_urlVersion=0&_userid=443835&md5=99b0bf86863c646c13a3445a37c46db4.

Cirelli, C. and Tononi, G. (2004). Locus ceruleus control of state-dependent gene expression. *J. Neurosci.* **24**(23); 5410–5419. Available from http://www.ncbi.nlm.nih.gov/entrez/query.fcgi?cmd=Retrieve&db=PubMed&dopt=Citation&list_uids=15190114.

Cirelli, C. and Tononi, G. (2008). Is sleep essential? *PLoS Biol.* **6**(8); e216. Available from http://www.ncbi.nlm.nih.gov/entrez/query.fcgi?cmd=Retrieve&db=PubMed&dopt=Citation&list_uids=18752355.

Comery, T.A., Harris, J.B., Willems, P.J., Oostra, B.A., Irwin, S.A., Weiler, I.J. and Greenough, W.T. (1997). Abnormal dendritic spines in fragile x knockout mice: maturation and pruning deficits. *Proc. Natl. Acad. Sci. U S A* **94**(10); 5401–5404. Available from http://www.ncbi.nlm.nih.gov/entrez/query.fcgi?cmd=Retrieve&db=PubMed&dopt=Citation&list_uids=9144249.

Cowan, T.M. and Siegel, R.W. (1984). Mutational and pharmacological alterations of neuronal membrane function disrupt conditioning in *Drosophila*. *J. Neurogenet.* **1**(4); 333–344. Available from http://www.ncbi.nlm.nih.gov/entrez/query.fcgi?cmd=Retrieve&db=PubMed&dopt=Citation&list_uids=6100306.

Cowan, T.M. and Siegel, R.W. (1986). *Drosophila* mutations that alter ionic conduction disrupt acquisition and retention of a conditioned odor avoidance response. *J. Neurogenet.* **3**(4); 187–201. Available from http://www.ncbi.nlm.nih.gov/entrez/query.fcgi?cmd=Retrieve&db=PubMed&dopt=Citation&list_uids=2427679.

Crocker, A. and Sehgal, A. (2008). Octopamine regulates sleep in *Drosophila* through protein kinase a-dependent mechanisms. *J. Neurosci.* **28**(38); 9377–9385. Available from http://www.ncbi.nlm.nih.gov/entrez/query.fcgi?cmd=Retrieve&db=PubMed&dopt=Citation&list_uids=18799671.

Crocker, A. and Sehgal, A. (2010). Genetic analysis of sleep. *Genes Dev.* **24**(12); 1220–1235. Available from http://www.ncbi.nlm.nih.gov/entrez/query.fcgi?cmd=Retrieve&db=PubMed&dopt=Citation&list_uids=20551171.

Crocker, A., Shahidullah, M., Levitan, I.B. and Sehgal, A. (2010). Identification of a neural circuit that underlies the effects of octopamine on sleep:wake behavior. *Neuron* **65**(5); 670–681. Available from http://www.ncbi.nlm.nih.gov/entrez/query.fcgi?cmd=Retrieve&db=PubMed&dopt=Citation&list_uids=20223202.

Czarnecki, A., Birtoli, B. and Ulrich, D. (2007). Cellular mechanisms of burst firing-mediated long-term depression in rat neocortical pyramidal cells. *J. Physiol.* **578**(Pt 2); 471–479. Available from http://www.ncbi.nlm.nih.gov/entrez/query.fcgi?cmd=Retrieve&db=PubMed&dopt=Citation&list_uids=17082228.

Dash, M.B., Douglas, C.L., Vyazovskiy, V.V., Cirelli, C. and Tononi, G. (2009). Long-term homeostasis of extracellular glutamate in the rat cerebral cortex across sleep and waking states. *J. Neurosci.* **29**(3); 620–629. Available from http://www.ncbi.nlm.nih.gov/entrez/query.fcgi?cmd=Retrieve&db=PubMed&dopt=Citation&list_uids=19158289.

Diekelmann, S. and Born, J. (2010). The memory function of sleep. *Nat. Rev. Neurosci.* **11**(2); 114–126. Available from http://www.ncbi.nlm.nih.gov/entrez/query.fcgi?cmd=Retrieve&db=PubMed&dopt=Citation&list_uids=20046194.

Donlea, J.M., Ramanan, N. and Shaw, P.J. (2009). Use-dependent plasticity in clock neurons regulates sleep need in Drosophila. *Science* **324**(5923); 105–108. Available from http://www.ncbi.nlm.nih.gov/entrez/query.fcgi?cmd=Retrieve&db=PubMed&dopt=Citation&list_uids=19342592.

Douglas, C.L., Vyazovskiy, V., Southard, T., Chiu, S.Y., Messing, A., Tononi, G. and Cirelli, C. (2007). Sleep in kcna2 knockout mice. *BMC Biol.* **5**, 42. Available from http://www.ncbi.nlm.nih.gov/entrez/query.fcgi?cmd=Retrieve&db=PubMed&dopt=Citation&list_uids=17925011.

Dudai, Y., Jan, Y.N., Byers, D., Quinn, W.G. and Benzer, S. (1976). Dunce, a mutant of *Drosophila* deficient in learning. *Proc. Natl. Acad. Sci. U S A* **73**(5); 1684–1688. Available from http://www.ncbi.nlm.nih.gov/entrez/query.fcgi?cmd=Retrieve&db=PubMed&dopt=Citation&list_uids=818641.

Feldman, D.E. (2009). Synaptic mechanisms for plasticity in neocortex. *Annu. Rev. Neurosci.* **32**, 33–55. Available from http://www.ncbi.nlm.nih.gov/entrez/query.fcgi?cmd=Retrieve&db=PubMed&dopt=Citation&list_uids=19400721.

Feng, Y., Gutekunst, C.A., Eberhart, D.E., Yi, H., Warren, S.T. and Hersch, S.M. (1997). Fragile x mental retardation protein: nucleocytoplasmic shuttling and association with somatodendritic ribosomes. *J. Neurosci.* **17**(5); 1539–1547. Available from http://www.ncbi.nlm.nih.gov/entrez/query.fcgi?cmd=Retrieve&db=PubMed&dopt=Citation&list_uids=9030614.

Foltenyi, K., Greenspan, R.J. and Newport, J.W. (2007). Activation of egfr and erk by rhomboid signaling regulates the consolidation and maintenance of sleep in *Drosophila*. *Nat. Neurosci.* **10**(9); 1160–1167. Available from http://www.ncbi.nlm.nih.gov/entrez/query.fcgi?cmd=Retrieve&db=PubMed&dopt=Citation&list_uids=17694052.

Franken, P., Gip, P., Hagiwara, G., Ruby, N.F. and Heller, H.C. (2003). Changes in brain glycogen after sleep deprivation vary with genotype. *Am. J. Physiol. Regul. Integr. Comp. Physiol.* **285**(2); R413–R419. Available from http://www.ncbi.nlm.nih.gov/entrez/query.fcgi?cmd=Retrieve&db=PubMed&dopt=Citation&list_uids=12730076.

Franken, P., Gip, P., Hagiwara, G., Ruby, N.F. and Heller, H.C. (2006). Glycogen content in the cerebral cortex increases with sleep loss in c57bl/6j mice. *Neurosci. Lett.* **402**(1-2); 176–179. Available from http://www.ncbi.nlm.nih.gov/entrez/query.fcgi?cmd=Retrieve&db=PubMed&dopt=Citation&list_uids=16644123.

Ganguly-Fitzgerald, I., Donlea, J. and Shaw, P.J. (2006). Waking experience affects sleep need in Drosophila. *Science* **313**(5794); 1775–1781. Available from http://www.ncbi.nlm.nih.gov/entrez/query.fcgi?cmd=Retrieve&db=PubMed&dopt=Citation&list_uids=16990546.

Gilestro, G.F., Tononi, G. and Cirelli, C. (2009). Widespread changes in synaptic markers as a function of sleep and wakefulness in Drosophila. *Science* **324**(5923); 109–112. Available from http://www.ncbi.nlm.nih.gov/entrez/query.fcgi?cmd=Retrieve&db=PubMed&dopt=Citation&list_uids=19342593.

Graves, L.A., Hellman, K., Veasey, S., Blendy, J.A., Pack, A.I. and Abel, T. (2003). Genetic evidence for a role of CREB in sustained cortical arousal. *J. Neurophysiol.* **23**, 23. Available from http://www.ncbi.nlm.nih.gov/htbin-post/Entrez/query?db=m&form=6&dopt=r&uid=12711709.

Gronke, S., Beller, M., Fellert, S., Ramakrishnan, H., Jackle, H. and Kuhnlein, R.P. (2003). Control of fat storage by a *Drosophila* pat domain protein. *Curr. Biol.* **13**(7); 603–606. Available from http://www.ncbi.nlm.nih.gov/entrez/query.fcgi?cmd=Retrieve&db=PubMed&dopt=Citation&list_uids=12676093.

Gronke, S., Mildner, A., Fellert, S., Tennagels, N., Petry, S., Muller, G., Jackle, H. and Kuhnlein, R.P. (2005). Brummer lipase is an evolutionary conserved fat storage regulator in *Drosophila*. *Cell Metab.* **1**(5); 323–330. Available from http://www.ncbi.nlm.nih.gov/entrez/query.fcgi?cmd=Retrieve&db=PubMed&dopt=Citation&list_uids=16054079.

Han, P.L., Levin, L.R., Reed, R.R. and Davis, R.L. (1992). Preferential expression of the *Drosophila* rutabaga gene in mushroom bodies, neural centers for learning in insects. *Neuron* **9**(4); 619–627. Available from http://www.ncbi.nlm.nih.gov/entrez/query.fcgi?cmd=Retrieve&db=PubMed&dopt=Citation&list_uids=1382471.

Han, K.A., Millar, N.S. and Davis, R.L. (1998). A novel octopamine receptor with preferential expression in *Drosophila* mushroom bodies. *J. Neurosci.* **18**(10); 3650–3658. Available from http://www.ncbi.nlm.nih.gov/entrez/query.fcgi?cmd=Retrieve&db=PubMed&dopt=Citation&list_uids=9570796.

Hanlon, E.C., Faraguna, U., Vyazovskiy, V.V., Tononi, G. and Cirelli, C. (2009). Effects of skilled training on sleep slow wave activity and cortical gene expression in the rat. *Sleep* **32**(6); 719–729. Available from http://www.ncbi.nlm.nih.gov/entrez/query.fcgi?cmd=Retrieve&db=PubMed&dopt=Citation&list_uids=19544747.

Harbison, S.T., Mackay, T.F. and Anholt, R.R. (2009). Understanding the neurogenetics of sleep: progress from *Drosophila*. *Trends Genet.* **25**(6); 262–269. Available from http://www.ncbi.nlm.nih.gov/entrez/query.fcgi?cmd=Retrieve&db=PubMed&dopt=Citation&list_uids=19446357.

Harley, C. (1991). Noradrenergic and locus coeruleus modulation of the perforant path-evoked potential in rat dentate gyrus supports a role for the locus coeruleus in attentional and memorial processes. *Prog. Brain Res.* **88**, 307–321. Available from http://www.ncbi.nlm.nih.gov/entrez/query.fcgi?cmd=Retrieve&db=PubMed&dopt=Citation&list_uids=1687619.

Hendricks, J.C., Finn, S.M., Panckeri, K.A., Chavkin, J., Williams, J.A., Sehgal, A. and Pack, A.I. (2000). Rest in *Drosophila* is a sleep-like state. *Neuron* **25**(1); 129–138. Available from http://www.ncbi.nlm.nih.gov/cgi-bin/Entrez/referer?http://www.neuron.org/cgi/content/full/25/1/129.

Hendricks, J.C., Williams, J.A., Panckeri, K., Kirk, D., Tello, M., Yin, J.C. and Sehgal, A. (2001). A non-circadian role for camp signaling and creb activity in *Drosophila* rest homeostasis. *Nat. Neurosci.* **4**(11); 1108–1115. Available from http://www.ncbi.nlm.nih.gov/entrez/query.fcgi?cmd=Retrieve&db=PubMed&dopt=Citation&list_uids=11687816.

Hewes, R.S. (2008). The buzz on fly neuronal remodeling. *Trends Endocrinol. Metab.* **19**(9); 317–323. Available from http://www.ncbi.nlm.nih.gov/entrez/query.fcgi?cmd=Retrieve&db=PubMed&dopt=Citation&list_uids=18805704.

Hinton, V.J., Brown, W.T., Wisniewski, K. and Rudelli, R.D. (1991). Analysis of neocortex in three males with the fragile x syndrome. *Am. J. Med. Genet.* **41**(3); 289–294. Available from http://www.ncbi.nlm.nih.gov/entrez/query.fcgi?cmd=Retrieve&db=PubMed&dopt=Citation&list_uids=1724112.

Ho, K.S. and Sehgal, A. (2005). *Drosophila melanogaster*: an insect model for fundamental studies of sleep. *Methods Enzymol.* **393**, 772–793. Available from http://www.ncbi.nlm.nih.gov/entrez/query.fcgi?cmd=Retrieve&db=PubMed&dopt=Citation&list_uids=15817324.

Huber, R., Ghilardi, M.F., Massimini, M. and Tononi, G. (2004a). Local sleep and learning. *Nature* **430** (6995); 78–81. Available from http://www.ncbi.nlm.nih.gov/entrez/query.fcgi?cmd=Retrieve&db=PubMed&dopt=Citation&list_uids=15184907.

Huber, R., Hill, S.L., Holladay, C., Biesiadecki, M., Tononi, G. and Cirelli, C. (2004b). Sleep homeostasis in *Drosophila melanogaster*. *Sleep* **27**(4); 628–639. Available from http://www.ncbi.nlm.nih.gov/entrez/query.fcgi?cmd=Retrieve&db=PubMed&dopt=Citation&list_uids=15282997.

Irwin, S.A., Idupulapati, M., Gilbert, M.E., Harris, J.B., Chakravarti, A.B., Rogers, E.J., Crisostomo, R.A., Larsen, B.P., Mehta, A., Alcantara, C.J., Patel, B., Swain, R.A., Weiler, I.J., Oostra, B.A. and Greenough, W.T. (2002). Dendritic spine and dendritic field characteristics of layer v pyramidal neurons in the visual cortex of fragile-x knockout mice. *Am. J. Med. Genet.* **111**(2); 140–146. Available from http://www.ncbi.nlm.nih.gov/entrez/query.fcgi?cmd=Retrieve&db=PubMed&dopt=Citation&list_uids=12210340.

Ishimoto, H. and Kitamoto, T. (2010). The steroid molting hormone ecdysone regulates sleep in adult *Drosophila melanogaster*. *Genetics* **185**(1); 269–281. Available from http://www.ncbi.nlm.nih.gov/entrez/query.fcgi?cmd=Retrieve&db=PubMed&dopt=Citation&list_uids=20215472.

Ishimoto, H., Sakai, T. and Kitamoto, T. (2009). Ecdysone signaling regulates the formation of long-term courtship memory in adult Drosophila melanogaster. *Proc. Natl. Acad. Sci. U S A* **106**(15); 6381–6386. Available from http://www.ncbi.nlm.nih.gov/entrez/query.fcgi?cmd=Retrieve&db=PubMed&dopt=Citation&list_uids=19342482.

Joiner, W.J., Crocker, A., White, B.H. and Sehgal, A. (2006). Sleep in Drosophila is regulated by adult mushroom bodies. *Nature* **441**(7094); 757–760. Available from http://www.ncbi.nlm.nih.gov/entrez/query.fcgi?cmd=Retrieve&db=PubMed&dopt=Citation&list_uids=16760980.

Jones, B.E. (2005). From waking to sleeping: neuronal and chemical substrates. *Trends Pharmacol. Sci.* **26** (11); 578–586. Available from http://www.ncbi.nlm.nih.gov/entrez/query.fcgi?cmd=Retrieve&db=PubMed&dopt=Citation&list_uids=16183137.

Jones, S., Pfister-Genskow, M., Benca, R.M. and Cirelli, C. (2008). Molecular correlates of sleep and wakefulness in the brain of the white-crowned sparrow. *J. Neurochem.* **105**(1); 46–62. Available from http://www.ncbi.nlm.nih.gov/entrez/query.fcgi?cmd=Retrieve&db=PubMed&dopt=Citation&list_uids=18028333.

Keene, A.C., Duboue, E.R., Mcdonald, D.M., Dus, M., Suh, G.S., Waddell, S. and Blau, J. (2010). Clock and cycle limit starvation-induced sleep loss in *Drosophila*. *Curr. Biol.* **20**(13); 1209–1215. Available from http://www.ncbi.nlm.nih.gov/entrez/query.fcgi?cmd=Retrieve&db=PubMed&dopt=Citation&list_uids=20541409.

Kemp, N. and Bashir, Z.I. (2001). Long-term depression: a cascade of induction and expression mechanisms. *Prog. Neurobiol.* **65**(4); 339–365.

Killgore, W.D. (2010). Effects of sleep deprivation on cognition. *Prog. Brain Res.* **185**, 105–129.

Knutson, K.L., Spiegel, K., Penev, P. and Van Cauter, E. (2007). The metabolic consequences of sleep deprivation. *Sleep Med. Rev.* **11**(3); 163–178. Available from http://www.ncbi.nlm.nih.gov/entrez/query.fcgi?cmd=Retrieve&db=PubMed&dopt=Citation&list_uids=17442599.

Koh, K., Joiner, W.J., Wu, M.N., Yue, Z., Smith, C.J. and Sehgal, A. (2008). Identification of sleepless, a sleep-promoting factor. *Science* **321**(5887); 372–376. Available from http://www.ncbi.nlm.nih.gov/entrez/query.fcgi?cmd=Retrieve&db=PubMed&dopt=Citation&list_uids=18635795.

Krashes, M.J., Keene, A.C., Leung, B., Armstrong, J.D. and Waddell, S. (2007). Sequential use of mushroom body neuron subsets during *Drosophila* odor memory processing. *Neuron* **53**(1); 103–115. Available from http://www.ncbi.nlm.nih.gov/entrez/query.fcgi?cmd=Retrieve&db=PubMed&dopt=Citation&list_uids=17196534.

Kuhl, D., Kennedy, T.E., Barzilai, A. and Kandel, E.R. (1992). Long-term sensitization training in aplysia leads to an increase in the expression of BIP, the major protein chaperon of the ER. *J. Cell Biol.* **119**(5); 1069–1076. Available from http://www.ncbi.nlm.nih.gov/entrez/query.fcgi?cmd=Retrieve&db=PubMed&dopt=Citation&list_uids=1360013.

Kume, K., Kume, S., Park, S.K., Hirsh, J. and Jackson, F.R. (2005). Dopamine is a regulator of arousal in the fruit fly. *J. Neurosci.* **25**(32); 7377–7384. Available from http://www.ncbi.nlm.nih.gov/entrez/query.fcgi?cmd=Retrieve&db=PubMed&dopt=Citation&list_uids=16093388.

Lesku, J.A., Vyssotski, A.L., Martinez-Gonzalez, D., Wilzeck, C. and Rattenborg, N.C. (2011). Local sleep homeostasis in the avian brain: convergence of sleep function in mammals and birds? *Proc. Biol. Sci.* Available from http://www.ncbi.nlm.nih.gov/entrez/query.fcgi?cmd=Retrieve&db=PubMed&dopt=Citation&list_uids=21208955

Li, X., Yu, F. and Guo, A. (2009). Sleep deprivation specifically impairs short-term olfactory memory in *Drosophila*. *Sleep* **32**(11); 1417–1424. Available from http://www.ncbi.nlm.nih.gov/entrez/query.fcgi?cmd=Retrieve&db=PubMed&dopt=Citation&list_uids=19928381.

Liu, Z.W., Faraguna, U., Cirelli, C., Tononi, G. and Gao, X.B. (2010). Direct evidence for wake-related increases and sleep-related decreases in synaptic strength in rodent cortex. *J. Neurosci.* **30**(25); 8671–8675. Available from http://www.ncbi.nlm.nih.gov/entrez/query.fcgi?cmd=Retrieve&db=PubMed&dopt=Citation&list_uids=20573912.

Lubenov, E.V. and Siapas, A.G. (2008). Decoupling through synchrony in neuronal circuits with propagation delays. *Neuron* **58**(1); 118–131. Available from http://www.ncbi.nlm.nih.gov/entrez/query.fcgi?cmd=Retrieve&db=PubMed&dopt=Citation&list_uids=18400168.

Mackiewicz, M., Shockley, K.R., Romer, M.A., Galante, R.J., Zimmerman, J.E., Naidoo, N., Baldwin, D.A., Jensen, S.T., Churchill, G.A. and Pack, A.I. (2007). Macromolecule biosynthesis—a key function of sleep. *Physiol. Genom* **31**(3); 441–457. Available from http://www.ncbi.nlm.nih.gov/entrez/query.fcgi?cmd=Retrieve&db=PubMed&dopt=Citation&list_uids=17698924.

Mackiewicz, M., Zimmerman, J.E., Shockley, K.R., Churchill, G.A. and Pack, A.I. (2009). What are microarrays teaching us about sleep? *Trends Mol. Med.* **15**(2); 79–87. Available from http://www.ncbi.nlm.nih.gov/entrez/query.fcgi?cmd=Retrieve&db=PubMed&dopt=Citation&list_uids=19162550.

Maret, S., Dorsaz, S., Gurcel, L., Pradervand, S., Petit, B., Pfister, C., Hagenbuchle, O., O'hara, B.F., Franken, P. and Tafti, M. (2007). Homer1a is a core brain molecular correlate of sleep loss. *Proc.*

Natl. Acad. Sci. U S A **104**(50); 20090–20095. Available from http://www.ncbi.nlm.nih.gov/entrez/query.fcgi?cmd=Retrieve&db=PubMed&dopt=Citation&list_uids=18077435.

Maret, S., Faraguna, U., Nelson, A., Cirelli, C. and Tononi, G. (2011). Sleep and wake modulate spine turnover in the adolescent mouse cortex. *Sleep* **34**(Suppl.); .

Mcbride, S.M., Choi, C.H., Wang, Y., Liebelt, D., Braunstein, E., Ferreiro, D., Sehgal, A., Siwicki, K. K., Dockendorff, T.C., Nguyen, H.T., Mcdonald, T.V. andJongens, T.A. (2005). Pharmacological rescue of synaptic plasticity, courtship behavior, and mushroom body defects in a *Drosophila* model of fragile x syndrome. *Neuron* **45**(5); 753–764. Available from http://www.ncbi.nlm.nih.gov/entrez/query.fcgi?cmd=Retrieve&db=PubMed&dopt=Citation&list_uids=15748850.

Misonou, H. and Trimmer, J.S. (2004). Determinants of voltage-gated potassium channel surface expression and localization in mammalian neurons. *Crit. Rev. Biochem. Mol. Biol.* **39**(3); 125–145. Available from http://www.ncbi.nlm.nih.gov/entrez/query.fcgi?cmd=Retrieve&db=PubMed&dopt=Citation&list_uids=15596548.

Mongrain, V., Hernandez, S.A., Pradervand, S., Dorsaz, S., Curie, T., Hagiwara, G., Gip, P., Heller, H.C. and Franken, P. (2010). Separating the contribution of glucocorticoids and wakefulness to the molecular and electrophysiological correlates of sleep homeostasis. *Sleep* **33**(9); 1147–1157. Available from http://www.ncbi.nlm.nih.gov/entrez/query.fcgi?cmd=Retrieve&db=PubMed&dopt=Citation&list_uids=20857860.

Naidoo, N., Casiano, V., Cater, J., Zimmerman, J. and Pack, A.I. (2007). A role for the molecular chaperone protein bip/grp78 in *Drosophila* sleep homeostasis. *Sleep* **30**(5); 557–565. Available from http://www.ncbi.nlm.nih.gov/entrez/query.fcgi?cmd=Retrieve&db=PubMed&dopt=Citation&list_uids=17552370.

Nikonova, E.V., Naidoo, N., Zhang, L., Romer, M., Cater,J.R., Scharf, M.T., Galante, R.J. and Pack, A.I. (2010). Changes in components of energy regulation in mouse cortex with increases in wakefulness. *Sleep* **33**(7); 889–900. Available from http://www.ncbi.nlm.nih.gov/entrez/query.fcgi?cmd=Retrieve&db=PubMed&dopt=Citation&list_uids=20614849.

Nithianantharajah, J. and Hannan, A.J. (2006). Enriched environments, experience-dependent plasticity and disorders of the nervous system. *Nat. Rev. Neurosci.* **7**(9); 697–709. Available from http://www.ncbi.nlm.nih.gov/entrez/query.fcgi?cmd=Retrieve&db=PubMed&dopt=Citation&list_uids=16924259.

Nitz, D.A., Van Swinderen, B., Tononi, G. and Greenspan, R.J. (2002). Electrophysiological correlates of rest and activity in *Drosophila melanogaster*. *Curr. Biol.* **12**(22); 1934–1940. Available from http://www.ncbi.nlm.nih.gov/htbin-post/Entrez/query?db=m&form=6&dopt=r&uid=12445387.

Pan, L. and Broadie, K.S. (2007). *Drosophila* fragile x mental retardation protein and metabotropic glutamate receptor a convergently regulate the synaptic ratio of ionotropic glutamate receptor subclasses. *J. Neurosci.* **27**(45); 12378–12389. Available from http://www.ncbi.nlm.nih.gov/entrez/query.fcgi?cmd=Retrieve&db=PubMed&dopt=Citation&list_uids=17989302.

Pan, L., Zhang, Y.Q., Woodruff, E. and Broadie, K. (2004). The *Drosophila* fragile x gene negatively regulates neuronal elaboration and synaptic differentiation. *Curr. Biol.* **14**(20); 1863–1870. Available from http://www.ncbi.nlm.nih.gov/entrez/query.fcgi?cmd=Retrieve&db=PubMed&dopt=Citation&list_uids=15498496.

Parisky, K.M., Agosto,J., Pulver, S.R., Shang, Y., Kuklin, E., Hodge,J.J., Kang, K., Liu, X., Garrity, P. A., Rosbash, M. and Griffith, L.C. (2008). Pdf cells are a GABA-responsive wake-promoting component of the *Drosophila* sleep circuit. *Neuron* **60**(4); 672–682. Available from http://www.ncbi.nlm.nih.gov/entrez/query.fcgi?cmd=Retrieve&db=PubMed&dopt=Citation&list_uids=19038223.

Park, J.H., Helfrich-Forster, C., Lee, G., Liu, L., Rosbash, M. and Hall, J.C. (2000). Differential regulation of circadian pacemaker output by separate clock genes in Drosophila. *Proc. Natl. Acad. Sci. U S A* **97**(7); 3608–3613. Available from http://www.ncbi.nlm.nih.gov/entrez/query.fcgi?cmd=Retrieve&db=PubMed&dopt=Citation&list_uids=10725392.

Perazzona, B., Isabel, G., Preat, T. and Davis, R.L. (2004). The role of camp response element-binding protein in *Drosophila* long-term memory. *J. Neurosci.* **24**(40); 8823–8828. Available from http://www.ncbi.nlm.nih.gov/entrez/query.fcgi?cmd=Retrieve&db=PubMed&dopt=Citation&list_uids=15470148.

Peschel, N. and Helfrich-Forster, C. (2011). Setting the clock—by nature: circadian rhythm in the fruitfly *Drosophila melanogaster*. *FEBS Lett.* **585**(10); 1435–1442. Available from http://www.ncbi.nlm.nih.gov/entrez/query.fcgi?cmd=Retrieve&db=PubMed&dopt=Citation&list_uids=21354415.

Pitman, J.L., Mcgill, J.J., Keegan, K.P. and Allada, R. (2006). A dynamic role for the mushroom bodies in promoting sleep in Drosophila. *Nature* **441**(7094); 753–756. Available from http://www.ncbi.nlm.nih.gov/entrez/query.fcgi?cmd=Retrieve&db=PubMed&dopt=Citation&list_uids=16760979.

Qiu, Y. and Davis, R.L. (1993). Genetic dissection of the learning/memory gene dunce of *Drosophila* melanogaster. *Genes Dev.* **7**(7B); 1447–1458. Available from http://www.ncbi.nlm.nih.gov/entrez/query.fcgi?cmd=Retrieve&db=PubMed&dopt=Citation&list_uids=7687228.

Rattenborg, N., Martinez-Gonzalez, D., Roth Ii, T., and Pravosudov, V., 2010. Hippocampal memory consolidation during sleep: a comparison of mammals and birds. *Biol. Rev. Camb. Philos. Soc*, http://www.ncbi.nlm.nih.gov/entrez/query.fcgi?cmd=Retrieve&db=PubMed&dopt=Citation&list_uids=21070585.

Restivo, L., Ferrari, F., Passino, E., Sgobio, C., Bock, J., Oostra, B.A., Bagni, C. and Ammassari-Teule, M. (2005). Enriched environment promotes behavioral and morphological recovery in a mouse model for the fragile x syndrome. *Proc. Natl. Acad. Sci. U S A* **102**(32); 11557–11562. Available from http://www.ncbi.nlm.nih.gov/entrez/query.fcgi?cmd=Retrieve&db=PubMed&dopt=Citation&list_uids=16076950.

Rogero, O., Hammerle, B. and Tejedor, F.J. (1997). Diverse expression and distribution of shaker potassium channels during the development of the *Drosophila* nervous system. *J. Neurosci.* **17**(13); 5108–5118. Available from http://www.ncbi.nlm.nih.gov/entrez/query.fcgi?cmd=Retrieve&db=PubMed&dopt=Citation&list_uids=9185548.

Rosanova, M. and Ulrich, D. (2005). Pattern-specific associative long-term potentiation induced by a sleep spindle-related spike train. *J. Neurosci.* **25**(41); 9398–9405. Available from http://www.ncbi.nlm.nih.gov/entrez/query.fcgi?cmd=Retrieve&db=PubMed&dopt=Citation&list_uids=16221848.

Rothman, D.L., Behar, K.L., Hyder, F. and Shulman, R.G. (2003). In vivo NMR studies of the glutamate neurotransmitter flux and neuroenergetics: implications for brain function. *Annu. Rev. Physiol.* **65**, 401–427.

Rubio, M.E. and Wenthold, R.J. (1999). Calnexin and the immunoglobulin binding protein (bip) coimmunoprecipitate with ampa receptors. *J. Neurochem.* **73**(3); 942–948. Available from http://www.ncbi.nlm.nih.gov/entrez/query.fcgi?cmd=Retrieve&db=PubMed&dopt=Citation&list_uids=10461883.

Sakai, T., Tamura, T., Kitamoto, T. and Kidokoro, Y. (2004). A clock gene, period, plays a key role in long-term memory formation in *Drosophila*. *Proc. Natl. Acad. Sci. U S A* **101**(45); 16058–16063. Available from http://www.ncbi.nlm.nih.gov/entrez/query.fcgi?cmd=Retrieve&db=PubMed&dopt=Citation&list_uids=15522971.

Sakurai, T. (2007). The neural circuit of orexin (hypocretin): maintaining sleep and wakefulness. *Nat. Rev. Neurosci.* **8**(3); 171–181. Available from http://www.ncbi.nlm.nih.gov/entrez/query.fcgi?cmd=Retrieve&db=PubMed&dopt=Citation&list_uids=17299454.

Saper, C.B., Fuller, P.M., Pedersen, N.P., Lu, J. and Scammell, T.E. (2010). Sleep state switching. *Neuron* **68**(6); 1023–1042. Available from http://www.ncbi.nlm.nih.gov/entrez/query.fcgi?cmd=Retrieve&db=PubMed&dopt=Citation&list_uids=21172606.

Sara, S.J. (2009). The locus coeruleus and noradrenergic modulation of cognition. *Nat. Rev. Neurosci.* **10**(3); 211–223. Available from http://www.ncbi.nlm.nih.gov/entrez/query.fcgi?cmd=Retrieve&db=PubMed&dopt=Citation&list_uids=19190638.

Schenck, A., Van De Bor, V., Bardoni, B. and Giangrande, A. (2002). Novel features of dfmr1, the *Drosophila* orthologue of the fragile x mental retardation protein. *Neurobiol. Dis.* **11**(1); 53–63. Available from http://www.ncbi.nlm.nih.gov/entrez/query.fcgi?cmd=Retrieve&db=PubMed&dopt=Citation&list_uids=12460546.

Schwaerzel, M., Monastirioti, M., Scholz, H., Friggi-Grelin, F., Birman, S. and Heisenberg, M. (2003). Dopamine and octopamine differentiate between aversive and appetitive olfactory memories in *Drosophila. J. Neurosci.* **23**(33); 10495–10502. Available from http://www.ncbi.nlm.nih.gov/entrez/query.fcgi?cmd=Retrieve&db=PubMed&dopt=Citation&list_uids=14627633.

Schwarz, T.L., Papazian, D.M., Carretto, R.C., Jan, Y.N. and Jan, L.Y. (1990). Immunological characterization of K+ channel components from the shaker locus and differential distribution of splicing variants in *Drosophila. Neuron* **4**(1); 119–127. Available from http://www.ncbi.nlm.nih.gov/entrez/query.fcgi?cmd=Retrieve&db=PubMed&dopt=Citation&list_uids=2310570.

Seol, G.H., Ziburkus, J., Huang, S., Song, L., Kim, I.T., Takamiya, K., Huganir, R.L., Lee, H.K. and Kirkwood, A. (2007). Neuromodulators control the polarity of spike-timing-dependent synaptic plasticity. *Neuron* **55**(6); 919–929. Available from http://www.ncbi.nlm.nih.gov/entrez/query.fcgi?cmd=Retrieve&db=PubMed&dopt=Citation&list_uids=17880895.

Seugnet, L., Suzuki, Y., Vine, L., Gottschalk, L. and Shaw, P.J. (2008). D1 receptor activation in the mushroom bodies rescues sleep-loss-induced learning impairments in *Drosophila. Curr. Biol.* **18**(15); 1110–1117. Available from http://www.ncbi.nlm.nih.gov/entrez/query.fcgi?cmd=Retrieve&db=PubMed&dopt=Citation&list_uids=18674913.

Shaw, P. (2003). Awakening to the behavioral analysis of sleep in *Drosophila. J. Biol. Rhythms* **18**(1); 4–11. Available from http://www.ncbi.nlm.nih.gov/htbin-post/Entrez/query?db=m&form=6&dopt=r&uid=12568240.

Shaw, P.J., Cirelli, C., Greenspan, R.J. and Tononi, G. (2000). Correlates of sleep and waking in Drosophila melanogaster. *Science* **287**(5459); 1834–1837. Available from http://www.ncbi.nlm.nih.gov/cgi-bin/Entrez/referer?http://www.sciencemag.org/cgi/content/full/287/5459/1834.

Shaw, P.J., Tononi, G., Greenspan, R.J. and Robinson, D.F. (2002). Stress response genes protect against lethal effects of sleep deprivation in Drosophila. *Nature* **417**(6886); 287–291. Available from http://www.ncbi.nlm.nih.gov/htbin-post/Entrez/query?db=m&form=6&dopt=r&uid=12015603.

Sheeba, V., Gu, H., Sharma, V.K., O'dowd, D.K. and Holmes, T.C. (2008). Circadian- and light-dependent regulation of resting membrane potential and spontaneous action potential firing of *Drosophila* circadian pacemaker neurons. *J. Neurophysiol.* **99**(2); 976–988. Available from http://www.ncbi.nlm.nih.gov/entrez/query.fcgi?cmd=Retrieve&db=PubMed&dopt=Citation&list_uids=18077664.

Singh, N., Lorbeck, M.T., Zervos, A., Zimmerman, J. and Elefant, F. (2010). The histone acetyltransferase elp3 plays in active role in the control of synaptic bouton expansion and sleep in *Drosophila. J. Neurochem.* **115**(2); 493–504. Available from http://www.ncbi.nlm.nih.gov/entrez/query.fcgi?cmd=Retrieve&db=PubMed&dopt=Citation&list_uids=20626565.

Skoulakis, E.M., Kalderon, D. and Davis, R.L. (1993). Preferential expression in mushroom bodies of the catalytic subunit of protein kinase a and its role in learning and memory. *Neuron* **11**(2); 197–208. Available from http://www.ncbi.nlm.nih.gov/entrez/query.fcgi?cmd=Retrieve&db=PubMed&dopt=Citation&list_uids=8352940.

Steriade, M. (2003). The corticothalamic system in sleep. *Front. Biosci.* **8**, D878–D899. Available from http://www.ncbi.nlm.nih.gov/htbin-post/Entrez/query?db=m&form=6&dopt=r&uid=12700074.

Terao, A., Greco, M.A., Davis, R.W., Heller, H.C. and Kilduff, T.S. (2003a). Region-specific changes in immediate early gene expression in response to sleep deprivation and recovery sleep in the mouse brain. *Neuroscience* **120**(4); 1115–1124. Available from http://www.ncbi.nlm.nih.gov/entrez/query.fcgi?cmd=Retrieve&db=PubMed&dopt=Citation&list_uids=12927216.

Terao, A., Steininger, T.L., Hyder, K., Apte-Deshpande, A., Ding, J., Rishipathak, D., Davis, R.W., Heller, H.C. and Kilduff, T.S. (2003b). Differential increase in the expression of heat shock protein family members during sleep deprivation and during sleep. *Neuroscience* **116**(1); 187–200. Available from http://www.ncbi.nlm.nih.gov/entrez/query.fcgi?cmd=Retrieve&db=PubMed&dopt=Citation&list_uids=12535952.

Terao, A., Wisor, J.P., Peyron, C., Apte-Deshpande, A., Wurts, S.W., Edgar, D.M. and Kilduff, T.S. (2006). Gene expression in the rat brain during sleep deprivation and recovery sleep: an affymetrix genechip study. *Neuroscience* **137**(2); 593–605. Available from http://www.ncbi.nlm.nih.gov/entrez/query.fcgi?cmd=Retrieve&db=PubMed&dopt=Citation&list_uids=16257491.

Thimgan, M.S., Suzuki, Y., Seugnet, L., Gottschalk, L. and Shaw, P.J. (2010). The perilipin homologue, lipid storage droplet 2, regulates sleep homeostasis and prevents learning impairments following sleep loss. *PLoS Biol.* **8**(8); . Available from http://www.ncbi.nlm.nih.gov/entrez/query.fcgi?cmd=Retrieve&db=PubMed&dopt=Citation&list_uids=20824166.

Tinklenberg, J.A., Segal, I.S., Guo, T.Z. and Maze, M. (1991). Analysis of anesthetic action on the potassium channels of the shaker mutant of *Drosophila. Ann. N. Y. Acad. Sci.* **625**, 532–539. Available from http://www.ncbi.nlm.nih.gov/entrez/query.fcgi?cmd=Retrieve&db=PubMed&dopt=Citation&list_uids=1905502.

Tononi, G. and Cirelli, C. (2003). Sleep and synaptic homeostasis: a hypothesis. *Brain. Res Bull.* **62**(2); 143–150. Available from http://www.ncbi.nlm.nih.gov/entrez/query.fcgi?cmd=Retrieve&db=PubMed&dopt=Citation&list_uids=14638388.

Tononi, G. and Cirelli, C. (2006). Sleep function and synaptic homeostasis. *Sleep Med. Rev.* **10**(1); 49–62. Available from http://www.ncbi.nlm.nih.gov/entrez/query.fcgi?cmd=Retrieve&db=PubMed&dopt=Citation&list_uids=16376591.

Ungless, M.A., Whistler, J.L., Malenka, R.C. and Bonci, A. (2001). Single cocaine exposure in vivo induces long-term potentiation in dopamine neurons. *Nature* **411**(6837); 583–587. Available from http://www.ncbi.nlm.nih.gov/entrez/query.fcgi?cmd=Retrieve&db=PubMed&dopt=Citation&list_uids=11385572.

Van Swinderen, B., Nitz, D.A. and Greenspan, R.J. (2004). Uncoupling of brain activity from movement defines arousal states in *Drosophila. Curr. Biol.* **14**(2); 81–87. Available from http://www.ncbi.nlm.nih.gov/entrez/query.fcgi?cmd=Retrieve&db=PubMed&dopt=Citation&list_uids=14738728.

Vandenberghe, W., Nicoll, R.A. and Bredt, D.S. (2005). Interaction with the unfolded protein response reveals a role for stargazin in biosynthetic ampa receptor transport. *J. Neurosci.* **25**(5); 1095–1102. Available from http://www.ncbi.nlm.nih.gov/entrez/query.fcgi?cmd=Retrieve&db=PubMed&dopt=Citation&list_uids=15689545.

Vecsey, C.G., Baillie, G.S., Jaganath, D., Havekes, R., Daniels, A., Wimmer, M., Huang, T., Brown, K. M., Li, X.Y., Descalzi, G., Kim, S.S., Chen, T., Shang, Y.Z., Zhuo, M., Houslay, M.D. and Abel, T. (2009). Sleep deprivation impairs camp signalling in the hippocampus. *Nature* **461**(7267); 1122–1125. Available from http://www.ncbi.nlm.nih.gov/entrez/query.fcgi?cmd=Retrieve&db=PubMed&dopt=Citation&list_uids=19847264.

Vyazovskiy, V.V., Cirelli, C., Pfister-Genskow, M., Faraguna, U. and Tononi, G. (2008). Molecular and electrophysiological evidence for net synaptic potentiation in wake and depression in sleep. *Nat. Neurosci.* **11**(2); 200–208. Available from http://www.ncbi.nlm.nih.gov/entrez/query.fcgi?cmd=Retrieve&db=PubMed&dopt=Citation&list_uids=18204445.

Vyazovskiy, V.V., Olcese, U., Lazimy, Y.M., Faraguna, U., Esser, S.K., Williams, J.C., Cirelli, C. and Tononi, G. (2009). Cortical firing and sleep homeostasis. *Neuron* **63**(6); 865–878. Available from http://www.ncbi.nlm.nih.gov/entrez/query.fcgi?cmd=Retrieve&db=PubMed&dopt=Citation&list_uids=19778514.

Waddell, S. (2010). Dopamine reveals neural circuit mechanisms of fly memory. *Trends Neurosci.* **33**(10); 457–464. Available from http://www.ncbi.nlm.nih.gov/entrez/query.fcgi?cmd=Retrieve&db=PubMed&dopt=Citation&list_uids=20701984.

Weber, B., Schaper, C., Bushey, D., Rohlfs, M., Steinfath, M., Tononi, G., Cirelli, C., Scholz, J. and Bein, B. (2009). Increased volatile anesthetic requirement in short-sleeping *Drosophila* mutants. *Anesthesiology* **110**(2); 313–316. Available from http://www.ncbi.nlm.nih.gov/entrez/query.fcgi?cmd=Retrieve&db=PubMed&dopt=Citation&list_uids=19164958.

Werk, C.M. and Chapman, C.A. (2003). Long-term potentiation of polysynaptic responses in layer v of the sensorimotor cortex induced by theta-patterned tetanization in the awake rat. *Cereb. Cortex* **13**(5); 500–507. Available from http://www.ncbi.nlm.nih.gov/entrez/query.fcgi?cmd=Retrieve&db=PubMed&dopt=Citation&list_uids=12679296.

Werk, C.M., Klein, H.S., Nesbitt, C.E. and Chapman, C.A. (2006). Long-term depression in the sensorimotor cortex induced by repeated delivery of 10 Hz trains in vivo. *Neuroscience* **140**(1); 13–20. Available from http://www.ncbi.nlm.nih.gov/entrez/query.fcgi?cmd=Retrieve&db=PubMed&dopt=Citation&list_uids=16530972.

Wu, M.N., Koh, K., Yue, Z., Joiner, W.J. and Sehgal, A. (2008). A genetic screen for sleep and circadian mutants reveals mechanisms underlying regulation of sleep in *Drosophila. Sleep* **31**(4); 465–472. Available from http://www.ncbi.nlm.nih.gov/entrez/query.fcgi?cmd=Retrieve&db=PubMed&dopt=Citation&list_uids=18457233.

Wu, M.N., Joiner, W.J., Dean, T., Yue, Z., Smith, C.J., Chen, D., Hoshi, T., Sehgal, A. and Koh, K. (2010). Sleepless, a ly-6/neurotoxin family member, regulates the levels, localization and activity of shaker. *Nat. Neurosci.* **13**(1); 69–75. Available from http://www.ncbi.nlm.nih.gov/entrez/query.fcgi?cmd=Retrieve&db=PubMed&dopt=Citation&list_uids=20010822.

Yamazaki, D., Horiuchi, J., Nakagami, Y., Nagano, S., Tamura, T. and Saitoe, M. (2007). The *Drosophila* dco mutation suppresses age-related memory impairment without affecting lifespan. *Nat. Neurosci.* **10**(4); 478–484. Available from http://www.ncbi.nlm.nih.gov/entrez/query.fcgi?cmd=Retrieve&db=PubMed&dopt=Citation&list_uids=17322874.

Yao, W.D. and Wu, C.F. (2001). Distinct roles of camkii and pka in regulation of firing patterns and k(+) currents in *Drosophila* neurons. *J. Neurophysiol.* **85**(4); 1384–1394. Available from http://www.ncbi.nlm.nih.gov/entrez/query.fcgi?cmd=Retrieve&db=PubMed&dopt=Citation&list_uids=11287463.

Yin, J.C., Del Vecchio, M., Zhou, H. and Tully, T. (1995). Creb as a memory modulator: induced expression of a dcreb2 activator isoform enhances long-term memory in *Drosophila. Cell* **81**(1); 107–115. Available from http://www.ncbi.nlm.nih.gov/entrez/query.fcgi?cmd=Retrieve&db=PubMed&dopt=Citation&list_uids=7720066.

Yu, D., Ponomarev, A. and Davis, R.L. (2004). Altered representation of the spatial code for odors after olfactory classical conditioning; memory trace formation by synaptic recruitment. *Neuron* **42**(3); 437–449. Available from http://www.ncbi.nlm.nih.gov/entrez/query.fcgi?cmd=Retrieve&db=PubMed&dopt=Citation&list_uids=15134640.

Yu, D., Keene, A.C., Srivatsan, A., Waddell, S. and Davis, R.L. (2005). *Drosophila* dpm neurons form a delayed and branch-specific memory trace after olfactory classical conditioning. *Cell* **123**(5); 945–957. Available from http://www.ncbi.nlm.nih.gov/entrez/query.fcgi?cmd=Retrieve&db=PubMed&dopt=Citation&list_uids=16325586.

Yu, D., Akalal, D.B. and Davis, R.L. (2006). *Drosophila* alpha/beta mushroom body neurons form a branch-specific, long-term cellular memory trace after spaced olfactory conditioning. *Neuron* **52**(5); 845–855. Available from http://www.ncbi.nlm.nih.gov/entrez/query.fcgi?cmd=Retrieve&db=PubMed&dopt=Citation&list_uids=17145505.

Yuan, L.L. and Chen, X. (2006). Diversity of potassium channels in neuronal dendrites. *Prog. Neurobiol.* **78**(6); 374–389. Available from http://www.ncbi.nlm.nih.gov/entrez/query.fcgi?cmd=Retrieve&db=PubMed&dopt=Citation&list_uids=16716489.

Yuan, Q., Joiner, W.J. and Sehgal, A. (2006). A sleep-promoting role for the *Drosophila* serotonin receptor 1a. *Curr. Biol.* **16**(11); 1051–1062. Available from http://www.ncbi.nlm.nih.gov/entrez/query.fcgi?cmd=Retrieve&db=PubMed&dopt=Citation&list_uids=16753559.

Zhang, S., Yin, Y., Lu, H. and Guo, A. (2008). Increased dopaminergic signaling impairs aversive olfactory memory retention in *Drosophila. Biochem. Biophys. Res. Commun.* **370**(1); 82–86. Available from http://www.ncbi.nlm.nih.gov/entrez/query.fcgi?cmd=Retrieve&db=PubMed&dopt=Citation&list_uids=18342622.

Zimmerman, J.E., Rizzo, W., Shockley, K.R., Raizen, D.M., Naidoo, N., Mackiewicz, M., Churchill, G.A. and Pack, A.I. (2006). Multiple mechanisms limit the duration of wakefulness in *Drosophila* brain. *Physiol. Genomics* **27**(3); 337–350. Available from http://www.ncbi.nlm.nih.gov/entrez/query.fcgi?cmd=Retrieve&db=PubMed&dopt=Citation&list_uids=16954408.

Zimmerman, J.E., Naidoo, N., Raizen, D.M. and Pack, A.I. (2008). Conservation of sleep: insights from non-mammalian model systems. *Trends Neurosci.* **31**(7); 371–376. Available from http://www.ncbi.nlm.nih.gov/entrez/query.fcgi?cmd=Retrieve&db=PubMed&dopt=Citation&list_uids=18538867.

Subject Index

E

F

G

N

O

P

T

U

V

CONTENTS OF RECENT VOLUMES

Volume 39

Volume 40

Volume 41

Volume 42

Volume 43

Volume 44

Volume 45

Volume 46

Volume 47

Volume 48

Volume 49

Volume 50

Volume 51

Volume 52

Volume 53

Volume 54

Volume 55

Volume 56

Volume 57

Volume 58

Volume 59

Volume 60

Volume 61

Volume 62

Volume 63

Volume 64

Volume 65

Volume 66

Volume 67

Volume 68

Volume 69

Volume 70

Volume 71

Volume 72

Volume 73

Volume 74

Volume 75

Volume 76

Volume 77

Volume 78

Volume 79

Volume 80

Volume 81

Volume 82

Volume 83

Volume 84

Volume 88

Volume 89

Volume 90

Volume 91

Volume 92

Volume 93

Volume 94

Volume 95

Volume 96

Volume 97

Volume 98

FSC
www.fsc.org
MIX
Papier | Fördert gute Waldnutzung
FSC® C083411

Kolibri 360 GmbH
Ferdinand-Jühlke-Straße 7
99095 Erfurt, Deutschland
produktsicherheit@kolibri360.de